Composites Science and Technology

Series Editor

Mohammad Jawaid, Laboratory of Biocomposite Technology, Universiti Putra Malaysia, INTROP, Serdang, Malaysia

This book series publishes cutting edge research monographs comprehensively covering topics in the field of composite science and technology. The books in this series are edited or authored by top researchers and professional across the globe. The series aims at publishing state-of-the-art research and development in areas including, but not limited to:

- Conventional Composites from natural and synthetic fibers
- Advanced Composites from natural and synthetic fibers
- Chemistry and biology of Composites and Biocomposites
- Fatigue damage modelling of Composites and Biocomposites
- Failure Analysis of Composites and Biocomposites
- Structural Health Monitoring of Composites and Biocomposites
- Durability of Composites and Biocomposites
- Biodegradability of Composites and Biocomposites
- Thermal properties of Composites and Biocomposites
- Flammability of Composites and Biocomposites
- Tribology of Composites and Biocomposites
- Applications of Composites and Biocomposites

Review Process

The proposal for each volume is reviewed by the main editor and/or the advisory board. The chapters in each volume are individually reviewed single blind by expert reviewers (at least two reviews per chapter) and the main editor.

Ethics Statement for this series can be found in the Springer standard guidelines here - https://www.springer.com/us/authors-editors/journal-author/journal-author-helpdesk/before-you-start/before-you-start/1330#c14214

Tabrej Khan · Mohammad Jawaid
Editors

Green Hybrid Composite in Engineering and Non-Engineering Applications

Springer

Editors

Tabrej Khan
Department of Engineering Management
College of Engineering
Prince Sultan University
Riyadh, Saudi Arabia

Mohammad Jawaid
Laboratory of Biocomposite Technology
Institute of Tropical Forestry and Forest Products (INTROP)
Universiti Putra Malaysia
Serdang, Malaysia

ISSN 2662-1819 ISSN 2662-1827 (electronic)
Composites Science and Technology
ISBN 978-981-99-1582-8 ISBN 978-981-99-1583-5 (eBook)
https://doi.org/10.1007/978-981-99-1583-5

This Springer imprint is published by the registered company Springer Nature Singapore Pte Ltd.
The registered company address is: 152 Beach Road, #21-01/04 Gateway East, Singapore 189721, Singapore

Dr. Tabrej Khan, Editor, is honored to dedicate this book to his Mother Miss. Daraksa Khan.

She supported me to pursue a Ph.D. degree to fulfill my dreams; otherwise, I would not be in this position without her support, inspiration, and motivation.

Preface

Manufacturing green hybrid composites in engineering and non-engineering applications is one of the major advances in cellulosic materials and has become a topic of great interest over the last decade. Natural hybrid composites, biodegradable character, economy, high aspect ratio, low weight, and sustainability are an impetus for this growing interest. There are currently no limits to the possible uses of natural hybrid composites. However, the main application of natural hybrid composites is their inclusion into polymer matrixes to produce composites with high mechanical characteristics. Given the growing number of scientific publications on natural hybrid composites and the tremendous advances in processing capabilities to make compounds based on natural hybrid composites, a book that summarizes and updates all of these innovations is ideal for researchers and students involved in manufacturing. Interest in and use of natural hybrid composites in innovative materials. The basic objective of this book is to present the development and application of natural hybrid composites as reinforcement in compounds. This book is written by leading experts and aims to cover a wide range of properties and uses of natural hybrid composites, including the manufacture of natural hybrid composites from various biomass resources and the utility of natural hybrid composites as reinforcement of polymers; great challenges for successful scaling of production. The chapters will provide cutting-edge research on the use of natural hybrid composite reinforcements in polymer composites to achieve material properties and significant improvements in physical, mechanical, and thermal properties. The book provides an up-to-date overview of the most important innovations in the field of natural hybrid composites and provides reference materials for future research on natural hybrid composites, which are in high demand due to their sustainability, recyclability, and environmental friendliness.

We thank all the authors who contributed to this book and who made our idea come true. We also thank the Springer Nature Singapore support team, especially Mr. Balbir Singh, for helping us during the preparation of this book.

Riyadh, Saudi Arabia — Tabrej Khan
Serdang, Malaysia — Mohammad Jawaid

Contents

Editors and Contributors

About the Editors

Dr. Tabrej Khan was born on April 22, 1989, in Jamui, Bihar, India. He received his primary education in sector 4 "E" high school, Bokaro steel city, Jharkhand, India (1992–2006), high school education in N. J. S. Inter College Jainamore, Bokaro, Jharkhand (2001–2006), higher secondary education in the same place, college in Bokaro, Jharkhand (2006–2008), and completed her bachelor's degree in Aerospace Engineering in (2009–2012) at The Indian Institute Aeronautics and Information Technology (IIAEIT) Pune, Maharashtra, India. During his bachelor's, he worked on unmanned aerial vehicle using Coanda effect hovercraft, balloon satellite, airport system, airport layout, internal combustion engine, and analysis of aircraft winglet design. He continued his master's in Aerospace Engineering in Faculty of Engineering, Department of Aerospace Engineering, Universiti Putra Malaysia. During his candidature, he had attended various seminars and workshops organized by the School of Graduate Studies and conferences (2013–2015). He received his Ph.D. from the Universiti Putra Malaysia, (2020). His area of research is Hybrid Composites, Advanced Materials, Structural Health Monitoring and Impact Studies. His research interest area includes Hybrid Composites, Advanced Materials, Structural Health Monitoring and Impact Studies, Damage Detections and Repairs, Impact Studies, Signal Processing and Instrumentations, and Non-Destructive Testing and Destructive Testing.

Dr. Mohammad Jawaid is currently working as High Flyer Fellow (Professor) at Biocomposite Technology Laboratory, Universiti Putra Malaysia, Malaysia. He is working in the composite field since 2008 and having ten years of experience in teaching and research in the field of composites. His area of research interests includes Hybrid Reinforced/Filled Polymer Composites, Advanced Materials: Graphene/Nanoclay/Fire Retardant, Lignocellulosic Reinforced/Filled Polymer Composites, Modification and Treatment of Lignocellulosic Fibers and Solid Wood, Nanocomposites, and Nanocellulose Fibers. Presently, he is supervising

12 Ph.D. students and 5 master's students in hybrid composites, green composites, nanocomposites, natural fiber reinforced composites, etc. About 26 Ph.D. and 10 master's students were graduated under his supervision in 2014–2020. He has several research grants at university/national/international level on polymer composites and its related research of around RM 6 million (USD 1.5 million). So far, he has published 40 books, 60 book chapters, and more than 350 international journal papers in reputed journal in composites and allied field. He has published five review papers under top 25 hot articles in science direct from 2014 to 2020. His H-index is 63 (Google) and 53 (Scopus).

Contributors

Kamarul Arifin Ahmad Department of Aerospace Engineering, Faculty of Engineering, Universiti Putra Malaysia (UPM), Serdang, Selangor Darul Ehsan, Malaysia;
Faculty of Engineering, Aerospace Malaysia Research Centre, Universiti Putra Malaysia (UPM), Serdang, Selangor Darul Ehsan, Malaysia

Tauseef Ahmad Department of Mechanical Engineering, Universiti Teknologi PETRONAS, Seri Iskandar, Perak, Malaysia

Mohammad Azad Alam Department of Mechanical Engineering, Universiti Teknologi PETRONAS, Seri Iskandar, Perak, Malaysia

Fairuz Alias Faculty of Mechanical and Manufacturing Engineering, University Tun Hussein Onn Malaysia, Johor, Malaysia

Tibor Alpár Faculty of Wood Engineering and Creative Industry, University of Sopron, Sopron, Hungary

Adriano Amaral Textiles Technologies Study Group (GETTEX), Laboratory of Knitting, Textile Engineering Laboratory, Department of Textile Engineering, Federal University of Rio Grande Do Norte, Natal, Rio Grande Do Norte, Brazil

Marcos Aquino Textiles Technologies Study Group (GETTEX), Laboratory of Knitting, Textile Engineering Laboratory, Department of Textile Engineering, Federal University of Rio Grande Do Norte, Natal, Rio Grande Do Norte, Brazil

D. Aravind University Science Instrumentation Centre, Madurai Kamaraj University, Madurai, Tamil Nadu, India;
Department of Mechanical Engineering, Kalasalingam Academy of Research and Education, Krishnankoil, Tamil Nadu, India

Ahmad Hamdan Ariffin Faculty of Mechanical and Manufacturing Engineering, Universiti Tun Hussein Onn Malaysia (UTHM), Parit Raja, Batu Pahat, Johor Darul Takzim, Malaysia;

Research Centre for Unmanned Vehicle (ReCUV), Faculty of Mechanical and Manufacturing Engineering, Universiti Tun Hussein Onn Malaysia (UTHM), Parit Raja, Batu Pahat, Johor Darul Takzim, Malaysia;
Green Design and Manufacture Research Group, Center of Excellence Geopolymer and Green Technology (CEGeoGTech), Universiti Malaysia Perlis, Kangar, Perlis, Malaysia

Mochamad Asrofi Department of Mechanical Engineering, University of Jember, Jember, East Java, Indonesia

M. R. M. Asyraf Engineering Design Research Group (EDRG), Faculty of Mechanical Engineering, Universiti Teknologi Malaysia, Johor Bahru, Johor, Malaysia;
Centre of Advanced Composite Materials (CACM), Universiti Teknologi Malaysia, Johor Bahru, Johor, Malaysia

Mohammad Azeem Department of Mechanical Engineering, Universiti Teknologi PETRONAS, Seri Iskandar, Perak, Malaysia

S. G. Borisade Department of Metallurgical and Materials Engineering, Federal University of Technology, Akure, Ondo State, Nigeria;
Department of Materials and Metallurgical Engineering, Federal University Oye-Ekiti, Oye-Ekiti, Ekiti State, Nigeria

Edson Cocchieri Botelho UNESP—Sao Paulo State University, SP, Brazil

Luiz Filipe Castro Department of Production Engineering, Federal University of Rio Grande do Norte, Natal, Rio Grande do Norte, Brazil

Maria Eduarda Cavalcante Textile Engineering Laboratory, Department of Textile Engineering, Federal University of Rio Grande do Norte, Natal, Rio Grande do Norte, Brazil;
Laboratory of Knitting, Textiles Technologies Study Group (GETTEX), Federal University of Rio Grande do Norte, Natal, Rio Grande do Norte, Brazil

Ivana Cesarino UNESP—Sao Paulo State University, SP, Brazil

M. Chandrasekar Department of Aeronautical Engineering, Hindustan Institute of Technology & Science, Padur, Kelambakkam, Chennai, Tamil Nadu, India

Muthukumar Chandrasekar School of Aeronautical Sciences, Hindustan Institute of Technology and Science, Chennai, India

Milena Chanes UNESP—Sao Paulo State University, SP, Brazil

Shuai Chen Department of Biochemistry and Molecular Biology, College of Basic Medical Sciences, Guizhou Medical University, Guiyang, China

Mariana Dias Textiles Technologies Study Group (GETTEX), Laboratory of Knitting, Textile Engineering Laboratory, Department of Textile Engineering, Federal University of Rio Grande Do Norte, Natal, Rio Grande Do Norte, Brazil

Otavio Augusto Titton Dias University of Toronto, Toronto, Canada

Rubens Fonseca Textiles Technologies Study Group (GETTEX), Laboratory of Knitting, Textile Engineering Laboratory, Department of Textile Engineering, Federal University of Rio Grande Do Norte, Natal, Rio Grande Do Norte, Brazil

Shahrul Malek Faizsal Bin Shahrul Hairi Faculty of Mechanical and Manufacturing Engineering, Universiti Tun Hussein Onn Malaysia (UTHM), Parit Raja, Batu Pahat, Johor Darul Takzim, Malaysia

K. M. Faridul Hasan Faculty of Wood Engineering and Creative Industry, University of Sopron, Sopron, Hungary

Hussain Hasanuthin Faculty of Mechanical and Manufacturing Engineering, University Tun Hussein Onn Malaysia, Johor, Malaysia

Péter György Horváth Faculty of Wood Engineering and Creative Industry, University of Sopron, Sopron, Hungary

Emad K. Hussein Mechanical Power Engineering Department, Mussaib Technical College, Al Furat Al Awsat Technical University, Babil, Iraq

Emad Kamil Hussein Al Furat Al Awsat Technical University, Kufa, Iraq

R. A. Ilyas Faculty of Chemical and Energy Engineering, Universiti Teknologi Malaysia, Johor Bahru, Johor, Malaysia

A. R. Irfan Faculty of Mechanical Engineering Technology, Universiti Malaysia Perlis, Kampus Tetap Pauh Putra, Arau, Perlis, Malaysia;
Green Design and Manufacture Research Group, Center of Excellence Geopolymer and Green Technology (CEGeoGTech), Universiti Malaysia Perlis, Kangar, Perlis, Malaysia

M. R. Ishak Department of Aerospace Engineering, Faculty of Engineering, Universiti Putra Malaysia, UPM Serdang, Selangor, Malaysia

Mohammad Jawaid Institute of Tropical Forestry and Forest Products (INTROP), Universiti Putra Malaysia, Putra, Malaysia

Ashwini Karrupaswamy Department of Biotechnology (DDE), Madurai Kamaraj University, Madurai, Tamil Nadu, India

Rehan Khan College of Electrical and Mechanical Engineering, National University of Sciences and Technology, Rawalpindi, Pakistan

Tabrej Khan Department of Engineering Management, College of Engineering, Prince Sultan University, Riyadh, Saudi Arabia

Alcides Lopes Leao UNESP—Sao Paulo State University, SP, Brazil

M. Manikandan Department of Aeronautical and Automobile Engineering, Manipal Institute of Technology, Manipal Academy of Higher Education, Manipal, Karnataka, India

Mohamad Norani Mansur Faculty of Mechanical and Manufacturing Engineering, University Tun Hussein Onn Malaysia, Johor, Malaysia

Faisal Masood Department of Electrical Engineering, University of Engineering and Technology Taxila, Rawalpindi, Pakistan

Katia M. Melo Textile Engineering Laboratory, Department of Textile Engineering, Federal University of Rio Grande do Norte, Natal, Rio Grande do Norte, Brazil;
Laboratory of Knitting, Textiles Technologies Study Group (GETTEX), Federal University of Rio Grande do Norte, Natal, Rio Grande do Norte, Brazil

Mohammad Sukri Mustapa Faculty of Mechanical and Manufacturing Engineering, University Tun Hussein Onn Malaysia, Johor, Malaysia

Eddie Yin Kwee Ng School of Mechanical and Aerospace Engineering, College of Engineering, Nanyang Technological University, Singapore, Singapore

M. N. F. Norrahim Research Center for Chemical Defence, Universiti Pertahanan Nasional Malaysia (UPNM), Kuala Lumpur, Malaysia

N. M. Nurazzi Bioresource Technology Division, School of Industrial Technology, Universiti Sains Malaysia, Pulau Pinang, Malaysia

G. S. Ogunwande Department of Metallurgical and Materials Engineering, Federal University of Technology, Akure, Ondo State, Nigeria

I. O. Oladele Department of Metallurgical and Materials Engineering, Federal University of Technology, Akure, Ondo State, Nigeria;
Centre for Nanomechanics and Tribocorrosion, School of Metallurgy, Chemical and Mining Engineering, University of Johannesburg, Johannesburg, South Africa

Zamri Bin Omar Faculty of Mechanical and Manufacturing Engineering, Universiti Tun Hussein Onn Malaysia (UTHM), Parit Raja, Batu Pahat, Johor Darul Takzim, Malaysia

L. N. Onuh Department of Metallurgical and Materials Engineering, Federal University of Technology, Akure, Ondo State, Nigeria

Raghuvir Pai Department of Mechanical and Manufacturing Engineering, Manipal Institute of Technology, Manipal Academy of Higher Education, Manipal, Karnataka, India

M. Rafidah Department of Civil Engineering, Universiti Putra Malaysia, UPM Serdang, Selangor, Malaysia

N. Rajini Department of Mechanical Engineering, Kalasalingam Academy of Research and Education, Krishnankoil, Tamil Nadu, India

M. R. Razman Research Centre for Sustainability Science and Governance (SGK), Institute for Environment and Development (LESTARI), Universiti Kebangsaan Malaysia, UKM Bangi, Selangor, Malaysia

Siti Juita Mastura Binti Mohd Saleh Research Centre for Unmanned Vehicle (ReCUV), Faculty of Mechanical and Manufacturing Engineering, Universiti Tun Hussein Onn Malaysia (UTHM), Parit Raja, Batu Pahat, Johor Darul Takzim, Malaysia

Jeyaguru Sangilimuthukumar Department of Automobile Engineering, Kalasalingam Academy of Research and Education, Krishnankoil, Tamil Nadu, India

M. R. Sanjay Natural Composite Research Group Lab, Department of Materials and Production Engineering, The Sirindhorn International Thai-German Graduate School of Engineering (TGGS), King Mongkuts University of Technology North Bangkok (KMUTNB), Bangkok, Thailand

Caroliny M. Santos Textiles Technologies Study Group (GETTEX), Laboratory of Knitting, Textile Engineering Laboratory, Department of Textile Engineering, Federal University of Rio Grande Do Norte, Natal, Rio Grande Do Norte, Brazil

Thiago F. Santos Textiles Technologies Study Group (GETTEX), Laboratory of Knitting, Textile Engineering Laboratory, Department of Textile Engineering, Federal University of Rio Grande Do Norte, Natal, Rio Grande Do Norte, Brazil

S. M. Sapuan Laboratory of Biocomposite Technology, Institute of Tropical Forestry and Forest Products, Universiti Putra Malaysia, UPM, Serdang, Selangor, Malaysia;
Department of Mechanical and Manufacturing Engineering, Universiti Putra Malaysia, UPM, Serdang, Selangor, Malaysia

Siti Noorbaini Sarmin Biocomposite Technology and Design Unit, Institute of Tropical Forestry & Forest Products, Universiti Putra Malaysia, UPM, Serdang, Selangor, Malaysia;
Department of Wood Technology, Faculty of Applied Sciences, Universiti Teknologi MARA, Bandar Pusat Jengka, Pahang, Malaysia

T. Senthil Muthu Kumar Department of Mechanical Engineering, Kalasalingam Academy of Research and Education, Krishnankoil, Tamil Nadu, India

Thiagamani Senthil Muthu Kumar Department of Automobile Engineering, Kalasalingam Academy of Research and Education, Krishnankoil, Tamil Nadu, India

K. Senthilkumar Department of Mechanical Engineering, PSG Institute of Technology and Applied Research, Coimbatore, Tamil Nadu, India

Krishnasamy Senthilkumar Departmet of Mechanical Engineering, PSG Institute of Technology and Applied Research, Coimbatore, Tamil Nadu, India

Suchart Siengchin Department of Materials and Production Engineering, The Sirindhorn International Thai-German Graduate School of Engineering (TGGS), King Mongkut's University of Technology North Bangkok (KMTNB), Bangkok, Thailand;

Institute of Plant and Wood Chemistry, Technische Universität Dresden, Tharandt, Germany

Balbir Singh Department of Aerospace Engineering, Faculty of Engineering, Universiti Putra Malaysia (UPM), Serdang, Selangor Darul Ehsan, Malaysia;
Department of Aeronautical and Automobile Engineering, Manipal Institute of Technology, Manipal Academy of Higher Education, Manipal, Karnataka, India

Jayavel Sridhar Department of Biotechnology (DDE), Madurai Kamaraj University, Madurai, Tamil Nadu, India

Mohamed Thariq Bin Hameed Sultan Department of Aerospace Engineering, Faculty of Engineering, Universiti Putra Malaysia, UPM Serdang, Selangor Darul Ehsan, Putra, Malaysia;
Laboratory of Biocomposite Technology, Institute of Tropical Forestry and Forest Products (INTROP), Universiti Putra Malaysia, Selangor Darul Ehsan, Serdang, Seri Kembangan, Malaysia;
Aerospace Malaysia Innovation Centre (944751-A), Prime Minister's Department, MIGHT Partnership Hub, Selangor Darul Ehsan, Cyberjaya, Malaysia

K. Tabrej Department of Engineering Management, College of Engineering, Prince Sultan University, Riyadh, Saudi Arabia

György Török Faculty of Wood Engineering and Creative Industry, University of Sopron, Sopron, Hungary

Liu Xiaoyi School of Public Health, the key, Laboratory of Environmental Pollution Monitoring and Disease Control, Ministry of Education, Guizhou Medical University, Guiyang, China

H. H. Ya Department of Mechanical Engineering, Universiti Teknologi PETRONAS, Seri Iskandar, Perak, Malaysia

Noorfaizal Yidris Department of Aerospace Engineering, Faculty of Engineering, Universiti Putra Malaysia (UPM), Serdang, Selangor Darul Ehsan, Malaysia

Mohammad Yusuf Department of Petroleum Engineering, Universiti Teknologi PETRONAS, Seri Iskandar, Malaysia

Lucas Zilio Textiles Technologies Study Group (GETTEX), Laboratory of Knitting, Textile Engineering Laboratory, Department of Textile Engineering, Federal University of Rio Grande Do Norte, Natal, Rio Grande Do Norte, Brazil

The Challenges of Natural Fiber in Manufacturing, Material Selection and Technology Application

Tabrej Khan, Mohamed Thariq Bin Hameed Sultan, and Ahmad Hamdan Ariffin

Abstract In this chapter, we will address prior research on the characteristics and uses of natural fiber composites in the automotive and aerospace industries. Due to its advantages over current artificial fiber composites, such as their biodegradability, lightweight design, and low cost, natural fiber composites are a preferable replacement. Natural fibers have been employed for the preparation of buildings, baskets, ropes, clothing, and many other things since the dawn of civilization. Natural fibers including jute, kenaf, sisal, hemp, and flax have more recently been utilized in the engineering manufacturing sector. The aerospace and automobile sectors are increasingly using natural fiber composites. To determine the best materials for engineering areas, researchers are now comparing natural fiber composites with synthetic composites. Natural fibers are also attracting increasing attention from

T. Khan (✉)
Engineering Management Department, College of Engineering, Prince Sultan University, Riyadh, Saudi Arabia
e-mail: tkhan@psu.edu.sa

M. T. B. H. Sultan
Department of Aerospace Engineering, Faculty of Engineering, Universiti Putra Malaysia, UPM Serdang, Selangor Darul Ehsan, Seri Kembangan 43400, Putra, Malaysia

Laboratory of Biocomposite Technology, Institute of Tropical Forestry and Forest Products (INTROP), Universiti Putra Malaysia, Selangor Darul Ehsan, 43400 Serdang, Seri Kembangan, Malaysia

Aerospace Malaysia Innovation Centre (944751-A), Prime Minister's Department, MIGHT Partnership Hub, Selangor Darul Ehsan, Jalan Impact, 63000 Cyberjaya, Malaysia

A. H. Ariffin
Faculty of Mechanical and Manufacturing Engineering, University Tun Hussein Onn Malaysia, 86400 Batu Pahat, Johor, Malaysia

Research Centre for Unmanned Vehicle (ReCUV), University Tun Hussein Onn Malaysia, 86400 Batu Pahat, Johor, Malaysia

Green Design and Manufacture Research Group, Center of Excellence Geopolymer and Green Technology (CEGeoGTech), Universiti Malaysia Perlis, 01000 Kangar, Perlis, Malaysia

T. Khan and M. Jawaid (eds.), *Green Hybrid Composite in Engineering and Non-Engineering Applications*, Composites Science and Technology,
https://doi.org/10.1007/978-981-99-1583-5_1

researchers because of their biodegradability and affordable manufacture. The possibility for adopting natural fiber composites in place of the materials used in airplane components and panel structures has been evaluated.

1 Introduction

Over the last few decades, it has been discovered that natural fiber composites have applications in a wide range of technical sectors [1]. This is most likely caused by the polymers' biodegradability. Natural fiber's importance stems from how simple and inexpensive it is to process. Natural fiber polymer composites outperform conventional materials when compared to their individual qualities [2]. Numerous other aerospace applications and allied industries make use of these types of green composites [3].

The use of natural fiber composites has a number of advantages, but there are also some drawbacks that have an impact on applications in the aerospace and automotive industries [4]. These drawbacks include progressive moisture absorption, reduced fire resistance, microbe infection, low temperature limitations, poor mechanical properties, and, most importantly, price fluctuations during the annual harvest [5]. Previous research has demonstrated that the quality of natural fibers may be enhanced for better fiber matrix bonding by applying chemical treatments, such as surface modification [6]. Additionally, natural fiber composites can suit human demands and offer appealing environmental and financial perspectives. Green fiber composites have a lot of potential applications in the automotive and aerospace industries [7].

1.1 Natural Fibers

One of the most able to demonstrate the ability in engineering is a natural fiber. Natural fibers can be found as long strands, woven materials, and more. Additionally, they may be shaped into mats for use in goods like paper and felt [8]. There are two categories of fibers: those that are created artificially or naturally. Directly from nature sourced fibers fall into one of three categories: those derived from plants, animals, or minerals [9]. Figure 1 displays a thorough categorization of fibers.

From a variety of fruit and vegetable species, natural fibers may be produced. The majority of these fibers are readily available in nature. Wool and silk are examples of natural fibers that may be obtained from animals. Some minerals, like asbestos, include inorganic natural fiber [4, 10]. Several of the significant fibers used in the strengthening of natural fiber composites, with their classes and roots, are listed in Table 1. A general composite application spectrum of fiber reinforcements covers almost all kinds of progressive engineering structures.

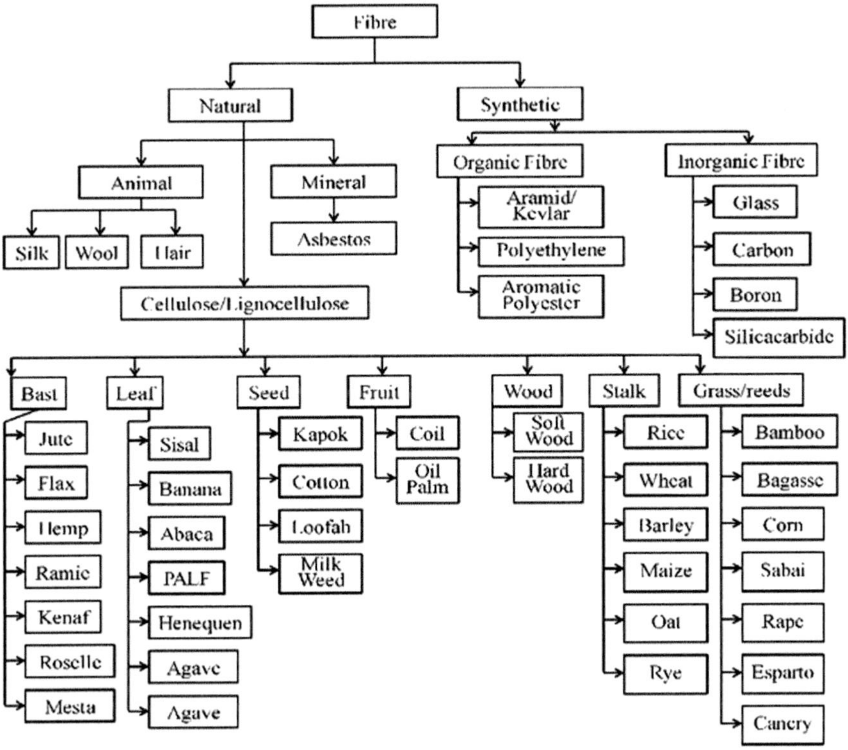

Fig. 1 Classification of Natural fibers [1]

Physical and mechanical properties of natural fibres

Some mechanical qualities of natural fiber, such as tensile strength and Young's modulus, rise exponentially as cellulose content increases. The strength of the natural fibers is dependent on the micro-fibrillar angle [11]. Natural fibre plants are more flexible if the micro-fibrils are flexuous in relation to the natural fibre axis [12]. If the micro-fibrils are parallel to the fibber's axis, the fibers will be stiffer and more inflexible, with greater mechanical qualities such as the tensile strength shown in Table 2.

1.2 A Comparison of Natural Fiber with Artificial Fiber

A wide range of values for natural fibers' mechanical and physical qualities may be found in the literature because of the considerable variability in their characteristics and testing procedures [13]. The highest values for the same attributes were taken into account for the assessment. Bast natural fiber offers high mechanical strength,

Table 1 List of important natural fibers [1]

Sr. No	Natural fiber types	Species	Origin
1	Abaca	Musa textile	Leaf
2	Bamboo	(1250 Species)	Grass
3	Banana	Musa indica	Leaf
4	Coir	Cocos nucifera	Fruit
5	Cotton	Gossypium sp.	Seed
6	Curaua	Anasnas erectifoluis	Leaf
7	Flax	Linum usiatissimum	Stem
8	Hemp	Cannabis sativa	Stem
9	Henequen	Agava fourcroydes	Leaf
10	Jute	Corchorus capsularis	Steam
11	Kenaf	Hibiscus cannabinus	Steam
12	Oil	Elaeis guineensis	Fruit
13	Pineapple	Annanus comosus	Leaf
14	Ramie	Boehmeria nivea	Steam
15	Sisal	Agava sisilana	Leaf
16	Wood	(>10 000 species)	Steam

Table 2 Physical and mechanical characteristics of different natural fibers [1]

Fiber name	Density [g cm^{-3}]	Diameter [μm]	Tensile strength [Mpa]	Specific strength [S/ρ]	Tensile modulus [Mpa]	Specific Modulus [E/ρ]	Elongation at break [%]
Abaca	1.5	–	400	267	12	8	3–12
Bamboo	1.1	240–330	500	454	35.91	32.6	1.40
Banana	1.35	50–250	600	444	17.85	13.6	3.36
Coconut	1.15	100–450	500	435	2.5	2.17	3.36
Coir	1.2	–	175	146	4–6	3.3–5	30
Cotton	1.6	–	287–597	179–373	5.5–12.6	3.44–7.9	7–8
Curaua	1.4	170	158–729	113–521	–	–	5
Flax	1.5	–	800–1500	535–1000	27.6–80	18.4–53	1.2–3.2
Hemp	1.48	–	550–900	372–608	70	47.3	2–4
Jute	1.46	40–350	393–938	269–548	10–30	6.85–20.6	1.5–1.8
Kenaf	1.45	70–250	930	641	53	36.55	1.6
Raime	1.5	50	220–938	147–625	44–128	29.3–85	2–3.8
Sisal	1.45	50–3000	530–64	366–441	9.4–22	6.5–15.2	3–7
Softwood	1.5	–	1000	667	40	26.67	4.4

making it ideal for aerospace and automotive applications [14]. In compared to other natural fibers, flax offers a wonderful mix of cheaper cost, lighter weight, and good mechanical qualities. Jute fiber is another often used natural fiber; however, it is not as strong as flax or bast [15]. In contrast to glass reinforced composites, natural fiber reinforced thermoset composites release energy at lower levels of stress and higher levels of strain, according to prior studies. The qualities of the resin, however, are what primarily contribute to the energy degradation of natural fiber reinforced thermoplastic composites [16].

1.3 Natural Fibre Composite Challenges

Glass fiber composites now account for 90% of the composites market. Due to the lesser interference between the reinforcements and the adhesive, fibers are inappropriate for use in interior components of airplanes and automobiles [17]. Through the use of research, technology in materials such chemical treatments, additives, and coatings, these characteristics can be improved. Compared to glass fiber composites, green composites have a significantly lower impact on the environment [18]. Glass fibers are hydrophobic by nature and humidity resistant, according to earlier research findings. Natural fibers may struggle to compete with glass fibers due to their hydrophilic nature, which might be a drawback [19]. Natural fibers absorb water from the environment under humid circumstances, causing fiber bumps within the composite and making it difficult to utilize for interior components of automobiles and airplanes [20].

The influence of hydrothermal and weathering conditions on the properties of jute fiber composites has been studied. The exposure period under varied moisture conditions was connected to the change in different humidity conditions [21]. The swelling of natural fiber jute fibers will increase the weight and thickness as humidity levels rise. The impact of water submersion on sisal and jute fiber reinforced epoxy composites was also tested and compared. It was discovered that sisal-based composites absorb more water than jute-reinforced composites [22]. Natural fibers with a higher cellulose content have a higher fiber volume, which increases their ability to accept a higher moisture percentage. Improved fiber properties can be obtained by carefully working on improving the weak ecological and dimensional reliability of natural fibers [23].

1.4 Natural Fibre Resistance

Natural fiber composites also have very low fire resistance due to their nature, which is a significant disadvantage for employing them in aviation components, the automotive sector, and many other production industries where inflammability is a vital consideration [24]. This disadvantage is one of the most significant problems for natural fibers, making it even more difficult for them to compete with synthetic fibers [25]. Natural fibers are not thermoplastic, and their decomposition temperature is extremely low when compared to other thermal characteristics such as melting and glass temperature. Natural fiber is mostly comprised of hemicellulose, cellulose, lignin, minerals, and waxes [26]. When the cellulose concentration of the fibers is high, the fire resistance is significantly lowered. The temperature range for cellulose decomposition is 260 to 350 °C, but hemicellulose decomposes at a lower temperature range of 2000 °C to 2600 °C and produces more non-combustible gases [27]. Lignin begins to degrade at 1600 °C and continues to decompose until 4000 °C. Flax fiber has a low lignin concentration, which results in higher decomposition temperatures but poor oxidation resistance [28]. Variable flammability characteristics can be caused by different chemical configurations of natural fibers. Higher lignin concentration usually improves surface morphology. The chemical composition of the natural fiber, as well as its structure and orientation, contribute to its flammability [29].

Previously, only a small series of researches on the flammability of green composite fibers were conducted, hence there is relatively little literature in this area. Researchers continue to face significant problems in finding new approaches and methods to improve the fire-resistant characteristics of natural fiber-based composites. When choosing natural fiber as a reinforcement for composites, other characteristics such as the fiber's mechanical properties must be addressed.

1.5 Natural Fibre Durability

Humidity, hydrothermal conditions, and general weathering can all have an impact on the mechanical characteristics and physical behavior of natural fiber composites. These are some of the major challenges that green fiber composites encounter [30]. The toughness of the fiber is clearly connected to the surface and interior adhesion of the composite. The current literature on natural fiber strength and what is required to select the best fibers for a certain application seems to be quite limited. The mechanical characteristics of natural fiber may alter as a result of changing environmental factors such as humidity and time exposure [31]. Previous study indicates that exposing the jute-reinforced composite to a 95% RH state reduces its tensile characteristics by 32% to 52%. Under the same conditions, flexural strength decreased by 11% to 57%. SEM investigation showed a few black dots and white patches on the external surface of the jute fiber composite; these spots and patches are fungal hyphae [32]. After 3 days of exposure to moisture, fungal hyphae were seen on the surface of flax and jute fibers. The application of a suitable coating and many fiber changes seems to slow the effect of weathering and fungal development [33].

In the aviation industry, aircraft interiors and many structural are made of aluminum or synthetic composite materials. Aluminum was traditionally used for these structures; however composite materials are currently being utilized due to their advanced and appropriate properties. Carbon fiber composites are being employed to reduce an aircraft's mass [34]. To counteract these negative effects, researchers have been looking for a suitable alternative. Natural composite materials are environmentally friendly, economical, and disposable. Bast fibers like kenaf, hemp, jute, flax, sisal, and others are widely utilized as reinforcements in composites [35].

1.6 Natural Fibre Jute

Natural jute fiber is classified into three main sections, based on their morphological construction, as follows:

(a) Bast fibers.
(b) Leaf fibers.
(c) Seed fibers.

The stems of several dicotyledonous plants contain bast stem fibers. It is a fast-growing plant that may be harvested within a year. The plant's bast fibers are subsequently harvested in cotton form. In a warm and humid weather, a jute plant may grow to a height of up to 3 m in 4–6 months, as shown in Fig. 3 [36] (Fig. 2).

Table 3 lists the chemical modifications required for jute fiber to overcome one of its major issues, specifically moisture absorption within the fibers, and make it more hydrophobic in nature. This is critical for limiting the degradation of the mechanical and physical properties of the composites, as shown in Table 3. Making it moisture resistant decreases the effect of delamination. The decrease in chemical properties

Fig. 2 Natural unlaminated Jute fabric [1]

Fig. 3 Kenaf woven mat [1]

could be attributed to inadequate interfacial bonding between the resin and the fiber [36]. It is critical to modify the external surface of the fiber to make it less hydrophilic. To achieve the optimal bonding between the matrix and the reinforcement, well-matched bonding compounds and matrices must be carefully chosen. Implantation copolymerization is a very successful and useful technique of natural fiber chemical modification [37]. The resulting co-polymer exhibits the properties of both fibroid cellulose and the attaching polymer. Acetylation or esterification of the natural fiber cellulose-OH by treating it with aceticanhydride is another well-known chemical modification. This reaction decreases the hydrophilic nature of the fiber and the swelling of the composites; Table 3 presents some mechanical properties of jute [38].

Table 3 Mechanical and Physical Properties of Kenaf and Jute Fiber [1]

Properties	Jute fiber
Density (g/cm^3)	1.3
Tensile strength (MPa)	393–773
Young's modulus	26.5
Elongation at break (%)	1.5–1.8
Cellulose content (%)	58–63
Hemicellulose content (%)	20
Lignin content (%)	12–14
Diameter	20–200

1.7 Natural Fiber Kenaf

Kenaf fiber is a very attractive natural fiber for many industries and researchers, and it is employed in a variety of polymers green and environmental benign manufacturing and government policies in many countries [39]. The strength of the polymeric composite is increased by reinforcing kenaf fiber in epoxy resin. investigated the moulding conditions, fiber matrix bonding, and mechanical characteristics for the production of kenaf-based composites They also observed that the shape, size, and strength of kenaf fiber are largely determined by the harvesting environment. Natural fibers are extremely beneficial to the environment because they absorb 1.5 times their weight in carbon dioxide [40]. These natural fiber plants have the highest carbon dioxide absorption rate. Kenaf fiber plants may reach heights of 3 to 4.5 m in 4 to 5 months and produce an annual fiber production of 6 to 10 tons per dry acre. Kenaf fiber offers superior mechanical properties when compared to other natural fibers such as flax and bamboo, which have previously been utilized in composites. Kenaf fiber takes relatively minimal water to grow. It has a 6-month growing cycle and an average growth rate of 1700 kg/ha.

Manufacturing in the aviation and automotive industries is one of the applications of kenaf fiber. Scientists are keen to use natural fibers like jute, kenaf, hemp, and flax into composite fabrications. This can result in the production of materials with extremely advanced specifications, as well as qualities such as recyclability and low density. They have several advantages, including being lightweight and non-toxic. These fibers are comparatively cheap in cost and easy to make, and they are sufficiently strong to be used in the construction of a wide range of products, as demonstrated by the Kenaf woven mat in Fig. 3. Kenaf fiber is an interesting and important material in engineering. The fiber has become more attractive as a natural source for the development of biodegradable parts for the aviation and car industries. They are also used in the manufacture of sports items. Natural fibers are also extremely advantageous in the packaging business, where they are used to make hygienic and non-toxic products. Especially compared to other natural fibers, kenaf fiber is a very competitive fiber since it has a short farm sequence, is less vulnerable to environmental conditions, and requires comparatively low quantities of insecticides and herbicides. Table 3 shows that the mechanical strength and thermal characteristics of kenaf fiber-based composites are superior to those of other natural fiber composites. Kenaf is therefore suitable for advanced performance engineering applications.

1.8 Natural Composites and Their Applications in Aerospace and Automobiles

Kenaf fibers, like some of the other natural fibers, have recently been used in the production of highly high-tech products, and it is now time to start reducing their use.

In the aerospace and automobile industries, a new green material is being created that is biodegradable and suitable for the current global requirement for environmental protection. Because of their low cost and weight, natural fiber composites are gaining popularity over glass fibers and carbon fibers in certain applications. Natural composites are being used in the aircraft industry by European countries to prepare interior components such as side walls. Natural fiber composites are also being used in the same capacity in automobiles such as Audi and Volvo.

In the aerospace industries, green composites are very impressive and appropriate materials, e.g., adhesives used for attaching Aeroplan components. Natural fiber sandwich composites are typically prepared as thermoset composites, like epoxy resin in the aerospace field.

1.9 Conclusion and Future Developments

1. The use of kenaf and jute in composite materials is extremely beneficial in engineering solutions for manufacturing new components in the aerospace and automobile industries. Green composite renewable resource alternatives include kenaf and jute. Natural fibers also grow extremely rapidly.
2. Meanwhile, there is a consistent supply of authentically sourced kenaf and jute fibers at very low manufacturing costs as compared to synthetic fibers. Furthermore, the tensile properties of jute and kenaf are used, and they may be used to remove or replace synthetic glass fibers in reinforced polymer matrices.
3. However, various characteristics and properties of jute and kenaf have an impact on the performance of natural jute and kenaf fiber composites. Appropriate natural fiber extraction procedures must be developed in order to optimize separation of the lining from the natural fiber surface of jute and kenaf fibers. This may assist in the enhancement of interfacial adhesion between natural fibers and matrices. The bonding agent's location and the natural fiber pre-treatment, both contributes to this bonding.
4. A high amount of fiber loading in composites may enhance mechanical properties. The potential of natural jute and kenaf fiber composites is enhanced by hybridization with synthetic and carbon fibers that have adequate tensile properties. Natural kenaf and jute composites have superior performance that can compete with synthetic materials.
5. The use of natural jute and kenaf fibers rather than synthetic fibers in the development of biodegradable products may assist reduce key experimental concerns, the most prominent of which are energy consumption and solid waste material treatment. However, further research is required in the future to give improved understanding about kenaf and jute natural fibers, as well as to enhance their potential for usage in modern technologies.
6. Kenaf and jute natural fiber composites have been employed in numerous manufacturing industries across the world, including interior design applications such as house paneling and furniture. Although natural fibers are attractive study areas

currently, the utilization of kenaf and jute composites in aviation and automotive applications is still limited. New research into kenaf and jute fibers has the ability to increase their potential not just in aviation and automotive applications, but also in a variety of other industries.

Acknowledgements The authors would like to thank Universiti Putra Malaysia (UPM) for supporting this research with the research grant GP-IPB-9490602. Special thanks to the Aerospace Manufacturing Research Centre (AMRC), Universiti Putra Malaysia, and also to the Laboratory of Biocomposite Technology, Institute of Tropical Forestry and Forest Products (INTROP), Universiti Putra Malaysia.

Conflict of Interests The authors declare that there is no conflict of interests regarding the publication of this paper.

References

1. Khan T, Hameed Sultan MTB, Ariffin AH (2018) The challenges of natural fiber in manufacturing, material selection, and technology application: A review, J Reinf Plast Compos, 37(11), pp 770–779. https://doi.org/10.1177/0731684418756762
2. George M, Chae M, Bressler DC (2016) Composite materials with bast fibres: Structural, technical, and environmental properties. Prog Mater Sci 83:1–23
3. Kopparthy SDS, Netravali AN (2021) Green composites for structural applications. Compos Part C: Open Access 6:100169
4. Keya KN, Kona NA, Koly FA, Maraz KM, Islam MN, Khan RA (2019) Natural fiber reinforced polymer composites: history, types, advantages and applications. Mater Eng Res 1(2):69–85
5. Raju A, Shanmugaraja M (2021) Recent researches in fiber reinforced composite materials: A review. Mater Today: Proc 46:9291–9296
6. Asim M, Jawaid M, Fouad H, Alothman O (2021) Effect of surface modified date palm fibre loading on mechanical, thermal properties of date palm reinforced phenolic composites. Compos Struct 267:113913
7. Shah SS, Shaikh MN, Khan MY, Alfasane MA, Rahman MM, Aziz MA (2021) Present status and future prospects of jute in nanotechnology: A review. Chem Rec 21(7):1631–1665
8. Mohamed S, Zainudin E, Sapuan S, Azaman M, Arifin A, (2018) Introduction to natural fiber reinforced vinyl ester and vinyl polymer composites, In Natural Fibre Reinforced Vinyl Ester And Vinyl Polymer Composites: Elsevier, pp 1–25
9. Jagadeesh P, Puttegowda M, Mavinkere Rangappa S, Siengchin S (2021) A review on extraction, chemical treatment, characterization of natural fibers and its composites for potential applications. Polymer Composites, 42(12) pp 6239–6264
10. Awais H, Nawab Y, Amjad A, Anjang A, Akil HM, Abidin MSZ (2021) Environmental benign natural fibre reinforced thermoplastic composites: A review. Composites Part C: Open Access 4:100082
11. Ashraf MA, Zwawi M, Taqi Mehran M, Kanthasamy R, Bahadar A (2019) Jute based bio and hybrid composites and their applications. Fibers, 7(9), p 77
12. Melelli A et al (2022) Anticipating global warming effects: A comprehensive study of drought impact of both flax plants and fibres. Ind Crops Prod 184:115011
13. Bui H, Sebaibi N, Boutouil M, Levacher D (2020) Determination and review of physical and mechanical properties of raw and treated coconut fibers for their recycling in construction materials. Fibers 8(6):37

14. Bhadra D, Dhar NR (2022) Selection of the natural fiber for sustainable applications in aerospace cabin interior using fuzzy MCDM model. Materialia 21:101270
15. Noryani M, Sapuan SM, Mastura MT, Zuhri MYM, Zainudin ES (2019) Material selection of natural fibre using a stepwise regression model with error analysis. J Market Res 8(3):2865–2879
16. Ahmed MM, Dhakal H, Zhang Z, Barouni A, Zahari R (2021) Enhancement of impact toughness and damage behaviour of natural fibre reinforced composites and their hybrids through novel improvement techniques: A critical review. Compos Struct 259:113496
17. Jayan JS, Appukuttan S, Wilson R, Joseph K, George G, Oksman K (2021) An introduction to fiber reinforced composite materials. In Fiber Reinforced Composites: Elsevier, pp 1–24
18. Ilyas RA et al (2022) Natural-Fiber-Reinforced Chitosan, chitosan blends and their nanocomposites for various advanced applications. Polymers 14(5):874
19. Megahed M, Ali-Eldin SS, Abd El Moezz SM, Abdalla W (2020) Synthesis of developed rice straw sheets and glass fiber-reinforced polyester composites. J Compos Mater 54(23):3381–3394
20. F. Bogard et al. (2021) A comparative review of Nettle and Ramie fiber and their use in biocomposites, particularly with a PLA matrix," J Nat Fibers, pp 1–25
21. Mayandi K et al (2020) An overview of endurance and ageing performance under various environmental conditions of hybrid polymer composites. J Market Res 9(6):15962–15988
22. Kannan G, Thangaraju R (2021) Recent progress on natural lignocellulosic fiber reinforced polymer composites: A review. J Nat Fibers, pp 1–32.
23. Sari NH et al (2020) The effect of water immersion and fibre content on properties of corn husk fibres reinforced thermoset polyester composite. Polym Testing 91:106751
24. Jamshaid H et al (2022) Natural cellulosic fiber reinforced concrete: influence of fiber type and loading percentage on mechanical and water absorption performance. Materials 15(3):874
25. Chauhan V, Kärki T, Varis J (2019) Review of natural fiber-reinforced engineering plastic composites, their applications in the transportation sector and processing techniques. J Thermoplast Compos Mater, p 0892705719889095
26. Thomason JL, Rudeiros-Fernandez JL (2021) Thermal degradation behaviour of natural fibres at thermoplastic composite processing temperatures. Polym Degrad Stab 188:109594
27. Wang F, Ouyang D, Zhou Z, Page SJ, Liu D, Zhao X (2021) Lignocellulosic biomass as sustainable feedstock and materials for power generation and energy storage. J Energy Chem 57:247–280
28. Yiga VA, Lubwama M, Pagel S, Benz J, Olupot PW, Bonten C (2021) Flame retardancy and thermal stability of agricultural residue fiber-reinforced polylactic acid: A Review. Polym Compos 42(1):15–44
29. Ma W, Elkin R (2021) Sandwich structural core materials and properties. In Sandwich Structural Composites: CRC Press, pp 1–72
30. Jawaid M, Chee SS, Asim M, Saba N, Kalia S (2022) Sustainable kenaf/bamboo fibers/clay hybrid nanocomposites: Properties, environmental aspects and applications. J Clean Prod 330:129938
31. Zaghloul MMY, Steel K, Veidt M, Heitzmann MT (2022) Wear behaviour of polymeric materials reinforced with man-made fibres: A comprehensive review about fibre volume fraction influence on wear performance. J Reinf Plast Compos 41(5–6):215–241
32. Karim MRA, Tahir D, Haq EU, Hussain A, Malik MS (2021) Natural fibres as promising environmental-friendly reinforcements for polymer composites. Polym Polym Compos 29(4):277–300
33. Jones M, Mautner A, Luenco S, Bismarck A, John S (2020) Engineered mycelium composite construction materials from fungal biorefineries: A critical review. Mater Des 187:108397
34. Mansor M, Nurfaizey A, Tamaldin N, Nordin M (2019) Natural fiber polymer composites: utilization in aerospace engineering. In Biomass, Biopolymer-Based Materials, and Bioenergy: Elsevier, 2019, pp 203–224
35. Sharma AK, Bhandari R, Sharma C, Dhakad SK, Pinca-Bretotean C (2022) Polymer matrix composites: A state of art review. In: Materials Today: Proceedings

36. Pathak S, Saha GC, Abdul Hadi MB, Jain NK (2021) Engineered nanomaterials for aviation industry in COVID-19 context: A time-sensitive review. Coatings, 11(4), p 382
37. Fulmali AO, Sen B, Nayak BA, Prusty RK (2021) Effect of repeated hydrothermal cycling on the durability of glass fiber/epoxy composites with and without carbon nanotube reinforcement. Polym Compos 42(11):6160–6172
38. Kumar R, Sharma RK, Singh AP (2017) Cellulose based grafted biosorbents-Journey from lignocellulose biomass to toxic metal ions sorption applications-A review. J Mol Liq 232:62–93
39. Kumar PSS, Allamraju KV (2019) A review of natural fiber composites [Jute, Sisal, Kenaf]. Materials Today: Proceedings 18:2556–2562
40. Klemm D, Philpp B, Heinze T, Heinze U, Wagenknecht W (1998) Comprehensive cellulose chemistry. Volume 1: Fundamentals and analytical methods. Wiley-VCH Verlag GmbH

Advanced Natural/Synthetic Polymer Hybrid Composites

Siti Noorbaini Sarmin

Abstract This chapter provides an overview of the advancements in natural/synthetic hybrid composites, which have become the researchers' eye candy. In today's market, components require exceptional mechanical, thermal, and tribological qualities, which can be accomplished by combining natural and synthetic fibers. Due to their appealing material properties, hybridization fiber-reinforced composites have produced a variety of captivating properties at a low cost of the finished product. They've worked in a variety of industries, including aerospace, automobiles, civil infrastructure, and marine. Therefore, the current discussion is a summary of the advancement of hybrid fiber-based polymer composites, with a focus on their progression.

1 Introduction

Natural fiber has been used as a reinforcement material in polymer composites for the past decade. According to Gholampour and Ozbakkaloglu [1], natural fibers are environmentally benign, have a lower carbon footprint, are biodegradable, have a low cost of manufacture, and are less expensive than synthetic fibers. Reusing waste natural fiber is an environmentally friendly solution. Although these characteristics make natural fibers a better substitute for synthetic fibers in practical applications, their affinity for water and inappropriate matrix interface bonding are key drawbacks to their usage in many applications [2–4]. The usage of synthetic fiber is necessary for the reasons stated above [5–7]. Synthetic fibers avoid these issues; however, they are non-biodegradable and more expensive than natural fibers. As a result, much work

S. N. Sarmin (✉)
Biocomposite Technology and Design Unit, Institute of Tropical Forestry & Forest Products, Universiti Putra Malaysia, UPM, 43400 Serdang, Selangor, Malaysia
e-mail: noorbaini.sarmin@upm.edu.my; baini@uitm.edu.my

Department of Wood Technology, Faculty of Applied Sciences, Universiti Teknologi MARA, 26400 Bandar Pusat Jengka, Pahang, Malaysia

T. Khan and M. Jawaid (eds.), *Green Hybrid Composite in Engineering and Non-Engineering Applications*, Composites Science and Technology,
https://doi.org/10.1007/978-981-99-1583-5_2

has been devoted on hybrid composites research and development. These hybrid composites combine natural and synthetic fibers to achieve the best results possible.

Polymer hybrid composites are materials formed by mixing two or different category of fillers in a common polymer matrix [8]. Hybrid materials are part of the most rapidly evolving types of novel materials, resulting in a slew of technical breakthroughs. Hybrid composites with a wide range of material qualities are produced by combining two or more fillers (natural/natural, natural/synthetic, and micro/nano) in a single matrix [6]. Hybridization of two distinct fibers has shown to be a successful way for creating materials that are suitable for a variety of applications.

The addition of natural/synthetic fibers to a polymer has been shown to significantly affect the physical and mechanical properties of composites, and in some situations, the hybrid's strength exceeds the rule of mixes (synergistic strengthening). The unique qualities that develop as a result of the arrangement and interactions between the various constituents open a world of possibilities for improved material technologies [7, 8]. One of the major advantages of composite materials is the concept of hybridization, which allows the design engineer to tailor the material properties according to the requirements.

In this respect, this chapter briefly reviews the progress on natural/synthetic polymer hybrid composites, including proposed progression hybridization and their particularities, as well as providing suitable sources for additional research. The chapter is structured into three sections: (i) designation of natural/synthetic composites, (ii) hybrid composites strategy, and (iii) evolution of natural/synthetic fiber reinforced hybrid composites.

2 Designation and Approach of Natural/Synthetic Composites

Fiber-reinforced polymer (FRP) composites have found several uses in sophisticated sectors during the last few decades. FRP composite structures are often designed to withstand a variety of difficult operational circumstances, and studies of loading rate and strain-dependency are critical throughout the design stage [9, 10]. For the next generation of FRP composite structures, hybridization of natural and synthetic fibers appears to be a viable option.

As the use of FRP grows, so does the need for more stringent quality control criteria for the materials, processes, and design methods that surround them. The occurrence of faults is a typical problem that affects both the production and design elements of industrial components [6]. As a result, determining fault criticality and structural reliability of FRP has become a major study topic. In actuality, FRP parts can be prone to a variety of faults that arise throughout the production process. Variability can come from a variety of places and can occur at any point throughout the fabrication process [8].

According to Bari et al. [7] and Bhardwaj et al. [11] FRP composites are classified based on their filler and base materials. Matrix materials are the base materials that grasp the filler material in structures. Fillers serve as a dispersion medium in the composite; they are frequently spread and disseminated throughout the matrix, strengthening or reinforcing the structure [12]. When fillers are introduced into the matrix phase, a complex interphase structure is produced, in which the configuration and interaction of the fillers and the matrix dictate the composite's properties. Both the filler and matrix phases can work together to produce a composite with improved characteristics. The parts that follow go through the designation and approach of natural fiber-reinforced polymer composites (NFRPC), synthetic fiber-reinforced polymer composite (SFRPC) and hybrid composites made of synthetic and natural fibres (HFRPC).

3 Natural Fiber-Reinforced Polymer Composites (NFRPC)

Natural fibres have a number of preferences over traditional reinforcing synthetic fibres (such as glass and carbon), including low density, inexpensive, durability, adequate specific strength, renewability, biodegradability, convenience of separation, lower processing energy requirements, and global availability. Natural fibres are classified into three categories based on their origin (Fig. 1). They are categorized as follows:

Fully and partially green composites are the two types of NFRPC. In the partially and totally green composites, petrochemical and bio-based resins are employed as matrix, and natural fibres are used in both. Although all biopolymers are compostable in principle, not all biodegradable and compostable plastics are made of biopolymers; some are made wholly of non-biodegradable elements. A petrochemical-based matrix is a chemical made from petroleum, which is sourced from fossil fuels such as coal and natural gas. Thermoplastics and thermosets are the two main forms of petrochemical-based matrix utilized in green composites. Thermoplastic matrices include polyethylene, polystyrene, polypropylene, and polyvinyl chloride (PVC), while thermoset matrices include epoxy, polyester, vinylester, and phenolic (phenol formaldehyde) [11, 14]. The primary distinction was that thermoplastics may be remelted and convert by adopting heat and shear after moulding, whereas thermosets cannot. Thermoset matrices, on the other hand, are stiffer and chemically stable [15]. This is why recycling them is more challenging. Polyester, vinyl ester, phenolic, amino, derived ester, and epoxy resins are the most common thermoset matrices used to make natural fibre composites [15]. Resin transfer moulding (RTM), sheet moulding compound (SMC), pultrusion, vacuum assisted resin transfer moulding (VARTM), and hand layup are all prevalent methods for processing thermoset composites [16]. All of these manufacturing techniques don't necessitate high-pressure reprocessing.

Natural fibres are hydrophilic and polar, which makes them incompatible with most polymer matrices, which are hydrophobic and non-polar [9, 11, 17]. Due to a lack of interfacial adhesion between the fibres and the matrix, this phenomenon

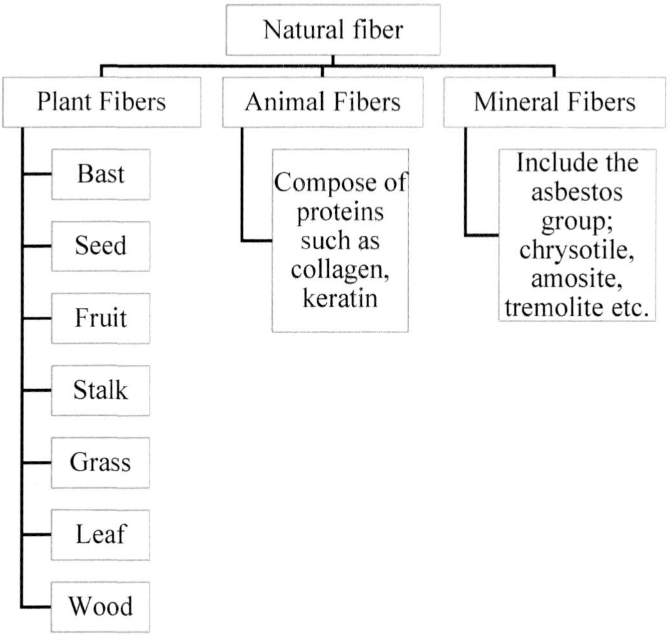

Fig. 1 Natural fibers classification [1, 13]

results in poor mechanical properties. Furthermore, even when inside a composite, the significant amount of hydroxyl groups available on the fibre surface increases water absorption [18–20]. Surface modification of the fibre, such as mercerization (treatment in sodium hydroxide solution to remove lignin and hemicellulose) [21, 22] followed by the addition of coupling agents [23–25], can fix these difficulties.

Recently, fibre treatment with a coupling agent in solution has been proposed. The addition of a coupling agent can also be combined with thermomechanical refinement [26, 27]. Copolymers having functional groups compatible with the fibres (hydroxyl groups) and the polymer matrix are commonly used as coupling agents [24, 28]. These chemical and physical interactions increase interfacial adhesion, resulting in enhanced mechanical characteristics and less water absorption [23, 25–27]. Coupling agents can be blended with the polymer matrix prior to fibre addition through extrusion, but they can also be introduced during composite compounding, which involves combining the matrix, fibre, and coupling agent all at the same time. Natural fibres can also be functionalized by treating them in solution with a coupling agent to improve compatibility with the polymer matrix [26, 29, 30]. Because natural fibres disintegrate at a lower temperature (150–275 °C) than most polymer matrices (350–460 °C), fibre mercerization and coupling agent addition have been proven to improve the thermal stability of the fibres and, as a result, the thermal stability of the final composites.

4 Synthetic Fiber-Reinforced Polymer Composite (SFRPC)

For the production of SFRPC, a variety of fibres are available on the market. Because of their strength, ease of manufacture, extended lifespan, and availability, synthetic fibres account for the majority of the fibres. Organic and inorganic fibres are the two main types of man-made fibres [31]. Carbon and glass fibres are the most well-known fibres on the planet. They do, however, come in a variety of forms; for instance, glass fibre can be established in E-Glass, S-Glass, and other varieties. Synthetic fibres can be made with a variety of qualities depending on their intended use (Fig. 2). They are categorized as follows [31–33].

Synthetic fibre performance has improved as an outcome of continued exploration and advance efforts in fibre materials and fibre processing. Structures made of composite materials including synthetic fibres (whether short staple fibres, whiskers, or fibres in extended filament form, such as roving or textiles) have a lot of design versatility [34, 35]. As a result, composites of low-cost class fibres incorporated in a plastic matrix have been widely employed in medium-to-high-volume applications by the transportation, building, and recreation industries for more than 40 years in applications such as auto body, boat hulls, and chemical tanks [32, 33].

Today's designer can acquire structural features that neither material displays on its own by spreading fibres or particles of one material in a matrix of alternative material. For instance, a metal alloy chosen for its high temperature resistance but with low creep resistance at usage temperature can be reinforced with fibrous inorganic oxide fibres to improve creep resistance while being stable at high temperatures [36]. The insertion of reinforcing fibres to offer equivalent mechanical capabilities at a much-decreased weight is frequently cited as a major benefit of composites over

Fig. 2 Synthetic fibers classification [1, 13]

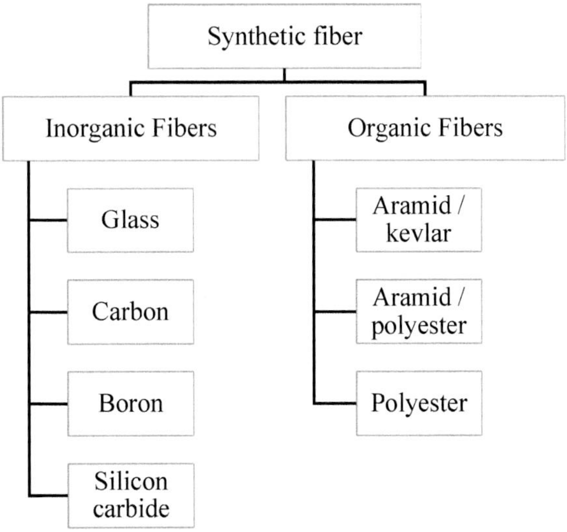

traditional structural materials. Additionally, important factor to consider is substituting easily existing materials for crucial elements that are in lessened supply or are only obtainable from foreign sources. Composite materials derived from readily available domestic materials like carbon, polymers, ceramics, or ordinary metals can often outperform these imported key materials.

In summary, the following are some of the special superiority of synthetic fiber-reinforced composites [33–37]:

i. Greater specific modulus and strength values
ii. Lightweight
iii. Governable ("tailorable") properties
    a. Toughness
    b. Electrical and thermal conductivities
    c. Thermal expansion
    d. Stiffness
iv. Barrier to corrosion
v. Better resistance to creep at high temperatures
vi. Substitution for critical or strategic materials
vii. Creation of multipurpose or "smart" structures

5 Hybrid Fiber-Reinforced Polymer Composite (HFRPC)

Natural and synthetic fibres can be mixed in the same matrix to create hybrid composites with a variety of qualities that are impossible to achieve with just one type of reinforcement. Since 2013, there has been an increasing interest in hybridizing components in composites in a common matrix. It was previously recognized that hybrid composites contained two or more fibres in a single matrix [8–10, 38]. Common particle fibres, woven and non-woven fibres, and nano-scale filler are all examples of fibre. Hybridization allows for a more balanced composite construction with more appealing qualities and cost, which is difficult to achieve with a single type of reinforcement [39]. Hybridization opens us new possibilities for enhancing the toughness and impact resistance of composite materials, especially in advanced applications. Hybrid composites also offer more design flexibility than monohybrid composites, resulting in a synergetic effect that no single material can achieve alone. The synergetic effect can be produced by a variety of factors, including fibre selection, fibre combination, and fibre interaction in the hybrid system [38–40].

Hybridization can be thought of as a cost-effective way to mitigate the drawbacks of one type of reinforcement. When natural-fibre reinforced polymer composites (NFRPC) are made, the bulk of the hybridization is done with glass fibre, which is an artificial fibre embedded in a matrix that could be partially bio-based or synthetic. Natural fibre hybridization could jeopardize NFRPC's major benefits, including as biodegradability, simplicity of processing, and reduced health hazards [41]. Hybridization, on the other hand, lowers the cost of raw materials, the overall weight of composites, and the environmental impact of synthetic fibre composites [42]. Hybrid composites can also be made by blending different natural fibres, which can demonstrate the advantages of employing hybrid composites [5–7, 43]. According to Arman, Chen and Ahmad [18], hybrid composites can also be made by hybridizing the matrix, which could increase the material's efficiency; however, most current research focuses solely on reinforcement hybridization. Autohybridization, on the other hand, is defined as the use of filler after combining more than one polymer matrix and hybridizing that filler type with the same filler of a different particle size.

Polymer science and engineering, energy sources, mechanics, materials technology, metallurgy, electrochemistry, and other related topics of engineering are all covered by hybridization approaches [3, 4, 39–41]. Regardless of the field, the fundamental goal is to combine more than three reinforcement materials with superior qualities for an intended application when compared to the separate components. Atmakuri et al. [41] mentioned that hybridization offers a one-of-a-kind opportunity to broaden the usability of a composite material, especially in more advanced applications. The materials employed, the preparation procedure, and the mixing of the fillers with the matrix are the three key aspects that have a considerable impact on the final hybrid composite's features.

6 Evolution in Natural/Synthetic Fiber Reinforced Hybrid Composites

Different elements, such as fibre and matrix type, as well as fibre surface treatment, influence the ultimate properties of hybrid composites [44]. In general, mixing natural fibres with inorganic reinforcements improves heat stability and impact strength while also increasing flexural and tensile moduli [45]. Furthermore, when two natural fibres from distinct sources are combined, or natural fibres with inorganic reinforcements are used in hybrid composites based on thermoplastic matrix, water uptake is reduced [46]. Many ways have been developed to improve the properties of composites, including polymer and fibre changes.

A new universe of interesting characteristics in polymer composites can be unlocked for multiple sophisticated applications in several industries by increasing global contributions of research results on the manufacturing of natural/synthetic polymer hybrid composites. The sections that follow discuss the most recent advancements in hybrid composites of common synthetic fibres (aramid, basalt, carbon, and glass) and natural fibres. Table 1 outline some of the most vital properties of hybrid fiber-reinforced polymer composite with result of fiber and matrix type and manufacturing process.

Table 1 Recent research finding of hybrid natural/synthetic polymer composites

Matrix	Fibers	Manufacturing process	Treatment	Finding description	References
Bio-Epoxy	Kenaf/Glass	Compression Molding	–	Hybridized samples demonstrated improved progressive crushing capabilities by combining local buckling, delaminate, and brittle fracturing as progressive crushing modes	[43]
Epoxy	*Grewia optiva*/Glass	Hand lay-up + compression molding	Addition of dolomite	Increased dolomite content improves density, void content, impact energy, and hardness, while flexural and tensile strength decreases	[47]
Epoxy	Kenaf/Kevlar	Hand lay-up + compression molding	–	The hybridization of Kenaf with Kevlar fiber improved the mechanical characteristics of epoxy composites	[48]
Epoxy	*Prosopis juliflora* bark fiber/E-Glass/Carbon	Hand lay-up + compression molding	–	Hybridization with the addition of E-glass and carbon textiles improved the characteristics of reinforced epoxy composites	[10]

(continued)

Table 1 (continued)

Matrix	Fibers	Manufacturing process	Treatment	Finding description	References
Epoxy	Jute/Glass Fabric	Hand lay-up + compression molding	–	When compared to pure synthetic adherend joints, the hybrid 3-layer adherend joints had a lower weight, allowing for a reduction in mass and an increase in sustainability without sacrificing joint efficiency	[50]
Epoxy	Flax/Carbon	Vacuum Assisted Resin Transfer Molding (VARTM)	Amine cross linker	Intra-ply hybridization of natural and synthetic fibers results in composites that are stable under both static and dynamic loading	[51]
Epoxy	Flax/Kenaf/Carbon/Glass	Hand lay-up + compression molding	Silane as coagulating agent	The results show that hybridizing flax fibre with carbon fibre produces the best hybrid effect and improves its flexural strength, modulus, and Izod impact strength	[52]

(continued)

Table 1 (continued)

Matrix	Fibers	Manufacturing process	Treatment	Finding description	References
Epoxy	Hemp/Sisal/Silica	Compression Molding	–	The composites with 2 wt% silica nanoparticles have the highest tensile strength, impact strength, and hardness, while the composite with 3 wt% silica nanoparticles has the highest flexural strength	[6]
Epoxy/bio-epoxy	Flax/Basalt/Carbon	Hand lay-up + compression molding	–	Based on the findings, the composites created in this study can be used in a variety of engineering medium load structural applications in the future	[53]
Epoxy/Bio-Epoxy	Coir/Basalt/Titanum Carbide (TiC)	Compression molding	–	A significant improvement in mechanical and thermal properties, with the effect of the highest load transfer between the fillers and matrix materials. Thermal stability analysis revealed that the newly developed epoxy hybrid composites are more resistant to temperature changes than the pure polymer sample	[55]

(continued)

Table 1 (continued)

Matrix	Fibers	Manufacturing process	Treatment	Finding description	References
Polyester	Jute/Glass	Hand lay-up + compression molding	–	Glass fiber hybridization significantly affects the properties	[5]
Polyester	Rice straw/glass fiber mats	Hand lay-up + compression molding	–	When compared to other fabricated composites, hybrid composites with high areal density on the outer surfaces have a significant increase in flexural-specific strength and hardness	[49]
Polyester	Jute/Carbon	Compression Molding	–	The study's findings revealed that the combination of jute and carbon fibre improved mechanical properties, reduced moisture absorption rate, and had good interfacial bonding with polyester, as confirmed by scanning electron microscopy	[54]
Unsaturated polyester	Flax/Sisal/Carbon/Glass	Hand lay-up + compression molding	–	After hydro ageing, hybrid composites with natural fiber at the outer layers show a decrease in tensile strength, whereas hybrid composites with synthetic fiber at the outer layers show only a modest decrease	[9]

7 Conclusion

The emergence of hybrid fibre reinforced composites as an efficient substitute for pure composites including artificial fibre is the subject of much research. Hybrid fibre reinforced composite materials displayed higher modulus values and strength at elevated temperatures when compared to pure natural fibre composites. It may also exhibit properties similar to synthetic fibre reinforced composites, which are dependent on fibre orientation and loading methods. This chapter demonstrates how suitable fibre hybridization can be used to modify the sturdiness of such composite materials, allowing for optimal performance at a low cost while also increasing the material's Eco friendliness. There's a good probability that a large amount of natural fibre reinforcements might be incorporated into traditional man-made fibre composites, which would be a tremendous step forward from an environmental and economic standpoint.

References

1. Gholampour A, Ozbakkaloglu T (2020) A review of natural fiber composites: properties, modification and processing techniques, characterization, applications. J Mater Sci 55(3):829–892
2. Praveen Kumar A, Nalla Mohamed M (2018) A comparative analysis on tensile strength of dry and moisture absorbed hybrid kenaf/glass polymer composites. J Ind Text 47(8):2050–2073
3. Sood M, Dwivedi G (2018) Effect of fiber treatment on flexural properties of natural fiber reinforced composites: a review. Egypt J Petrol 27(4):775–783
4. Li M, Pu Y, Thomas VM, Yoo CG, Ozcan S, Deng Y, Nelson K, Ragauskas AJ (2020) Recent advancements of plant-based natural fiber–reinforced composites and their applications. Compos Part B: Eng 108254
5. Das SC, Paul D, Grammatikos SA, Siddiquee MA, Papatzani S, Koralli P, Islam JM, Khan MA, Shauddin SM, Khan RA, Vidakis N (2021) Effect of stacking sequence on the performance of hybrid natural/synthetic fiber reinforced polymer composite laminates. Compos Struct 276:114525
6. Singh T, Gangil B, Singh B, Verma SK, Biswas D, Fekete G (2019) Natural-synthetic fiber reinforced homogeneous and functionally graded vinylester composites: effect of bagasse-Kevlar hybridization on wear behavior. J Market Res 8(6):5961–5971
7. Bari E, Morrell JJ, Sistani A (2019) Durability of natural/synthetic/biomass fiber–based polymeric composites: laboratory and field tests. Durability and life prediction in biocomposites, fibre-reinforced composites and hybrid composites, pp 15–26
8. Nurazzi NM, Asyraf MR, Fatimah Athiyah S, Shazleen SS, Rafiqah S, Harussani MM, Kamarudin SH, Razman MR, Rahmah M, Zainudin ES, Ilyas RA (2021) A review on mechanical performance of hybrid natural fiber polymer composites for structural applications. Polymers 13(13):2170
9. El-Wazery MS, El-Kelity AM, Elsad RA (2020) Effect of water absorption on the tensile characteristics of natural/synthetic fabrics reinforced hybrid composites. Int J Eng 33(11):2339–2446
10. Madhu P, Mavinkere Rangappa S, Khan A, Al Otaibi A, Al-Zahrani SA, Pradeep S, Gupta MK, Boonyasopon P, Siengchin S (2020) Experimental investigation on the mechanical and morphological behavior of *Prosopis juliflora* bark fibers/E-glass/carbon fabrics reinforced hybrid polymeric composites for structural applications. Polym Compos 41(12):4983–4993

11. Bhardwaj D, Gupta A, Chaudhary V, Gupta S (2021) Hybridization of natural fibers to develop the polymeric composite materials: a review. Adv Eng Mater 355–363
12. Samal S (2020) Effect of shape and size of filler particle on the aggregation and sedimentation behavior of the polymer composite. Powder Technol 366:43–51
13. Vigneshwaran S, Sundarakannan R, John KM, Johnson RD, Prasath KA, Ajith S, Arumugaprabu V, Uthayakumar M (2020) Recent advancement in the natural fiber polymer composites: a comprehensive review. J Cleaner Prod 124109
14. Obande W, Brádaigh CM, Ray D (2021) Continuous fibre-reinforced thermoplastic acrylic-matrix composites prepared by liquid resin infusion–a review. Compos Part B: Eng 108771
15. Mallick PK (2021) Thermoplastics and thermoplastic–matrix composites for lightweight automotive structures. In: Materials, design and manufacturing for lightweight vehicles. Woodhead Publishing, pp 187–228
16. Bogiatzidis C, Zoumpoulakis L (2021) Thermoset polymer matrix composites of epoxy, unsaturated polyester, and Novolac resin embedding construction and demolition wastes powder: a comparative study. Polymers 5:737
17. Zaini ES, Azaman MD, Jamali MS, Ismail KA (2020) Synthesis and characterization of natural fiber reinforced polymer composites as core for honeycomb core structure: a review. J Sandwich Struct Mater 3:525–550
18. Arman NS, Chen RS, Ahmad S (2021) Review of state-of-the-art studies on the water absorption capacity of agricultural fiber-reinforced polymer composites for sustainable construction. Constr Build Mater 302:124174
19. Sekar S, Suresh Kumar S, Vigneshwaran S, Velmurugan G (2020) Evaluation of mechanical and water absorption behavior of natural fiber-reinforced hybrid biocomposites. J Nat Fibers 1–11
20. Reddy BM, Reddy RM, Reddy BC, Reddy PV, Rao HR, Reddy YM (2020) The effect of granite powder on mechanical, structural and water absorption characteristics of alkali treated cordia dichotoma fiber reinforced polyester composite. Polym Testing 91:106782
21. Alsubari S, Zuhri MY, Sapuan SM, Ishak MR, Ilyas RA, Asyraf MR (2021) Potential of natural fiber reinforced polymer composites in sandwich structures: a review on its mechanical properties. Polymers 13(3):423
22. Hassan MZ, Roslan SA, Sapuan SM, Rasid ZA, Mohd Nor AF, Md Daud MY, Dolah R, Mohamed Yusoff MZ (2020) Mercerization optimization of bamboo (bambusa vulgaris) fiber-reinforced epoxy composite structures using a box–behnken design. Polymers 12(6):1367
23. Anbupalani MS, Venkatachalam CD, Rathanasamy R (2020) Influence of coupling agent on altering the reinforcing efficiency of natural fibre-incorporated polymers–a review. J Reinf Plast Compos 39(13–14):520–544
24. Avci A, Eker AA, Bodur MS (2020) Effect of coupling agent and alkali treatment on mechanical, thermal and morphological properties of flax-fiber-reinforced PLA composites. Green Mater 11:1–4
25. Gao Y, Chen Y, Gao J, Gao J, Yang L, Zhang J, Nan R, Cai M, Zhang J (2021) Properties of Arenga Engleri Becc palm fiber particles with silane coupling agent KH570 treatments for application in polymer/cement composites. J Nat Fibers 14:1–5
26. Liew FK, Hamdan S, Rahman MR, Rusop M, Khan A 92020) Thermo-mechanical properties of jute/bamboo/polyethylene hybrid composites: the combined effects of silane coupling agent and copolymer. Polym Compos 41(11):4830–4841
27. Orue A, Anakabe J, Zaldua-Huici AM, Eceiza A, Arbelaiz A (2021) The importance of coupling agent on tensile and thermomechanical performance of annealed composites based on poly (lactic acid)/poly (methyl methacrylate) matrix and sisal fiber bundles. J Nat Fibers 23:1–1
28. Kenned JJ, Sankaranarayanasamy K, Kumar CS (2021) Chemical, biological, and nanoclay treatments for natural plant fiber-reinforced polymer composites: a review. Polym Polym Compos 29(7):1011–1038
29. Sahai RS, Pardeshi RA (2021) Comparative study of effect of different coupling agent on mechanical properties and water absorption on wheat straw-reinforced polystyrene composites. J Thermoplast Compos Mater 34(4):433–450

30. Tanasă F, Zănoagă M, Teacă CA, Nechifor M, Shahzad A (2020) Modified hemp fibers intended for fiber-reinforced polymer composites used in structural applications—a review. I. Methods of modification. Polym Compos 41(1):5–31
31. Liu L, Yan F, Liu H, Zhang M, Jin L, Xiao L, Ao Y (2021) Construction of hierarchical "organic-inorganic" structure by self-assembly to enhance the interfacial properties of epoxy composites. Compos Sci Technol 210:108802
32. Liu Y, Liu P, Jiang Q, Jiang F, Liu J, Liu G, Liu C, Du Y, Xu J (2021) Organic/inorganic hybrid for flexible thermoelectric fibers. Chem Eng J 405:126510
33. Sandomierski M, Voelkel A (2021) Diazonium modification of inorganic and organic fillers for the design of robust composites: a review. J Inorg Organomet Polym Mater 1:1–21
34. Jiang Q, Yang J, Hing P, Ye H (2020) Recent advances, design guidelines, and prospects of flexible organic/inorganic thermoelectric composites. Mater Adv 1(5):1038–1054
35. Sun G, Liu Q, Si K, Zhang J, Xiong R, Liu K, Chen Y (2020) Flexible composite materials—between inorganic fibers and organic polymers. Fibers Polym 21(3):628–635
36. Li D, Gao X, Huang X, Liu P, Xiong W, Wu S, Hao F, Luo HA (2020) Preparation of organic-inorganic chitosan@ silver/sepiolite composites with high synergistic antibacterial activity and stability. Carbohyd Polym 249:116858
37. Sun H, Yang Y, Ge X (2020) Effect of various fibers on the ablation resistance of poly (diaryloxyphosphazene) elastomer. J Appl Polym Sci 137(14):48534
38. Abidin NM, Sultan MT, Basri EI, Basri AA, Shah AU (2021) Validation of experimental hybrid natural/synthetic composite laminate specimen using finite element analysis for UAV wing application. Impact Stud Compos Mater 143–160
39. Savran M, Aydin L (2021) Material selection for hybrid natural fiber laminated composites in vibration using a stochastic optimization method. Hybrid Nat Fiber Compos 281–307
40. Freire RT, dos Santos JC, Panzera TH, da Silva LJ (2021) Recent research and developments in hybrid natural fiber composites. Hybrid Nat Fiber Compos 91–112
41. Atmakuri A, Palevicius A, Vilkauskas A, Janusas G (2020) Review of hybrid fiber based composites with nano particles—material properties and applications. Polymers 12(9):2088
42. Kumar A, Biswal M, Mohanty S, Nayak SK (2021) Recent developments of lignocellulosic natural fiber reinforced hybrid thermosetting composites for high-end structural applications: a review. J Polym Res 28(12):1–37
43. Supian AB, Sapuan SM, Zuhri MY, Zainudin ES, Ya HH (2020) Crashworthiness performance of hybrid kenaf/glass fiber reinforced epoxy tube on winding orientation effect under quasi-static compression load. Defence Technol 16(5):1051–1061
44. Rodriguez-Castellanos W, Rodrigue D (2016) Production and characterization of hybrid polymer composites based on natural fibers. Compos Renew Sustain Mater 30:273–302
45. Nguyen H, Zatar W, Mutsuyoshi H (2017) Mechanical properties of hybrid polymer composite. Hybrid Polym Compos Mater 1:83–113
46. Sinha AK, Narang HK, Bhattacharya S (2020) Mechanical properties of hybrid polymer composites: a review. J Braz Soc Mech Sci Eng 42(8):1–3
47. Verma SK, Gupta A, Singh T, Gangil B, Jánosi E, Fekete G (2019) Influence of dolomite on mechanical, physical and erosive wear properties of natural-synthetic fiber reinforced epoxy composites. Mater Res Express 6(12):125704
48. Ramasamy M, Daniel AA, Nithya M, Kumar SS, Pugazhenthi R (2021) Characterization of natural–synthetic fiber reinforced epoxy-based composite–hybridization of kenaf fiber and kevlar fiber. Mater Today: Proc 37:1699–1705
49. Megahed M, Ali-Eldin SS, Abd El Moezz SM, Abdalla WS (2020) Synthesis of developed rice straw sheets and glass fiber-reinforced polyester composites. J Compos Mater 54(23):3381–3394
50. de Queiroz HF, Banea MD, Cavalcanti DK (2021) Adhesively bonded joints of jute, glass and hybrid jute/glass fibre-reinforced polymer composites for automotive industry. Appl Adhes Sci 9(1):1–4
51. Islam MZ, Amiri A, Ulven CA (2021) Fatigue behavior comparison of inter-ply and intra-ply hybrid flax-carbon fiber reinforced polymer matrix composites. J Compos Sci 5(7):184

52. Saroj S, Nayak RK (2021) Improvement of mechanical and wear resistance of natural fiber reinforced polymer composites through synthetic fiber (glass/carbon) hybridization. Trans Indian Inst Met 12:1–8
53. Yg TG, Kushvaha V, MR S, Siengchin S (2021) A new study on flax-basalt-carbon fiber reinforced epoxy/bioepoxy hybrid composites. Polym Compos 42(4):1891–1900
54. RaviKumar P, Rajeshkumar G, Prakash Maran J, Al-Dhabi NA, Karuppiah P (2021) Evaluation of mechanical and water absorption behaviors of jute/carbon fiber reinforced polyester hybrid composites. J Nat Fibers 31:1–3
55. Arshad MN, Mohit H, Sanjay MR, Siengchin S, Khan A, Alotaibi MM, Asiri AM, Rub MA (2021) Effect of coir fiber and TiC nanoparticles on basalt fiber reinforced epoxy hybrid composites: physico–mechanical characteristics. Cellulose 28(6):3451–3471

Tensile Properties of Kenaf Reinforced with Polypropylene Polymer Under Ultraviolet Light Exposure

Hussain Hasanuthin, Ahmad Hamdan Ariffin, Tabrej Khan, Fairuz Alias, Mohamad Norani Mansur, Mohammad Sukri Mustapa, and A. R. Irfan

Abstract This study focuses on the tensile properties of Kenaf under Ultraviolet concentration exposure. The specimens consist of Kenaf powder mixture, polypropylene, and Rice husk Ash Silica mixed via roll mill mixer. Then, the specimens in a dog-bone shape were fabricated through an injection moulding process. The specimens were exposed to Ultraviolet (UV) light in a UV machine for a different period. The tensile test was conducted using a Universal Testing Machine (UTM). The result shows that the specimen at longer time exposure presents a weaker strength compared to the specimen with UV in a shorter time.

H. Hasanuthin · A. H. Ariffin · F. Alias · M. N. Mansur · M. S. Mustapa
Faculty of Mechanical and Manufacturing Engineering, University Tun Hussein Onn Malaysia, Batu Pahat, 86400 Johor, Malaysia
e-mail: ad180185@siswa.uthm.edu.my

M. N. Mansur
e-mail: mnbm@uthm.edu.my

M. S. Mustapa
e-mail: sukri@uthm.edu.my

A. H. Ariffin (✉)
Research Centre for Unmanned Vehicle (ReCUV), University Tun Hussein Onn Malaysia, Batu Pahat, 86400 Johor, Malaysia
e-mail: ahmadh@uthm.edu.my

T. Khan
Department of Engineering Management, College of Engineering, Prince Sultan University, Riyadh 11586, Saudi Arabia
e-mail: tkhan@psu.edu.sa

A. R. Irfan
Faculty of Mechanical Engineering Technology, Universiti Malaysia Perlis, Kampus Tetap Pauh Putra, 02600 Arau, Perlis, Malaysia
e-mail: irfanrahim@unimap.edu.my

A. H. Ariffin · A. R. Irfan
Green Design and Manufacture Research Group, Center of Excellence Geopolymer and Green Technology (CEGeoGTech), Universiti Malaysia Perlis, 01000 Kangar, Perlis, Malaysia

T. Khan and M. Jawaid (eds.), *Green Hybrid Composite in Engineering and Non-Engineering Applications*, Composites Science and Technology,
https://doi.org/10.1007/978-981-99-1583-5_3

Keywords Kenaf fiber · Ultraviolet concentration · Mechanical properties · Geometry structure

1 Introduction

Kenaf Fiber is a natural fibre plant used to plant in the east-central of Africa. During the old days, Kenaf fibre was planted as food and fibre. Nowadays, the Kenaf fibre is planted in home gardens and sometimes are used as herbs. Kenaf fibre also can be used to make clothing and to make bags. Kenaf fibre can also be used as the source of raw materials in making papers and tissues. Besides all these applications in various industries, Kenaf fibre is also suitable for application in the aviation industry [1–3]. We can observe that most of the materials used in aircraft making are synthetic materials. Although these synthetic materials can be patterned to be strong and lightweight, they are dangerous to the environment. These synthetic materials can emit carbon dioxide that can harm mother nature. Besides that, synthetic materials are difficult to dispose of and turn into toxic material if not handled properly [4]. Because we want to apply the usage of this substance in the manufacture of aircraft, mainly in the window shade, we need to testify the endurance of this material. We choose to treat this material under UV concentration because when flights are flying at high altitude, they were exposed to a high amount of sunlight, especially during long haul flights.

The component composition of natural fibre will determine the characteristic of the natural. The three essential components of natural plant fibre are cellulose, hemicellulose, and lignin [5, 6]. The three-component compositions content can influence the mechanical properties of the fibres. Some studies treated the fibres to improve the component composition contents. Several researchers used an alkaline solution to increase the tensile strength of the fibres, in which the treatment partly extracts all hemicellulose and lignin, which serves as a cellulose binder. As a consequence, the lower the lignin content, the higher the cellulose content, and hence the higher the tensile strength of the fibre [7, 8].

As an example for the kenaf fibre, cellulose was the most dominant content in the natural fibre. However, some researchers treat the fibres with alkaline treatment to improve the composition content. The comparison of component composition between untreated and treated kenaf fibre was summarised in Table 1.

Ultraviolet (UV) radiation is a kind of energy emitted by the sun. UV radiation has a shorter wavelength than visible light, but it is harmless to the naked eye. The human skin, on the other hand, can detect UV radiation. The sun emits a wide range of ultraviolet radiation classified as UV-A, UV-B, and UV-C. The ozone layer will completely filter the UV-C before entering the earth surface as this ray is very dangerous to the human body. The UV-B could penetrate the earth surface and cause sunburn if human skin were exposed to this ray. Besides that, when human skin is exposed to UV-B for a long time, it can cause skin diseases such as skin cancer and damage the human body's DNA. However, only a small percentage of UV-B enter

Table 1 Component composition of untreated and treated kenaf fibre

Treatment	Cellulose (%)	Hemicellulose (%)	Lignin (%)	References
Untreated	44.00–57.00	21.00	15.00–19.00	[9]
Untreated	60.46–66.24	12.60–19.91	14.67–19.24	[10]
Untreated	45.00–57.00	21.50	8.00–13.00	[11]
Untreated	72.00	20.30	9.00	[12]
Untreated	70.00	19.00	3.00	[13]
Treated	74.10	12.20	6.30	[7]
Treated	79.30	9.69	7.22	[8]

the earth surface as 95% of UV-B is filtered by our ozone layer [14]. As for UV-A, it is the least dangerous of UV radiation. UV-A has a longer wavelength than UV-B and UV-C, thus less affecting the human skin. It only causes skin ageing if we're exposed to UV-A [15].

Generally, several types of research were conducted to study the effect of UV radiation on composite materials. The material, such as tensile strength and glass transition, was dropped after exposure to UV radiation. Table 2 summarises the finding conducted by several researchers on the UV effect on the composite.

Most of the research conducted is on synthetic fibres. There is still a need for an investigation on the natural fibre, especially under the UV radiation effect. Therefore, further effort is required to investigate the application of natural fibre in aviation and structure application. Therefore, this study aims to investigate kenaf's tensile properties when exposed to different UV period times.

Table 2 The effect of UV radiation on the composite materials

Material	UV expose duration	Performance	References
Polyurethane composite materials	50 h, 100 h	The mechanical properties performance is decreased	[16]
Glass Reinforced Polymer (GRP) composites	Exposure to Ultraviolet radiation. Increased temperature for about 1000 h	Proposed model of synergistic aging under UV & water condensation	[17]
Glass fiber/epoxy composites	15 days, 35 days, 45 days	The degradation effect is clearly presented	[18]
Styrene-based shape memory polymers	UV wavelength in the range of 325–400 nm	The mechanical properties and the glass transition temperature (Tg), considerably drop	[19]

2 Materials and Methods

Three different periods are employed for this study: Four days, one week, and two weeks. Two studied conditions are with UV treated conditions and the second one with the untreated condition. These treated and untreated specimens underwent a tensile test. The material employed in this research were kenaf fibre powder, Rice Husk Silica type NC A and Polypropylene. The NC A type contains 95% of silica. The Rice Husk Silica will be heated in a furnace for 6 h at a temperature of 700 °C and produce Rice Hush Ash (RHA). Polypropylene will be used as the polymer composite. The NC A is divided into several batches, as stated in Table 3.

Figure 1 shows the raw Rice Husk Silica before and after the heating process in the furnace. Figure 2 shows the other materials for sample fabrication: the fine-sized raw Kenaf Fiber and the polypropylene.

The mixing process will involve three materials: Rice Husk Silica (NC A), Kenaf Fiber, and Polypropylene. All these three materials will be mixed in the Roll Mill Mixer machine. Figure 3 shows the Roll Mill Mixer machine located at the Polymer Lab in the Faculty of Mechanical and Manufacturing Engineering, University Tun Hussein Onn Malaysia. The machine setup at a temperature of 190°C, a speed of 350 RPM, 30 min operation to obtain the size of the mixture of 63 Micron. The NC A, Kenaf Fiber, and Polypropylene will be weighted first before inserting them into

Table 3 The ratio of Kenaf Fiber powder (KF) and Rice Husk Silica Reinforced Polypropylene for four batch

Kenaf Fiber powder (KF)	Polypropylene (PP)	Rice Hush Ash (RHA)
30 wt%	60 wt%	0%
30 wt%	60 wt%	3%
30 wt%	60 wt%	6%
30 wt%	60 wt%	9%

(a) (b)

Fig. 1 Raw Rice Husk Silica in the furnace. **a** condition before the burning process, **b** condition after the burning process (Rice Hush Ash)

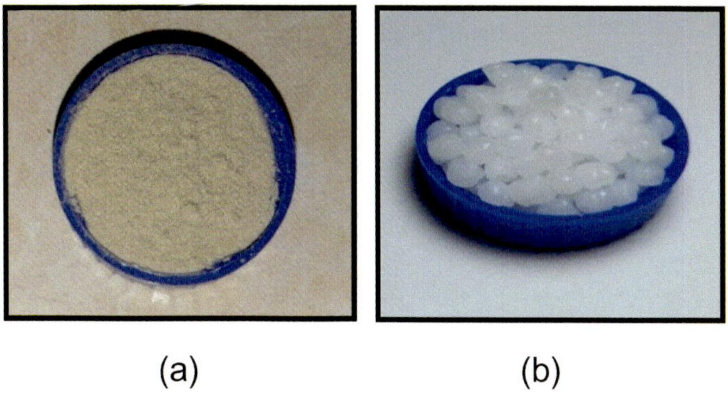

Fig. 2 The materials for sample fabrication, **a** The fine-sized Kenaf Fiber powder, **b** Polypropylene in the form of crystalline

Fig. 3 Roll mill mixer of sample material

the Roll Mill Mixer machine. The NC A will be inserted first. Then it will be put in a separate container. After that, polypropylene will be inserted gently to let it melt first and then the mixture of Kenaf Fiber and Rice Husk Silica will be poured. The mixture will be allowed to mix well for 20 min. Then, the mixture will be taken out from the machine to be cooled down.

After Rich Husk Silica, Kenaf Fiber and Polypropylene underwent a mixing process, the mixture was crushed. The mixture was crushed because the sample needs to be in pallet size to ensure the sample fit perfectly in the Injection Moulding machine. First, the mixture was cleaned from foreign substances to avoid polluted samples and get precise results. So, the combination will be inserted into Plastic Granulator Crusher Machine. Figure 4 shows the Plastic Granulator Crusher Machine. The mixture was inserted carefully to prevent any branch got stuck in the granulator machine during the operation. After the mixtures have undergone the crushing process, they produce a pallet form and are separated into containers.

The injection moulding machine was employed to fabricate a specimen from the pallets (shown in Fig. 5). The injection moulding machine temperature was setup

Fig. 4 Plastic Granulator Crushing Machine

properly to obtain the perfect shape. The pallets were melted down during the injection moulding process to fill the mould cavity to produce the specimen shape. The melted polymer composite was then compressed using the injection moulding process with mould shape of dog bone. The specimen was fabricated for the tensile test with ISO 527 standard. During this whole process, the Injection Moulding machine was setup at 190°C. The cooldown time for the specimen after this process was 10 min. Figure 5 shows the Injection Moulding machine.

3 UV Irradiation Test

The specimens formed during the Injection Moulding process were exposed under the UV concentration using the UV irradiation accelerated weathering tester, model: LUV, as shown in Fig. 6. This machine is located at the Polymer Lab in the Faculty of Mechanical and Manufacturing Engineering, University Tun Hussein Onn Malaysia. This machine can imitate the sunlight and rain by exposing UV radiation with moisture and high temperature. The temperature was set at 100°C based on cycle A in ASTM D4329-99 standard. The polymer composite specimen was exposed under

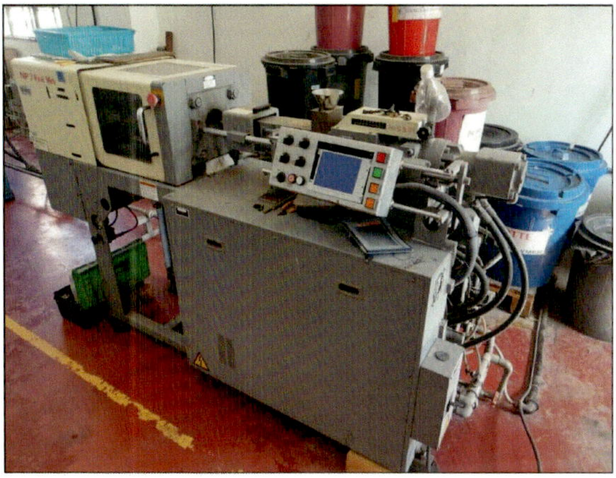

Fig. 5 Injection Moulding machine

Fig. 6 UV Accelerated Machine

UV radiation A 340 light with three separate exposure times, which are: four (4) days (96 h), one (1) week (168 h), and two (2) weeks (336 h).

Fig. 7 Universal Testing Machine

4 Tensile Test

The specimen underwent a tensile test on a Universal Testing Machine. Figure 7 shows the Universal Testing Machine employed in this testing following the ISO 527 standard. It is located at the Polymer Lab, Faculty of Mechanical and Manufacturing Engineering, University Tun Hussein Onn Malaysia. Ultimate Tensile Strength means the highest stress that the specimen can handle before approaching the limit and eventually breaking. The machine pulls the specimen until it breaks. The force and displacement data were recorded automatically in the computer. The Ultimate Tensile Strength of the specimen is presented in this paper.

5 Results and Discussion

Figure 8 shows data for the sample with 30% KF, 70% PP, and 3% RHA treated with different periods. The Ultimate Tensile Strength of the specimens shows a decreased trend. The untreated specimen has the highest Ultimate Tensile Strength, which is 22.6 MPa. The Ultimate Tensile Strength (22.4 MPa) starts to fall when the sample has been treated with UV radiation for four days. The value kept decreasing when treated with UV radiation for one week (22.08 MPa) and then two weeks (21.12 MPa). After two weeks (336 h) of exposure, the decreasing rate is 6%. This progress is agreed with the previous research conducted by F. K. Gorbunov [16] and S. Lohani [18]. The specimen is degraded due to the effect of UV over the time.

Figure 9 shows that the trend of the Ultimate Tensile Strength is decreasing. The 30% KF, 70% PP, and 6% of RHA untreated specimen have the highest Ultimate Tensile Strength, 21.62 MPa. The Ultimate Tensile Strength (21.33 MPa) specimen of four days UV radiation starts to decrease compared to the untreated specimen. The value kept decreasing when treated with UV radiation for one week (20.28 MPa)

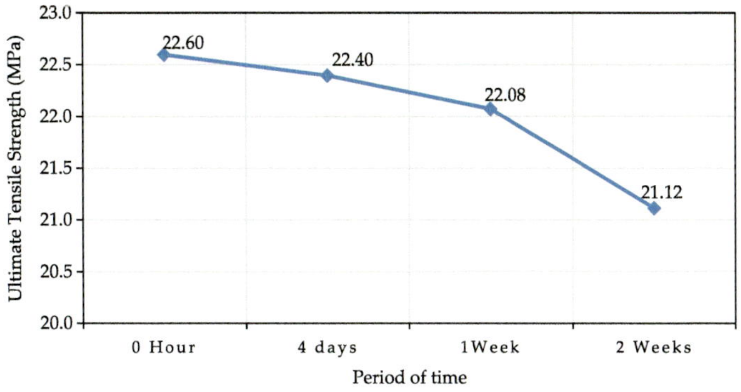

Fig. 8 Data for the sample with 30% KF, 70% PP, and 3% of RHA treated with different period of UV radiation

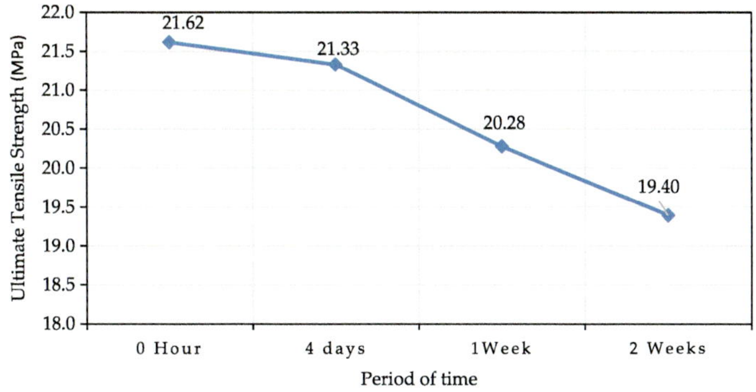

Fig. 9 Data for the sample with 30% KF, 70% PP and 6% of RHA treated with different period UV radiation

and then two weeks (19.40 MPa). The decreasing rate until two weeks of exposure compared to untreated is 10%. The UV radiation affected the Ultimate Tensile Strength of the specimen. Besides that, the increase of RHA value also cause the reduction of Ultimate Tensile Strength of the specimen as compared to 3% of RHA. This progress also agreed with the previous research conducted by F. K. Gorbunov [16] and S. Lohani [18].

Figure 10 shows the data for samples with 30% KF, 70% PP, and 9% of RHA treated with different periods of UV radiation. The trend of the graph is decreasing. The UV radiation decreases the Ultimate Tensile Strength of the sample for a longer period. The untreated specimen has the highest Ultimate Tensile Strength, which is 21.06 MPa. The Ultimate Tensile Strength of four days of exposure decreased to 20.76 MPa. The value kept falling when treated with UV radiation for one (1) week (20.69 MPa) and then two (2) weeks (19.15 MPa). The decreasing rate until two weeks of exposure as compared to untreated is 9%. This progress also agrees with the previous research conducted by F. K. Gorbunov [16] and S. Lohani [18].

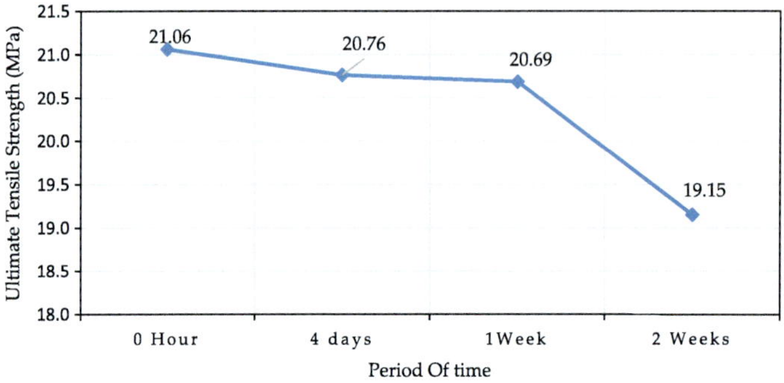

Fig. 10 Data for sample with 30% KF, 70% PP and 9% of RHA treated with different period of UV radiation

Table 4 Data for the sample with 30% KF, 70% PP and different percentage of RHA treated for two (2) weeks of UV radiation

Sample	Percentage of RHA	Ultimate tensile strength (MPa)
30% KF + 70% PP	3% RHA	21.12
	6% RHA	19.40
	9% RHA	19.15

Besides that, the increase of RHA value to 9% cause the reduction of Ultimate Tensile Strength of the specimen as compared to 6% and 3% of RHA.

Table 4 tabulates the data for the samples with 30% KF, 70% PP and different percentage of RHA treated for two (2) weeks of UV radiation. The value of Ultimate Tensile Strength was decreased when we kept the period of UV treatment for two weeks at the constant variable. The Ultimate Tensile Strength records decreasing trends as it exposed to UV radiation at a longer time has been recorded. Table 4 shows that the sample with 30% KF, 70% PP, and 3% RHA has the highest ability to withstand maximum stress before failure compared to the sample with 6% RHA and 9% RHA. Therefore, the selection of the optimum value of RHA is vital to obtain the best strength of a material.

6 Conclusions

From this research, the ratio of 30% kenaf, 70% Polypropylene obtained the optimum Ultimate Tensile Strength at the percentage of RHA at 3% as compared to 6% and 9% RHA. After two weeks of exposure, the UV radiation reduces the strength of the specimen. The reduction rate is 6%, 10%, and 95 for RHA at 3%, 6%, and 9%, respectively.

Acknowledgements This research was supported by Universiti Tun Hussein Onn Malaysia (UTHM) through TIER 1 (Vot: H773)

Conflicts of Interest The authors declare no conflict of interest.

Author Contributions All authors have read and agreed to the published version of the manuscript. Contribution and work management in this project can be presented; Hussain Hasanuthin, Ahmad Hamdan Ariffin, and Fairuz Alias experimented. Hussain Hasanuthin wrote the manuscript with support from Ahmad Hamdan Ariffin and Tabrej Khan. Hussain Hasanuthin and Fairuz Alias fabricated the specimen. Mohammad Sukri Mustapa, A.R Irfan and Ahmad Hamdan Ariffin conceived the original idea. Mohamad Norani Mansur expert in project management. Mohamad Norani Mansur, Tamer Sebaey and Mohammad Sukri Mustapa supervised the project.

Funding This research was funded by Universiti Tun Hussein Onn Malaysia (UTHM) through TIER 1 (Vot: H773). The authors would like to acknowledge the support of Prince Sultan University for paying the Article Processing Charges (APC) of this publication.

References

1. Hamdan A, Mustapha F, Ahmad KA, et al (2016) The Effect of Customized Woven and Stacked Layer Orientation on Tensile and Flexural Properties of Woven Kenaf Fibre Reinforced Epoxy Composites. Int J Polym Sci. https://doi.org/10.1155/2016/6514041
2. Suriani MJ, Ilyas RA, Zuhri MYM, et al (2021) Critical Review of Natural Fiber Reinforced Hybrid Composites: Processing, Properties, Applications and Cost. Polym. 13
3. Nor AFM, Sultan MTH, Hamdan A, et al (2018) Hybrid Composites Based on Kenaf, Jute, Fiberglass Woven Fabrics: Tensile and Impact Properties. Mater Today Proc 5:11198–11207. https://doi.org/10.1016/j.matpr.2018.01.144
4. Hamdan A, Sultan MTH, Mustapha F (2019) Structural health monitoring of biocomposites, fibre-reinforced composites, and hybrid composite. In: Structural Health Monitoring of Biocomposites, Fibre-Reinforced Composites and Hybrid Composites
5. Salit MS, Jawaid M, Bin YN, Hoque ME (2015) Manufacturing of natural fibre reinforced polymer composites. Springer, Cham
6. El-Shekeil YA, Sapuan SM, Abdan K, Zainudin ES (2012) Influence of fiber content on the mechanical and thermal properties of Kenaf fiber reinforced thermoplastic polyurethane composites. Mater Des 40:299–303. https://doi.org/10.1016/j.matdes.2012.04.003
7. Hassan A, Isa MRM, Ishak ZAM et al (2018) Characterization of sodium hydroxide-treated kenaf fibres for biodegradable composite application. High Perform Polym 30:890–899. https://doi.org/10.1177/0954008318784997
8. Guo A, Sun Z, Satyavolu J (2019) Impact of chemical treatment on the physiochemical and mechanical properties of kenaf fibers. Ind Crops Prod 141. https://doi.org/10.1016/j.indcrop.2019.111726
9. Alavudeen A, Rajini N, Karthikeyan S et al (2015) Mechanical properties of banana/kenaf fiber-reinforced hybrid polyester composites: Effect of woven fabric and random orientation. Mater Des 66:246–257. https://doi.org/10.1016/j.matdes.2014.10.067
10. Wang C, Bai S, Yue X et al (2016) Relationship between chemical composition, crystallinity, orientation and tensile strength of kenaf fiber. Fibers Polym 17:1757–1764. https://doi.org/10.1007/s12221-016-6703-5
11. Akil HM, Omar MF, Mazuki AAM et al (2011) Kenaf fiber reinforced composites: A review. Mater Des 32:4107–4121

12. Yusuff I, Sarifuddin N, Ali AM (2020) A review on kenaf fiber hybrid composites: Mechanical properties, potentials, and challenges in engineering applications. Prog Rubber, Plast Recycl Technol 147776062095343. https://doi.org/10.1177/1477760620953438
13. Millogo Y, Aubert JE, Hamard E, Morel JC (2015) How properties of kenaf fibers from Burkina Faso contribute to the reinforcement of earth blocks. Materials (Basel) 8:2332–2345. https://doi.org/10.3390/ma8052332
14. Ultraviolet Waves. https://science.nasa.gov/ems/10_ultravioletwaves
15. Anna Chien M, Heidi Jacobe M UV Radiation & Your Skin, The Facts,The Risks, How They Affect You. In: 2019. https://www.skincancer.org/risk-factors/uv-radiation/
16. Gorbunov FK, Berdnikova LK, Poluboyarov VA (2019) The influence of carbide modifiers on performance of polyurethane exposed to high temperature and UV radiation. In: Materials Today: Proceedings
17. Lu T, Solis-Ramos E, Yi YB, Kumosa M (2016) Synergistic environmental degradation of glass reinforced polymer composites. Polym Degrad Stab. https://doi.org/10.1016/j.polymdegradstab.2016.06.025
18. Lohani S, Shubham PRK, Ray BC (2021) Effect of ultraviolet radiations on interlaminar shear strength and thermal properties of glass fiber/epoxy composites. Mater Today Proc. https://doi.org/10.1016/j.matpr.2020.12.028
19. Al Azzawi W, Epaarachchi JA, Leng J (2018) Investigation of ultraviolet radiation effects on thermomechanical properties and shape memory behaviour of styrene-based shape memory polymers and its composite. Compos Sci Technol. https://doi.org/10.1016/j.compscitech.2018.07.001

Ecologically Enhanced Natural/Synthetic Polymer Hybrid Composites for Aviation-Interior and Secondary Structures

Alcides Lopes Leao, Ivana Cesarino, Milena Chanes, Edson Cocchieri Botelho, Otavio Augusto Titton Dias, and Mohammad Jawaid

Abstract The present paper reviews the latest development in composites application in the aviation industry, with emphasis in the bio-based and renewable components. With the environmental pressure over a sustainable mobility, the use of lighter and better materials are under worldwide research. The search for those materials goes through the bio-based components in composites. The recent advances in nanotechnology and in the applications for natural fibers can represent a future trend for the industry. Recent approaches such as the bioeconomy and circular economy aim to transition the current linear, economic system in the aviation industry to a more sustainable one, including bio-composites and recycling.

1 Introduction

The latest worldwide trend towards a bio-based and circular economy has opened new opportunities for a range of renewable and bio-based materials under the umbrella of sustainability and social responsibility [1]. The use of renewable resources must obey the natural production cycle, reducing its pressure which can be resulted of overexploitation of a linear economy [60]. Therefore, the use of bio-based materials, including the wastes, as a resource goes side by side with a sustainable and high-quality biomass production [18]. Renewable materials, most of them natural fibers, will play a key role in the bio-based or hybrid composites in several industry sectors, including aviation. Most of the aviation composites are mainly made from

A. L. Leao (✉) · I. Cesarino · M. Chanes · E. C. Botelho
UNESP—Sao Paulo State University, SP, Brazil
e-mail: alcideslopesleao@gmail.com

O. A. T. Dias
University of Toronto, Toronto, Canada

M. Jawaid
Institute of Tropical Forestry and Forest Products (INTROP), Universiti Putra Malaysia, Putra, Malaysia

T. Khan and M. Jawaid (eds.), *Green Hybrid Composite in Engineering and Non-Engineering Applications*, Composites Science and Technology,
https://doi.org/10.1007/978-981-99-1583-5_4

non-renewable, including Polyacrylonitrile-based Carbon Fibers (CF) [9]. A real closed-loop recycling, bio-based components and higher performance will improve the environmental profile those polymer systems and composites. The natural fibers are assumed environmentally and socially friendly, with a short carbon cycle and low greenhouse gas emissions when compared to its counterparts based on petroleum [36]. Not mentioning that the crop fibers and Agri-wastes does not compete with food production and contributes to the nutrients cycling in the soil.

Composites are versatile, being used in structural applications and components, including hot air balloon gondolas and gliders, commercial and military airplanes and helicopters, and the Space Shuttle. The first use of composites in the airplanes industry started in the 1950's with fiberglass (glass fibers embedded in a resin matrix) being used in the Boeing 707 passenger jet. Later in 1960s Rolls Royce made the compressor blades of the RB211 with carbon. The military started to use fibrous components in small amounts in the 1960's, which was followed by the transport planes in 1970's. Since later 1980s, composites had a steady growth in the airplanes industry with doubling every five years, with new materials and new application for the already developed ones [35].

With increasing fuel costs, the aviation industry is moving toward a weight reduction and the composites materials are the best option to be used. A good example is the Embraer E2 series with modern design and weight reduction resulting a specific fuel consumption, slightly taller landing gear and composite carbon fiber-based wing. The final result was a 24% reduction in per-seat fuel burn compared with the E195 model (Gubisch, 2019). This trend can also be observed by the Airbus company, which increased the use of composites in its aircraft and decreased the use of aluminum alloys, see Fig. 1 [53].

Nevertheless, the composites materials have more expensive components when compared to aluminum or most metal alloys and a higher fabrication cost, but this will not inhibit the composite growth for the next decades [59].

2 Airplane History

The aviation as we know nowadays started in later 18th Century. The beginning of the airplane's history is full of controversy. The invention is credited in most cases to two persons, the Americans Wright Brothers and the Brazilian Santos Dumont. Although several other claims were made, where one is particularly known, Gustave Whitehead that claimed to have made a sustained powered flight in a heavier-than-air machine on August 14, 1901:

- Clement Ader in the Avion III (1897);
- Gustave Whitehead in his No's 21 and 22 airplanes (1901–1903);
- Richard Pearse in his monoplane (1903–1904);
- Samuel Pierpont Langley's Aerodome A (1903);
- Karl Jatho in Jatho biplane (1903).

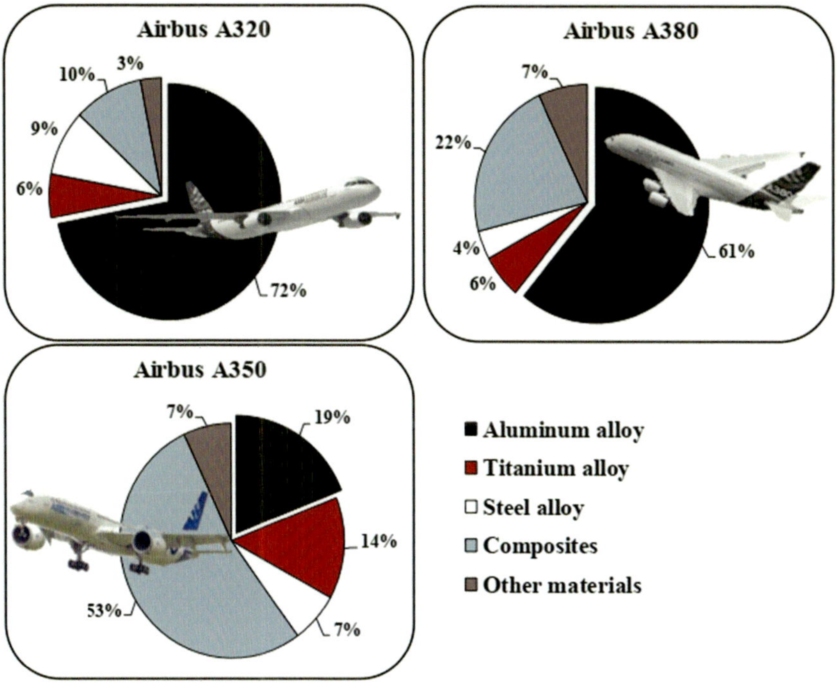

Fig. 1 Materials used in the manufacture of Airbus A320, Airbus A350 and Airbus A380 [53]

Orville and Wilbur Wright tried to develop a craft that would be light and with power enough to lift-off. This craft was named Flyer, a biplane made of wood and fabric with 12.3 m in length and with a wing area of 47.4 m^2. It is reported that on December 17, 1903, Orville Wright, after several trials, had a successful flight that lasted 59 s and covered 260 m. After a year of developing their aircraft, the Wright brothers performed the first circular flight of an engine powered airplane.

Santos-Dumont flew the 14-bis on October 23, 1906, and this was the first powered heavier-than-air flight in Europe to be certified by the Aro Club de France and the Fédération Aéronautique Internationale (FAI). The aircraft flew for 60 m at a height of about 5 m. It won the Deutsch-Archdeacon Prize for the first officially observed flight of more than 25 m, in comparison to the fact that Writes flight had no witnesses to their early accomplishments. Santos-Dumont's flight was the first public flight in the world; therefore, he was acclaimed as the inventor of the airplane across Europe. Another point in dispute was that the Wrights did not fulfill the conditions set up during this period to distinguish a true flight from a prolonged hop; on the flip side, Santos-Dumont took off unassisted and publicly flew a predetermined length in front of experts, in addition, he safely landed. Also, the Wright Flyer took off from a rail and, then later used a catapult, disqualifying the flight because there was no proof it could lift off on its own [8].

The airplanes had a strong development pushed by the military demand during the 1st World War, but continued in the years between the two world wars. During the Second World War, it was possible to notice an evolution in the models of airplanes, taking the planes to a vanguard level in military operations. Advances came mainly in designs, where there was a wide range of aircraft specifically suited to military operations, such as combat aircraft, strategic and photo-reconnaissance aircraft, bombers and attack aircraft, seaplanes, and transport and utility aircraft. After war the trend was a change in design with cone-shaped noses and sharp leading edges on the wings. The fuselage was also kept to a minimum cross-section, pressurized fuselages, windows and afforded flyers comfort and relative luxury [51].

3 Composite Materials

A composite material is composed of at least two materials with distinct phases, chemical and physical properties and are merged to create a material with properties unlike the individual elements, in a synergic effect resulting in properties superior to those of the individual constituents [16]. In addition, a biocomposite is a composite in which at least one component is bio-based [54].

Among the several types of composites, can be listed the fiber reinforced polymer (CFRP) composites, which can include carbon, glass, aramid, polymer or natural fibers embedded in a polymer matrix [51]. Other matrix materials can be used and composites may also contain fillers or nano-materials, like graphene, for example. The use of carbon nanotubes (CNT's) was another possibility considered to increase the performance of carbon and glass fibre reinforced composites, improving the interfacial shear strength, the interlaminar fracture toughness and the damage initiation [23].

Composites can be tailor-made resulting in an extremely versatile and efficient material, aiming to be in most of the cases, lighter, stronger, reliable, durable and sometimes cheaper than traditional materials.

The composites have nowadays a worldwide application in the airplanes industry, most of them in military aircrafts, including the bomber B2, fighters F (USA), Harrier (UK), Gripen (Sweden), Eurofighter, Rafalle, Mirage (Europe); Mig 29 and SU series (Russia). In the transport, planes can be listed: Airbus (A320, A340, A380); ATR (ATR42 and ATR70); Russian Tu204. Boeing 787 has about 80% by volume in composites. For the helicopters, the list includes V22, Eurocopter, Comanche, RAH66, BA609, EH101, Super Lynx 300, S92 and others [21].

4 Types of Matrix

A tough resin matrix reinforced with relatively strong and rigid fibers gives rise to composite materials. The lignocellulosic materials are natural composites: consists of cellulose fibers in a lignin matrix. Better known man-made composite materials, used in thc aerospace and other industries, are carbon- and glass-fiber-reinforced plastic (CFRP and GFRP respectively) that consist of carbon and glass fibers, for their density, both of which are stiff and strong, but brittle, in a polymer matrix, which is tough but neither particularly stiff nor strong. The aim is to combine materials with complementary properties resulting that a composite material with most or all of the benefits (high strength, stiffness, toughness and low density), thus this new material is obtained with few or none of the weaknesses of the individual components [50].

The request for more sustainable materials has led to the development and replacement of composites reinforced with synthetic fibers - carbon, glass, etc. - by composites reinforced with natural-based fibers, such as vegetable, animal, or mineral [48]. The matrix can typically be a ceramic, metal or polymer. Polymers can be classified into three classes: elastomer, thermoplastic and thermoset [16, 20, 54]. One example of green thermosetting material is Carvacrol, which is made from a renewable phenol. Carvacrol was converted to both a cyanate ester resin (CarvCy) and polycarbonate. The cured resin was unaffected by exposure to hot water, with good thermal stability. The polycarbonate prepared from carvacrol had better thermal stability than the cyanate ester. The results showed that the sustainable phenol carvacrol can be used as a platform molecule for the generation of high-performance polymers [26].

Thermoplastic materials currently dominate as matrices for vegetable-based fibers addition; the most commonly used thermoplastics for this purpose are polypropylene (PP), polyethylene (PE), polyvinyl chloride (PVC) and acrylonitrile butadiene styrene (ABS); on the other hand, the most commonly used thermosetting matrices are phenolic, epoxy and polyester resins [16, 34, 46].

Some advantages and disadvantages of thermoplastic matrices are:

- Polypropylene (PP): PP is widely used because it poses moderate dimensional stability, high temperature of thermal deformation, and flame resistance. In addition, recovered PP from the recycling process can be used to develop new natural-based fibers reinforced composites [61]. PP present good mechanical properties at ambient temperature, is absorbs a low amount of water, chemically stable, and maintains good insulating properties - even when moist - and is relatively cheap if compared with other polymers; however, its main disadvantages are low impact strength of some grades, low rigidity, ultraviolet (UV) irradiation sensitivity and poor creep behavior [14].
- Polyethylene (PE): This material has reasonable thermal and mechanical properties [61], associated with chemical inertness, good impact strength, low cost and easiness of processing [14]. PE is a non-biodegradable polymer and is normally produced from a fossil source. However, it is possible to produce bio-based-PE from sugar cane, sugar beet, starch plants, etc. [12]. Nevertheless, important

drawbacks for its utilization can be considered, such as low rigidity, limited UV resistance, and high shrinkage [14].

- Polystyrene (PS): PS is a material with significant demand because it has interesting properties such as transparency, fluidity, good electrical insulation, low cost, rigidity and dimensional stability [4, 14]. Also, PS matrix composites reinforced with vegetable-based fibers are less abrasive than traditional PS, and can be reused instead of discarded, for example it can have its calorific value recovered in a kiln or be composted [2, 61]. The main limitations of this material are its sensitivity to several conditions-UV irradiation, solvents, and heat-and flammability [14].
- Polycarbonate (PC): Owing to its characteristics, this material is used in engineering applications that require superior mechanical properties, such as tensile strength, impact resistance, fatigue and creep behavior, with the added benefit of transparency. On the one hand, is easy to mold and thermoform using wide temperature ranges [14, 61]. On the other hand, its production cost is higher, presenting sensitivity to light and weather, in addition to having high flammability [14].
- Polyvinyl chloride (PVC): in this case, the material exists in two conditions with different mechanical properties: rigid and flexible PVC. Composites usually are made with the rigid variation, since it has low cost, good durability and resistance to termites, good chemical and flame resistance, dimensional stability and good rigidity at ambient temperature; therefore, it shows suitable potential for use in building structures and construction works [14, 61]. Rigid PVC has also disadvantaged, namely: UV and heat sensitivity, low temperature brittleness, higher density than other polymers, and it is also more difficult to inject [14].
- Acrylonitrile butadiene styrene (ABS): is a terpolymer made by polymerizing acrylonitrile and styrene in the presence of polybutadiene. ABS composition can vary from 15 to 35% acrylonitrile, 5% to 30% butadiene and 40% to 60% styrene. It is stronger than pure polystyrene. The acrylonitrile contributes to chemical resistance, fatigue resistance, hardness, and rigidity, while increasing the heat deflection temperature [42].

Thermoset polymers can also be used as matrices in composites for structural applications subjected to lower mechanical stresses; also, these materials benefit from the fact that their processing methods are rather simple, due to the low viscosity of thermoset polymers. Fabrication processes include (but are not limited to) hand lay-up and spraying, compression, resin transfer and injection. The main disadvantage pertaining to this class is their impossibility of further processing after curing, which renders them unrecyclable [61].

Biocomposites can be composed of a polymeric matrix reinforced with different vegetable-based fibers - jute, coir, cotton, flax, wood, hemp, kenaf, luffa, bamboo, sisal, curaua, palm, pineapple, banana, etc. [3, 31, 34, 39, 46, 47]. In addition, another potential use of biobased materiais are the epoxy-curing agent, recycled and bio-based fibers [9]. Optimally, the use of a biodegradable renewable polymer as the matrix-natural rubber, polyhydroxybutyrate or polylactic acid - in association with natural fibers represents a complete eco-friendly green composite. These materials are of significant interest to governments and private research agencies [45].

5 Types of Fillers and Reinforcements

Aluminum wing with is well known characteristic of metal fatigue still are the dominant material, although carbon fiber is growing since has made important improvements in its properties. Also, the boron boron-reinforced (formed on a tungsten core) has been used successfully in the Advanced Tactical Fighter [28]. Aramid fibers have been used in honeycomb sheet form for bulkhead, fuel tanks, and floors, leading and trailing-edge wing components [15]. Nowadays, the principal choices for fiber reinforcement are glass, carbon, aramid and boron, from which the most used in composites are glass, carbon and aramid [28].

Advanced composites are made of high-strength stiff fibers as reinforcements blended in a matrix, in most of the cases thermosetting. Nowadays there are three in use: carbon fiber, glass, and aramid in epoxy matrix. Although others are reported, including boron-reinforced. Those reinforcement can fibrous and particulate and the final result is related to directionality of properties [44].

6 Fibrous

Polymer fibers exhibit a complex type of fibrous fracture. The manufacturing process of these fibers is done by a bundle of sub filaments or fibrils that are loosely bonded. Therefore, when comparing these fibers with ceramic fibers, polymeric fibers are relatively insensitive to failure. However, under compressive load, polymeric fibers can defibrillate, resulting in poor compression properties and other limitations [11].

Embraer had developed remarkable improvement on fiber polymer reinforcement together with aircraft design [5]. Fiber-reinforced polymer composite materials are fast gaining ground as preferred materials for construction of aircrafts and spacecrafts. In particular, the use of polymer fiber reinforced composite materials as primary structural materials in recent years has grown. The use of these materials in various frontline aerospace projects has demonstrated that the technology is viable around the world, ensuring greater trust and acceptance of these composites as primary materials for aerospace vehicles. Fiber polymer composites alongside aluminum alloys are the most frequently used materials in aircraft structures. The use of composites in civil aircraft, military fighters and helicopters has increased rapidly since the 1990s, and composites are now competing head-to-head with aluminum as the materials of choice in many airframe structures. In addition to weight reduction, the use of composites increases specific stiffness and strength, prolongs fatigue life and minimizes corrosion problems. Natural fiber composites present several advantages over man-made fiber composites such as low cost, light weight, high specific mechanical properties, non-hazardous nature, eco-friendliness, renewability and so on, and consequently their utilization in various industrial sectors including aerospace engineering is highly promising [10].

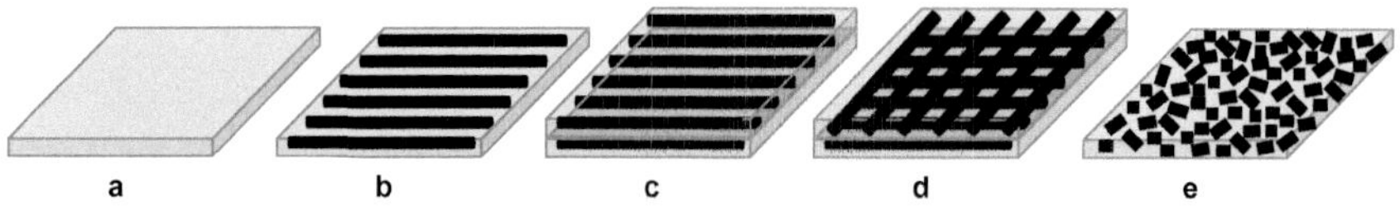

Fig. 2 Schematic of composite materials. (**a**) Isotropic material, (**b**) Anisotropic composite, (**c**) Unidirectional material, (**d**) composite with fibers oriented at different angles, and (**e**) Particulate composite (Elaborated by the authors)

7 Particulate

Particulate composites and conventional metallic materials are isotropic (Fig. 2a), that is, their strength and stiffness properties, for example, are the same in all directions. On the other hand, fibrous composites are anisotropic (Fig. 2b), and their properties vary according to the direction of the load—directly linked to fiber orientation. A unidirectional material (Fig. 2c), such as a sheet of wood, is much easier to bend and break along a line parallel to the fibers than perpendicular to the fibers. This anisotropy can be improved by stacking layers—just fractions of a millimeter thick—on top of each other, with the fibers oriented at different angles to form a laminate (Fig. 2d). CFRP and GFRP are fibrous composite materials, examples of such composites that can be manufactured with layers composed in different directions. Another category is particulate composites (Fig. 2e), such as metal matrix composites (MMC). CMMs are examples of particulate composites and are being developed especially for the aeronautical and aerospace industry. MMCs are manufactured, usually by non-metallic particles in a metallic matrix—silicon carbide particles combined with aluminum alloy, for example [62].

8 Natural Fibers

According to Puttegowda et al. [41], the natural fibers are represented by plant fibers, animal fibers, and mineral fibers. Natural fibers in general display properties which are similar to those of artificial fibers, but their use in polymeric composites is hindered by characteristics such as water absorption and insufficient thermal properties [49]. A comprehensive review about chemical composition, mechanical and physical properties of natural fibers was published by [33].

Natural Fiber Polymer Composites have low specific weight, relatively high strength, relatively low production cost, resistance to corrosion and fatigue, totally biodegradable, improving the surface finish of molded part composites. In addition to having relatively good mechanical properties, natural fibers also come from available and renewable sources, unlike synthetic fibers. Therefore, they can be used in electrical and electronic industries, aerospace, sports, recreation equipment, boats, machinery office products, etc. [37, 54]. The beginnings of natural fiber composites

(NFCs) in the aerospace industry started with the introduction of Gordon Aerolite—a flax roving impregnated with phenolic resin. Its mechanical properties—high tensile strength and stiffness—and the light weight—of Gordon Aerolite made it suitable for use as an aircraft material. This first NFC application was used for an experimental Bristol Blenheim bomber spar.

To ensure the mechanical properties of composites, sandwich panels are commonly used in aircraft interiors, mainly as galley walls, flooring, ceilings, cargo-hold liners and lavatories. Airbus in partnership with researchers from the Council for Scientific and Industrial Research (CSIR) in South Africa are involved in a project to develop natural fiber-reinforced thermoset sandwich panels from flax fabric phenolic skins and a Nomex honeycomb core for use in aircraft. The panels were manufactured by compression molding of pre impregnated composites (prepregs) at suitable temperatures to enable curing of the resin. As the flammability of the material for the aeronautical sector is very important, the research also developed aqueous-based flame-retardant treatments for flax fabric to ensure that the composite panels comply with Federal Aviation Administration (FAA) regulations. Besides being a non-fibrous natural silicate fire resistant material, the composite also contained primary flame retardant. The composite material was reported to exhibit superior flammability, smoke and toxicity properties [10, 29].

According to [31] and [21], the main plant fibers studied as reinforcement in composites are:

- Kenaf (*Hibiscus cannabinus*)
- Hemp (*Cannabis sativa*)
- Jute (*Corchorus capsularis*)
- Flax (*Linum usitatissimum*)
- Ramie (*Boehmeria nivea*)
- Nettle (*Urtica dioica*)
- Curaua (*Ananas erectifolius*)
- Pineapple Leaf (*Ananas comosus*)
- Sisal (*Agave sisalana*)
- Date Palm (*Phoenix dactylifera*)
- Cotton (Gossypium *herbaceum*)
- Coir Fiber (*Cocos nucifera*)
- Kapok (*Ceiba pentandra*)
- Bamboo (*Bambusoideae*)
- Silk (*Bombyx mori*)

The major problem identified with vegetable-based fibers are hydrophilic and thermoplastic and thermosetting matrices are hydrophobic [61]. The hydrophilic characteristic of vegetable-based fibers occurs because of the hydroxyl groups contained in the cellulose and hemicellulose molecular chains. Although some limitations in temperature processing PVC-Wood fiber composite were investigated. A blend with 20% wood fiber showed an increase in the tensile strength on further addition of the wood fiber a decrease was noticed. This decrease in tensile strength decreases the strain of the PVC-wood fiber composite and hence increases the modulus of elasticity of the PVC-wood fiber composite as the wood fiber is added [7]. Therefore, the major problem is the chemical incompatibility between hydrophilic natural fibers and hydrophobic polymeric matrices. This incompatibility can be treated by coupling agents, such as, Maleic Anhydride Grafted Polypropylene (PPgMA) to PP and methyl methacrylate (MMA) for PS [14, 55, 56].

Several surface treatments of natural fibers are conducted to improve their adhesion with different matrices. The main techniques are: alkali treatment—mercerization, acetylation, benzylation, permanganate treatment, silane treatment, peroxide treatment, enzyme treatment, isocyanate treatment, plasma treatment, esterification, TDI treatment, corona treatment, and others [14, 43, 47, 58, 61].

In addition to enhancing fiber properties before the composite fabrication, fiber treatments promote better adhesion between natural fiber and polymeric matrix improving the mechanical properties of these composites [13, 47, 58].

According to Puttegowda et al. [41], the main requirements of materials such as fiber-reinforced composites for aircraft structures are: Lightweight, high reliability, passenger safety, durability—fatigue and corrosion/degradation: Vacuum radiation thermal, aerodynamic performance, multi-role or functionality, fly-by-wire, stealth, and all-weather operation. Especially in the aeronautical sector, in commercial and military aircraft, composites reinforced with fibers, aramid, carbon, and glass are the most used because they provide better strength and stiffness-density ratio, along with superior physical properties. However, composites reinforced with natural fibers have advantages over synthetic ones, such as, biodegradability, low density, good processing flexibility, etc. Despite these advantages, there are still few developments of parts for the aircraft sector using composites reinforced with natural fibers—plant fibers, animal fibers, and mineral fibers.

Abstract Today, mainly man-made materials such as carbon and glass fibers are used to produce composite parts in aviation. Renewable materials such as natural fibers or bio-sourced resin systems have not found their way into aviation, yet. The project ECO-COMPASS aims to evaluate the potential applications of ecologically improved composite materials in the aviation sector in an international collaboration of Chinese and European partners. Natural fibers such as flax and ramie will be used for different types of reinforcements and sandwich cores. Furthermore, the bio-based epoxy resins to substitute bisphenol-A based epoxy resins in secondary structures are under investigation. Adapted material protection technologies to reduce environmental influence and to improve fire resistance are needed to fulfill the demanding safety requirements in aviation. Modeling and simulation of chosen eco-composites aims for an optimized use of materials while a life cycle assessment aims to prove the ecological advantages compared to synthetic state-of-the-art materials.

9 Composites Advantages

The primary reason composite materials are chosen for components is because of weight saving for its relative stiffness and strength also helps to increase acceleration or range in transport. The use of composites can allow 20–50 percent reduction in weight along with a lower production time and improved damage tolerance. For instance, composite reinforced with carbon-fiber has only 20% of 1020 grade steel and fivefold in strength. On the other hand aluminum (6061 grade) presents slight heavier weight to its counterpart carbon fiber, although the final composite showed twice the modulus and up to seven times the strength. Still aluminum is considered a very efficient material, thanks to its high specific strength and recyclability, specially under the umbrella of the circular economy [19].

In addition to the main benefit of reduced weight and formability, composite materials presents a good corrosion and fatigue resistance. Since they do not rust, the painting or coating is reduced, which is particularly important in marine and chemical environments. The need for maintenance and painting is reduced or eliminated. Combine the excellent fatigue resistance, and composites can increase product lifespan by several times in many applications.

Composites including wood itself are good insulators, thermal and acoustic, offering protection against heat, blasts and freezing. Other property is the electrical insulation, which is important to prevent radar detection. A conductive mesh or coating can be integrated if needed, e.g. to reflect radar or divert lightning. Sensors, electronics and cabling can be embedded.

Composites design allows for freedom of architectural form, allowing assembled parts, inserts, stiffeners. Also can be tailor-made for specific applications depending on its components and functionality. Another advantage is that, they can be formed into more complex shapes than their metallic counterparts, resulting in parts reduction, fasteners and joints with shorter assembly time.

Single-shell molded structures provide higher strength at lower weight. The high impact resistance, which is found in aramid-reinforced polymers reduces the damage to the engine pylons.

10 Airplanes Applications

The use of composites in airplanes is very dynamic and the changes occur in every new model or updates of old models. The use of composites in traditional parts started in 1980s, for secondary wing and tail components such as wing trailing edge panels and rudders. A summary of parts and models can be seen in Table 1.

The composite material can also be used by military forces to build aircraft, such as the B2 stealth bomber. The addition/exchange of conventional materials for materials that absorb radio energy is strategic to disguise radar aircraft and avoid detection [25].

Table 1 Components and parts in airplanes made of composites

Model	Part	Composite	Percent W/W
F14, F15	Initially: skins of the empennages (only in secondary structures) Posteriorly: wings and fuselages (primary structures)	Boron-reinforced epoxy composite	–
Eurofighter	Wing skins, forward fuselage, flaperons and rudder	Epoxy/carbon fiber-reinforced composite	40
Dassault's Rafael	–	–	26
Saab Gripen, EADS Mako	–	–	20–25
A300, A310	Rudder, vertical tail fin	–	–
A320	Rudder, vertical tail fin, the entire tail fuselage belly skins, fin/fuselage fairings, fixed leading- and trailing-edge bottom access panels and deflectors, trailing-edge flaps and flap-track fairings, spoilers, ailerons, wheel doors, main gear leg fairing doors, and nacelles, floor panels	Glass Fiber Reinforced Polymer (GFRP)	28
A340-500, A340-600	Rear pressure bulkhead, keel beam, fixed leading edge of the wing	Thermoplastic matrix composite	–
A330, A340	Spoilers	–	– -
A380	Upper deck floor beams and the rear pressure bulkhead	Carbon fiber reinforced plastic (CFRP); glass-fiber-reinforced aluminum alloy (GLARE)	20–22
A350-XWB	Wings, doors and surrounding door frames	Carbon fiber reinforced plastic (CFRP)	50

(continued)

Table 1 (continued)

Model	Part	Composite	Percent W/W
Boeing 777	Wing's fixed leading edge, trailing-edge panels, flaps and flaperons, spoilers, outboard aileron, floor beams, wing-to-body fairing, and the landing-gear doors	Carbon Fiber Reinforced Plastic (CFRP)	20
Boeing 787	Fuselage, wings, tail, doors, and interior	Carbon Fiber Reinforced Plastic (CFRP)	50
Embraer Super Tucano, ERJ 170, 190, 195, Legacy, Phenom	Rudder, elevator, vertical tail	Carbon Fiber Reinforced Plastic (CFRP)	–

Source Adapted by the authors from [43]

Composites are also used in the manufacture of helicopters primarily to reduce the number of components, hence the cost. The composites' excellent strength-to-weight ratio helps maximize payloads and overall performance. The first composite helicopter fairings were from the Boeing Vertol in the 1950s and rotor blades in the 1970s. Today, composites are used in the main structural elements of many modern helicopters. The V22 inclined rotor is about 50% composites by weight [43].

The gains due to an advanced design and lighter materials can be a significant point in the acquisition choice decision by the airlines, including purchasing cost, operating cost, maintenance cost, and recovery cost. Considering all the parameters the author has chosen the model EMBRAER E-195-E2, the best aircraft for regional jets [6].

11 Conclusions

With the growth and instability of fuel costs together with people awareness about the effects of climate change, the airplanes industry, since they are the most highly intensive fuel use in transportation must deliver aircrafts with improved performance, lower maintenance costs, and lighter. In this case, the composites are the most important pillar for the future. With the constant advance of composite technology and the development of new types of components in composites including graphene, nanotechnology, carbon nanotubes and basalt forms will allow new application in the aviation industry. Nevertheless, nowadays carbon fiber is the most used composite in the airplane industry.

Still, all the composites' benefits are not fully explored and the ongoing research in new materials result of improved products performance. New industry sectors can get the benefits of composite materials.

Acknowledgements We are thankful to CNPq—National Council for Research for Productivity Grant to the following authors: Leao, A. L.; Cesarino, I. and Botelho, I. C. Also, to CAPES—Coordination of Superior Level Staff Improvement, Program CAPES PrInt: scholarship grant to the author Souza, M. C.

References

1. Abad-Segura E et al (2021) Implications for sustainability of the joint application of bioeconomy and circular economy: a worldwide trend study. Sustainability 13:1–24
2. Abdulkareem, S. A. et al. (2019) Development of natural fibre reinforced polystyrene (NFRP) composites: Impact resistance study. In: IOP Conference. Series: Materials Science and Engineering, 640, pp 012059
3. Ali SZ et al (2018) Effect of fiber content and post stress on moisture absorption of jute polyester composite. IOP Conference Series Materials Science and Engineering 438:12024
4. Amin S, Amin M (2011) Thermoplastic elastomeric (TPE) materials and their use in outdoor electrical insulation. Rev Adv Mater Sci 29:15–30

5. Arakaki FK, Gonçalves WG (2007) EMBRAER composite material application. In: 16th International Conference on Composite Materials, Kyoto, Japan, 3 p 2007.
6. Ardil C (2020) Regional aircraft selection using preference analysis for reference ideal solution (PARIS). World Academy of Science, Engineering and Technology. In: International Scholarly and Scientific Research & Innovation, 14, 9, pp 378–388
7. Arthur DE et al (2021) Studies on some mechanical properties of PVC-wood fiber composite. Chem Rev Lett 4:85–91
8. Aviation Oil Outlet. Santos-dumont vs the wright brothers: who really invented the airplane? 2020
9. Bachmann J et al. (2021) Towards a circular economy in the aviation sector using eco-composites for interior and secondary structures. In: Results and recommendations from the EU/China project ECO-COMPASS 2018, Aerospace, 8, pp 131
10. Balakrishnan P et al. (2016) Natural fiber and polymer matrix composites and their applications in aerospace engineering. In: Sohel Rana and Raul Fangueiro editors. In: Advanced Composite Materials for Aerospace Engineering—Processing, Properties and Applications. Advanced Composite Materials for Aerospace Engineering, Chapter 12, pp. 365–383
11. Baker AA, Leong KH (2004) Fibers for polymer-matrix composites. Compos Mater Aircr Struct. Second Edition. Alan Baker; Stuart Dutton; Donald Kelly. Chapter 3, pp 55–80
12. Bazan P et al (2020) Biobased polyethylene hybrid composites with natural fiber: mechanical, thermal properties, and micromechanics. Materials 13:1–16
13. Bharath KN et al (2020) Alkaline effect on characterization of discarded waste of Moringa oleifera fiber as a potential eco-friendly reinforcement for biocomposites. J Polym Environ 2:2823–2836
14. Biron M (2013) Thermoplastic composites. Thermoplastics and thermoplastic composites: technical information for plastic users. 2nd ed Oxford: William Andrew. Chapter 6, pp 769–829
15. Botelho EC et al (2006) A review on the development and properties of continuous fiber/epoxy/aluminum hybrid composites for aircraft structures. Material Research 9:247–256
16. Callister Jr. WD, Rethwisch DG (2013) Composite Materials. Mater Sci Eng: Introd, 9th ed New Jersey: John Wiley & Sons, pp 627–671
17. Catalán J, Norppa H (2017) Safety aspects of bio-based nanomaterials. Bioengineering 4:94
18. Daystar J et al. (2019) An economic impact analysis of the U.S. biobased products industry. Rural Dev: US Dep Agric. pp 1–122
19. Devezas T (2020) Trends in aviation: rebound effect and the struggle composites x aluminum. Technol Forecast Soc Chang 160:120241
20. Erden S, Ho K (2017) Chapter 3—Fiber reinforced composites. Fiber Technol Fiber-Reinf Compos. Cambridge, UK: Woodhead Publishing, pp 51–79
21. Girijappa YG T et al. Natural fibers as sustainable and renewable resource for development of eco-friendly composites: a comprehensive review. Front. Mater. 6, pp 226
22. Gorbatikh L et al (2011) Nano-engineered composites: a multiscale approach for adding toughness to fibre reinforced composites. Procedia Eng 10:3252–3258
23. Gorbatikh L and Lomov S (2017) Nano-engineered carbon fibre-reinforced composites: Challenges and opportunities. https://doi.org/10.1007/978-3-319-46120-5_6
24. Gubisch M (February 2019) Embraer receives first production engines for E195-E2. Flightglobal
25. Harris T (2022) How stealth bombers work: Defenses against detection
26. Harvey BG et al (2016) Sustainable hydrophobic thermosetting resins and polycarbonates from turpentine. Green Chem 18:2416–2423
27. International Institute for Strategic Studies—IISS, The Military Balance: The annual Assessment of Global Military Capabilities and Defence Economics, Routledge, Taylor & Francis Group, pp 1–525
28. Kesarwan S (2017) Polymer composites in aviation sector - a brief review article. Int J Eng Res & Technol (IJERT) 6(06):518–525
29. Kumar R and Anandjiwala (2013) Compression molded flax fabric-reinforced polyfurfuryl alcohol bio-composites. Mechanical and thermal properties. J Therm Anal Calorim, 112(2):755–760

30. Layth M et al. (2015) A review on natural fiber reinforced polymer composite and its applications. Int J Polym Sci, pp 15
31. Leao AL et al. (2017) Curaua fibers—the queen of the fibers. Ryszard M. Kozlowski, Malgorzata Muzyczek (Org.). Natural Fibers—Properties, Mechanical Behavior, Functionalization and Applications. 1st Edition. Nova. Chapter 4, pp 83–106
32. Madhu P et al (2018) Potential of natural/synthetic hybrid composites for aerospace applications. Sustain Compos Aerosp Appl, Woodhead Publ Ser Compos Sci Eng, Chapter 15:315–351
33. Madhu P et al (2019) A review on synthesis and characterization of commercially available natural fibers: Part-I. J Nat Fibers 16:1132–1144
34. Malkapuram R et al (2008) Recent development in natural fiber reinforced polypropylene composites. J Reinf Plast Compos 28:1169–1189
35. Mrazova M (2013) Advanced composite materials of the future in aerospace industry. INCAS Bulletin. Bucharest, Ed 3, 5, pp 139–150
36. Mohanty AK et al. (2005) Natural fibers, biopolymers, and biocomposites. CRC Press, pp 274
37. Mohammed M et al. (2022) Improving hydrophobicity and compatibility between kenaf fiber and polymer composite by surface treatment with inorganic nanoparticles. Arab J Chem, 15(11):104233
38. Mukesh SS, Godara (2019) Effect of chemical modification of fiber surface on natural fiber composites: a review. In: Materials Today: Proceedings, 18, pp 3428–3434
39. Nandhakumar S et al (2021) Experimental investigations on natural fiber reinforced composites. Materials Today: Proceedings 37:2905–2908
40. Pereira PHF et al (2015) Vegetal fibers in polymeric composites: a review. Polymers 25:9–22
41. Puttegowda M et al. (2018) Potential of natural synthetic hybrid composites for aerospace applications. In: Sustainable composite for aerospace applications. Woodhead publishing series in composite science and engineering. pp. 315–351.
42. Pye A (2021) The basics of Acrylonitrile Butadiene Styrene (ABS)
43. Quilter A (2004) Composites in Aerospace Applications. Aviationpros
44. Rajak DK et al (2019) Recent progress of reinforcement materials: a comprehensive overview of composite materials. J Market Res 8:6354–6374
45. Rangappa SM et al (2020) Green-composites: ecofriendly and sustainability. Appl Sci Eng Prog 13:183–184
46. Rohit K, Dixit S (2016) A review—future aspect of natural fiber reinforced composite. Polym Renewable Resour 7:43–59
47. Salim S et al (2020) Enhanced mechanical properties of natural fiber bamboo/pineapple leaf/coconut husk reinforced composites for application in bio-board. Int J Geomate 19:168–174
48. Sanjay MR, Siengchin S (2021) Exploring the applicability of natural fibers for the development of biocomposites. Express Polym Lett 15:193
49. Sanjay MR et al (2018) Characterization and properties of natural fiber polymer composites: A comprehensive review. J Clean Prod 172:566–581
50. Shirvanimoghaddam K et al. (2021) Balancing the toughness and strength in polypropylene composites. Composites Part B: Engineering, 223
51. Soler M (2014) Fundamentals of Aerospace Engineering: An introductory course to aeronautical engineering
52. Soutis C et al (2019) How green composite materials could benefit aircraft construction. Sci China Tech Sci 62:1478–1480
53. Souza MC (2019) Essay proposal for defining the requirements of vegetable-based oils for biolubricants on Al 7050-T7451 and Ti-6Al-4V alloys and natural biocides. Doctor of Science Thesis in Materials, Manufacturing and Automation—Aeronautics Institute of Technology (ITA), pp 158
54. Souza MC et al (2022) A review of natural fibers reinforced composites for railroad applications. Appl Sci Eng Prog 15:1–12
55. Singha AS et al (2010) Natural fiber reinforced polystyrene matrix-based composites. Adv Mater Res 123–125:1175–1178

56. Singha AS, Rana RK (2012) Natural fiber reinforced polystyrene composites: Effect of fiber loading, fiber dimensions and surface modification on mechanical properties. Mater Des 41:289–297
57. Sivaraj S et al (2017) Mechanical behavior of saw wood dust filled polymer composites. Int J Sci Eng Res 5:3221–5687
58. Sreekumar PA et al (2012) Effect of chemical treatment on dynamic mechanical properties of sisal fiber-reinforced polyester composites fabricated by resin transfer molding. Compos Interfaces 15:263–279
59. Trzepieciński T et al (2021) New advances and future possibilities in forming technology of hybrid metal–polymer composites used in aerospace applications. Journal of Composites Science. 5:1–52
60. Wijkman A, Skanberg K (2017) The circular economy and benefits for society: jobs and climate clear winners in an economy based on renewable energy and resource efficiency. A study pertaining to the Czech Republic and Poland. The Club of Rome. pp 1–62
61. Yashas Gowda TG et al. (2018) Polymer matrix-natural fiber composites: an overview. Cogente Eng, 5, pp 1–13
62. Zweben C (2015) 10 Composite materials. Mech Engineers' Handb. I:1–35

Natural Fiber Reinforced Composites and Their Role in Aerospace Engineering

Balbir Singh, Kamarul Arifin Ahmad, M. Manikandan, Raghuvir Pai, Eddie Yin Kwee Ng, and Noorfaizal Yidris

Abstract Advanced natural hybrid composite materials were employed and are now being used in structural, mechanical, and high-end industrial applications. Composites offer various significant features, including the ability to resist fatigue, corrosion resistance, and the manufacturing of lightweight components with little compromise to dependability, among others. Natural Composites are a type of composite material that have significant mechanical properties compared to conventional composite materials. The use of composites in the aircraft industry now confronts a research deficit, with the major focus being on determining the future spectrum of use. The usage of appropriate composites so far is responsible for the majority of triumphs in the aviation sector. This chapter highlights the variety of available natural fiber hybrid composites, their general composition and properties, and their possible use in the Aerospace Industry.

B. Singh (✉) · K. A. Ahmad · N. Yidris
Department of Aerospace Engineering, Faculty of Engineering, Universiti Putra Malaysia (UPM), 43400 Serdang, Selangor Darul Ehsan, Malaysia
e-mail: balbir.s@manipal.edu

K. A. Ahmad
e-mail: aekamarul@upm.edu.my

B. Singh · M. Manikandan
Department of Aeronautical and Automobile Engineering, Manipal Institute of Technology, Manipal Academy of Higher Education, Manipal 576104, Karnataka, India

K. A. Ahmad
Faculty of Engineering, Aerospace Malaysia Research Centre, Universiti Putra Malaysia (UPM), 43400 Serdang, Selangor Darul Ehsan, Malaysia

R. Pai
Department of Mechanical and Manufacturing Engineering, Manipal Institute of Technology, Manipal Academy of Higher Education, Manipal 576104, Karnataka, India

E. Y. K. Ng
School of Mechanical and Aerospace Engineering, College of Engineering, Nanyang Technological University, 50, Nanyang Avenue, Singapore 639798, Singapore

T. Khan and M. Jawaid (eds.), *Green Hybrid Composite in Engineering and Non-Engineering Applications*, Composites Science and Technology,
https://doi.org/10.1007/978-981-99-1583-5_5

1 Introduction

Rapid expansion in the aerospace and manufacturing industry has necessitated the improvement of composite materials for their strength, stuffiness, reduced cost, and density with enhanced sustainability [1]. Natural composites have also advanced, as core materials with so much improvement in characteristics that they can be used now for a variety of purposes [2]. Composites in general are made up of more than two components, of which matrix and fibers are the primary ones [1]. Natural or synthetic fibers and their usage for the production of composite materials have given information on important applications in a wide range of areas, including construction and building, mechanical sectors, automotive, aerospace, marine and medical [3].

Composites are considered as an alternative to many traditional type materials for many applications and in research studies during the last twenty years. This is probably due to the substantial improvement in almost all the properties (structural, physical, mechanical, and tribological) of these fiber-reinforced composite (FRC) materials [4]. Even though composites have increased material endurance, there is still a great deal of worry about the environmental hazards and damage caused by accumulated plastic trash everywhere [5]. This issue has prompted researchers all around the globe to create ecologically friendly materials that are linked with cleaner manufacturing methods [6]. There are numerous methods developed for recycling composites to handle the hundreds of tons of waste from these composites that are produced annually. Smaller-sized recyclables are used as fillers in sheet molding compounds through a process known as pulverization, which is a type of mechanical recycling. Thermal recycling involves the combustion of composite materials with a high calorific value or the pyrolysis of composite waste to produce enormous amounts of heat energy [7]. There are additionally more effective techniques, including high-voltage fragmentation and solvolysis (chemical type recycling) [8]. Utilizing natural fillers in the polymer matrix, like natural fibers, cellulose- nanocrystals, and nano-fibrillated cellulose has improved material properties while reducing residue buildup [9]. Numerous authors have found benefits of cellulosic fibers, as they are very abundant in nature, non-toxic, renewable, and cost-effective, as well as providing sufficient polymer matrix bonding for more improvements in process parameters such as flexural capacity, ductility, rigidity, and wear resistance [10, 11]. Current practices use a variety of mineral additions to strengthen composite constructions, including brick powder, fly ash, limestone powder,, and others. Fly ash was added to a concrete based composites for structural purposes, which raises fracture toughness and prolonged the life of the material [12]. Now natural fibers are either plant-based, animal-based fibers, or mineral-based. Mineral-based fibers have not been extensively researched in terms of fiber-reinforced composite materials because the asbestos component is harmful to human health, whereas plant-based fibers offer desirable qualities like lower cost, biodegradability, and availability, as well as excellent physical and mechanical characteristics [13]. Sisal and abaca are examples of leaf fibers, along with bast, grass and reed, core, seed and all other varieties, which may also possibly include roots and wood. The natural and synthetic matrices used in polymer composites—including polypropylene (PP), polyester epoxy, and polyethylene (PE)—are all based on petrochemicals. The most recent research advances includes the hybrid developments made of synthetic and natural fibers. Hybrid composites are composite materials that contain multiple types of fiber. Techniques for combining these fibers include stacking layers of fibers, blending fibers, combining two different types of fibers to create a hybrid interaction, placing fibers selectively where they are requirements for great force, and orienting each one in a specific direction [1].

Stacking fibers is the simplest technique in composites, whereas the others present some problems in achieving a favorable hybridization result [10]. Many researchers have found success in optimizing these materials for effective usage in certain fields by changing fiber composition, orientation, their size, or development techniques [14]. To effectively apply FRCs, it is important to understand their characteristics. Because of their substantial mechanical characteristics, FRCs are now used in a variety of applications [15]. These polymer composites vary from their specified specifications from time to time due to faults such as manufacturing flaws that lead them to diverge from the predicted improvement in mechanical characteristics. Fiber misalignment, waviness, and occasionally delamination, breakage, debonding, and the development of holes (voids) in the composite matrices are some examples of these production flaws. With even 1% increase in their space content, tensile, flexural and interlaminar shear strength of these composites reduce by 10–20, 10, and 5–10% respectively. By changing the processing parameters used in the manufacturing process, it can be removed [16, 17]. Understanding and investigating different types of composite production processes is, therefore, necessary to implement the best practices that will reduce flaws and help produce self-sustaining and durable composites, effective for the intended area of application [18]. A number of newly developed automated composite fabrication techniques use robotic support for fabrication, which results in full automation and a huge boost in the productivity. Traditional fabrication methods for the production of composite materials have been in use for several decades [19].

2 Composite Materials (Natural and Synthetic)

In a composite, the base material with which the filler is bonded or fixed in the structure is called matrix or adhesive material. The filler is a plate material, fragment, particle, fiber, or whisker made of natural or synthetic materials. Modern fillers in aerospace are also comprised of certain nanomaterials. Figure 1 shows the composite materials and categories according to their structure. Composite materials can be categorized based on the fiber length that make up their matrix structure. Continuous FRC materials are those reinforced by long fibers, whereas discontinuous FRC materials are those reinforced by short fibers. A composite material that fortifies two or more than two fibers in a single matrix structure is called as a mixed fiber reinforced composite [20–22]. The fibers can be embedded into the continuous fiber composite

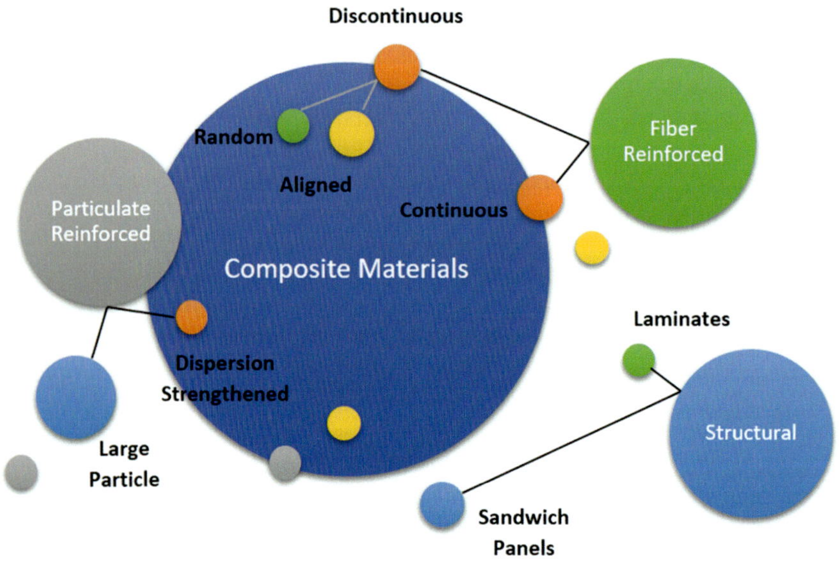

Fig. 1 Composite materials and their classification

material's matrix structure in one or both directions and they absorb the force from the fiber matrix in a very straightforward and efficient way. If the matrix is brittle, the discontinuous fibers need to be long enough to transfer loads effectively and to stop cracks from spreading. Otherwise, the material will fail. The performance as well as the structural behavior of these composite material are determined by the orientation and arrangements of the fiber [21, 23].

It is possible to determine improvements in qualities like impact strength and fatigue strength by using natural fibers that have undergone chemical treatment. The matrix structure of fiber-reinforced polymer (FRP) composite commonly uses glass, carbon, basalt, and aramid fibers in the dispersed phase [24]. Therefore, due to the current force on researchers to create environmentally friendly materials as a result of stringent environmental regulations, natural fiber polymer composites (NFPC) have potential modern industry applications [25].

The two main categories of fibers used in composite materials are natural fibers and synthetic fibers. Recent research has also demonstrated that the performance of the resulting hybrid composite material, formed by combining these two fibers with the matrix material, is unprecedented. Figure 2 depicts some natural and synthetic fibers. Table 1 shows the benefits and drawbacks of natural fiber composites (NFCs).

Advantages and disadvantages of NFCs. Source: https://doi.org/10.3390/polym13030423

Benefits	Drawback
Very good sound, acoustic, and electrical insulating properties	Moisture absorption, causing fibers to swell
Reactivity-materials provide sites for water absorption, and are also available for chemical modification	Restricted maximum processing temperature
Biodegradability: as a result of their tendency to absorb water, natural fibers will biodegrade under certain circumstances through the actions of fungi and\or bacteria	Lower durability, fiber treatments can improve this drawback
Combustibility: products can be disposed of through burning at the end of their useful service life, and energy can simultaneously be generated	Dimensional stability as a consequence of the hygroscopicity of fibers, products, and materials
Very good mechanical properties, especially tensile strength. In relation to their weight, the best fibers attain strength similar to Kevlar	Variability in quality, dependent on unpredictable variables such as weather
The abrasive nature of natural fibers is much lower compared to glass fibers, which leads to advantages in regards to the technical aspects, material recycling, or processing of composites materials	Less fire retardance
Plant fibers are renewable raw materials and their availability is unlimited	Lower strength properties, particularly impact strength

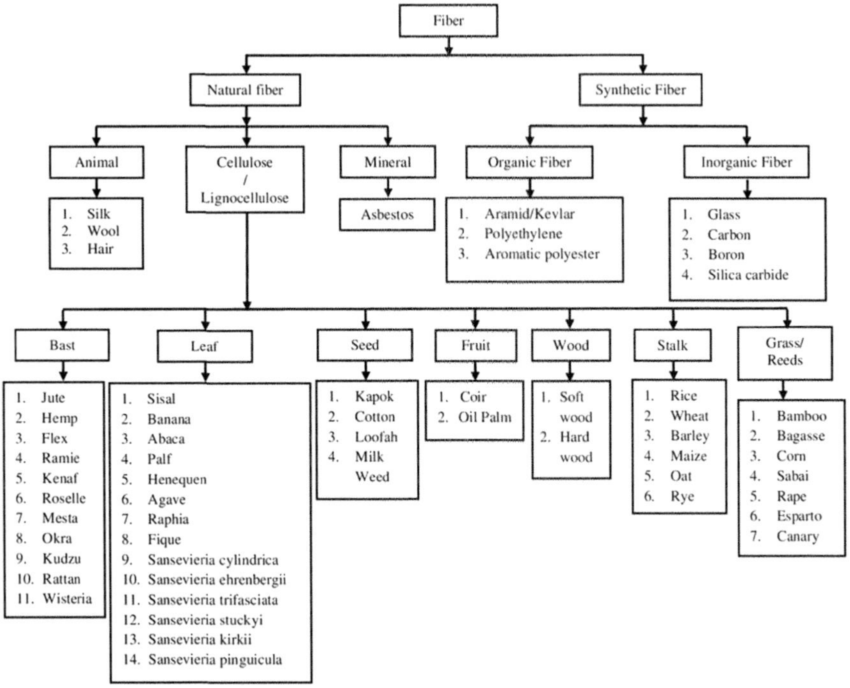

Fig. 2 Natural and synthetic fibers with their classification [26]

3 Available Natural Fiber-Based Hybrid Combinations and Their Properties

The available natural fiber-reinforced composites and their combinations are briefly described in this section, along with some of their key characteristics from existing literature. However, their use in the Aerospace sector is minimal as of now and is the grown area of research.

3.1 *The Bamboo/Glass Fiber Composites*

As far as the properties/characteristics of bamboo/GF reinforced composites is concerned [26] it clearly treated one has its tensile strength increased by a percentage of 5.7% with flexural strength, and modulus also increased to 23.5, and 32.3%, respectively. These bamboo/GF-reinforced unsaturated polyester composites have a substantial impact strength of 32 kJ/m^2 [27].

3.2 The Pineapple Leaves Fiber/Glass Fiber Composites

In these types of reinforced polyester composites, the Glass Fiber content is about 25% of the total fiber content. The tensile and flexural strengths are the highest when the GF content is 8.6 and 12.3%, respectively. In addition, a higher impact strength is obtained when the GF content is 12% by weight [28].

3.3 The Sisal/Glass Fiber Composites

The wool-sisal/fabric-GF-mat-reinforced polyester composite material has a fiber weight of 30%. Their break elongation depends on individual fiber reinforcements. The glass fiber loading varies from 0 to 8.5% by weight [29]. At 0% GF content, preliminary tensile and flexural strength of the composites are 69 and 99 MPa, respectively. Compare to sisal polyester composites, addition of GF usually increases both tensile and flexural strengths of sisal/glass hybrid composites, and it was observed that the maximum tensile and flexural properties were 5.7 and 2.8% of the GF load, respectively [30]. The bending strength also substantially rises by approximately 25%. Initial impact strength is 110 J/m at 0% glass fiber content. When the GF content is 8.5% (weight), the impact strength is increased by approximately 34%. Additionally, fiber plays a significant part in the magnitude of impact strength of composites. The fiber weight content of the sisal/glass hybrid composite material processed by alkali, cyanoethylation, and acetylation is 24.3% (sisal) and 5.7% (glass). Compared with other treated SF composite materials, the hybrid composite material treated with 5% alkali has better tensile strength. Tensile strength of glass hybrid composites based on cyanoethylation and acetylated SF is higher. The sisal/glass hybrid polyester composite with alkali treatment of 5% and acetylated SFs enhances the bending strength and boosts the impact strength by 8–6% [31]. The integration of GF in sisal-PP composites improves properties [32]. As compared with SFs, GFs have higher hardness and stiffness, so adding GFs to sisal-PP composites increases mechanical strength [33]. Due to the use of polyester resin, the tensile strength is increased to 176.20 MPa, but it is still lesser than the glass fiber-reinforced material [34].

3.4 The Roselle/Sisal Composites

These types are having random orientations, and are made with a constant weight ratio of 1:1 under different fiber content [35]. In these composites, the maximum flexural strength can be when the fiber length is 150 mm with 30% fiber content, and the maximum impact strength when the fiber length is 150 mm with 20% fiber content in a dry state than in a wet state [36].

3.5 The Sisal/Oil Palm Fiber Composites

For these hybrids, the mechanical properties in the longitudinal heading are better than in the transverse including higher tensile strength [37].

3.6 The Coir/Glass Fiber Composites

The properties like strength and modulus of these composites increase linearly as soon as the GF content is increased [38]. By increasing this content from 0 to 100%, there is a phenomenal increase in strength as well as the bending performance also increases The impact-bearing strength of untreated fiber composite material is 40 kJ/m^2 [39].

3.7 The Jute/Glass Fiber Composites

The researchers examined randomly oriented jute/GF composite materials. When sisal/glass and sisal/jute/GF composite materials are compared, the tensile strength was maximum of around 230 MPa. The jute/GF composite structures were made with a 14% weight content and a fiber length of 35 mm. The highest tensile strength was 62.99 MPa, which was lower than that of sisal/GF polymer composites [40]. The H2 laminate featured the best-woven jute/GF combination with the least amount of deflection, the highest peak load, and the best damage tolerance [41].

3.8 The Flax/Glass and Hemp/Glass Composites

These hybrids can have an impact strength of nearly 43 kJ/m^2 at 50% fiber without using any chemical treatment. At 41% fiber content, soybean oil as a matrix exhibited an impact strength of roughly 33.6 kJ/m^2. This will be 75 J/m^2 with 40% [42].

3.9 The Palmyra/Glass Fiber Composites

These reinforced roof lite composites with varying fiber lengths and 49–54% content can have improved properties and impact characteristics were obtained with 54% fiber content at 30 and 40 mm fiber length. With a fiber composition of 55%, all static mechanical characteristics were greater at 50 mm fiber length [43].

3.10 The Sisal/Silk Fiber Composites

In these unstructured polyester composites with fiber lengths of 1, 2, and 3 cm, the strength of 2 cm fiber-length composite is found greater than that of 1 and 3 cm. Furthermore, as compared to untreated composites, treated ones have higher values [43].

3.11 The Coir/silk Fiber Composites

Here also 2 cm fiber length possesses good properties and strength. Furthermore, treated fiber composite comes with good tensile strength [44]. This is attributed to mercerization which enhances surface-to-matrix adhesion because both lignin and hemicellulose get eliminated [45].

3.12 The Roystonea Regia/Glass Fiber Composites

Roystonea regia/GF reinforced composite's total fiber loading is kept constant at 20% by weight [46]. Mechanical parameters as well as impact strength generally, increased linearly with glass fiber loading. This was because GF had substantially greater strength and modulus than natural fiber. At 0% Regia and 20% glass fiber content, good properties and strength is attained [47].

3.13 The EFB/Jute Composites

Addition or loading of jute fiber is responsible for increase in tensile strength and modulus of these composites [48]. At the 1:4 fiber ratio, the highest tensile strength and tensile modulus can be observed. When compared to pure EFB composites, the hybridized OPEFB fiber, with woven jute fiber epoxy composites makes an improvement in both modulus and strength. The skin or core is basically a woven jute fiber mat was used as the skin or core for sandwich hybrid composites, and their tensile strength remains higher than that of pure EFB composite [49].

3.14 The Snake Grass/Banana and Snake Grass/Coir Fiber Composites

These composites at 20% fiber content show the highest tensile strength of snake grass combined with banana and coir fiber to be 48.6 and 44 MPa, respectively. Even the tensile modulus at the same fiber content in both banana and coir hybrid combinations was 710 and 660 MPa, respectively. At the same content, highest flexural strength of both the banana and coir hybrid combination was 88 MPa, and at 25% fiber content, it was 108 MPa [50].

3.15 The Regenerated Cellulose/Glass Fiber Composites

For this type of combination, the tensile stress parameter was defined as the ratio of maximal stress in the regenerated cellulose fiber to stress in the undamaged portion of the fiber. When the fiber/matrix contact is intact, the stress concentration factor is larger, and it decreases when they depend [51].

3.16 The Carbon/SiC Fiber/Boron Fiber Composites

Carbon/SiC fiber/boron-reinforced epoxy composites' compressive and flexural characteristics had the highest value with compressive strength of 991 MPa. SiC FRP achieved a maximum 2000 MPa and 220 GPa of flexural strength and modulus respectively at a fiber volume percentage of 51% [52].

3.17 The Carbon/GF-Reinforced Composites

One of the applications of this combination is hybrid composite tubes consisting of unidirectional type carbon/GF-reinforced plastic tubes [53]. The dynamic mechanical characteristics of fiber-reinforced composites are determined by the matrix material as well as the arrangement and orientations of reinforcing fibers [54]. The storage modulus rose as the weight percentage of fiber increased [55, 56].

4 Synthetic Versus Natural Composite Materials

Due to the significant variability in natural fiber qualities and testing methodologies, a broad range of values for their properties and strength may be found in the existing literature. The highest values for the same attributes were considered for the assessment [57]. Bast natural fiber offers high mechanical strength, making it ideal for aerospace and automotive applications. In comparison to other natural fibers, flax offers a wonderful mix of cheaper cost, lighter weight, and good mechanical qualities. Jute fiber is another extensively used natural fiber, however, its durability is less compared to flax or bast fiber [58]. Research shows that reinforced-type natural fiber thermosets lose energy at low values of stress and greater strain values than glass ones. However, the resin characteristics are the primary cause of energy deterioration. Natural fibers, such as jute, kenaf, flax, and hemp, possess low thicknesses and excellent mechanical qualities; and are better polymer reinforcements [59]. Glass fiber is typically less expensive than other synthetic fibers such as carbon, graphite, aramid, boron, and so on. Glass fiber composites now account for 90% of the composites market. Because of the lesser interference between the reinforcements and the binder, fibers are insufficient for use in the interiors of airplanes and cars. These techniques like chemical treatments, additives, and coatings can help improve the quality. Green composites have a substantially lower impact on the environment than glass fiber composites [60]. These composites are environmentally friendly, significantly decrease pollutant polymers, are lightweight, of good quality, and are biodegradable, resulting in lower carbon emissions when used for aerospace components. End-of-life natural fiber incineration yields good energy and carbon. Natural fibers are lighter and have better properties for aircraft and automobile industry [61]. Green fibers, on the other hand, have a variety of challenges in terms of achieving higher-quality mechanical qualities. Existing research investigations have revealed that glass fibers are damping resistant and very hydrophobic. In humid situations, natural fibers absorb environmental water, causing fiber irregularities within the composite and making it difficult to utilize for interior components of automobiles and airplanes [62]. Previous research has found that when a polymer composite is entirely immersed in water, its strength decreases by 13% to 31% when compared to 95% RH. The influence of hydrothermal and aging conditions on the properties of these fiber composites shows that the period of exposure is directly related to variations in humid levels [63]. Natural fibers with a higher cellulose content have a bigger fiber volume, which boosts their ability to tolerate a higher moisture percentage. Their environmental reliability shall be reduced for their use in the challenging aerospace sector [64].

5 Natural Reinforced Composites for Aerospace Applications

The eco-friendly, non-toxic, and biodegradable nature of natural fiber reinforced polymer composite materials has led to their increased use as a replacement for synthetic materials. Nonetheless, there are challenges that must be addressed to improve their properties and workability. https://doi.org/10.3390/polym13030423 FRPs are extensively used in the core aerospace industry to adopt materials that are with good durability, are thermal-resistant, light in weight for airplane construction and have superior properties. For example, natural fiber-reinforced thermoset and thermoplastic skins provide heat and flame resistance, simple recycling of materials, and material disposal that are less expensive and lighter than typical sandwich panels required for airplan, they are difficult to recycle. To address this, natural fiber/bio-composite materials have opened up new potential in the aerospace sector because they are biodegradable and low cost. Some fibers have conductive layers that eliminate the need for separate transmitter wires and power is transported across the fibers to particular electric devices. Over the last four decades, natural fiber and polymer matrix composite-based components have gained importance in the aerospace sector and with more emphasis on the use of green composites gained immense popularity. From small control surfaces to main lifting surfaces and the entire aircraft body, these composites are the leaders. Utmost care should be taken in their manufacturing and processing for specific purposes in the aerospace industry as per the standards. For example cavities and voids during impregnation etc. High fiber strength and stiffness paired with low density are apparent universal requirements for all sections of the airplane structure, that comes from these natural fiber based composites.

6 Conclusion and Outlook

Composite materials have shown many breakthroughs in several material qualities since their creation in the preceding century. Numerous research efforts have been made to generate optimal materials that will perform better for specific uses. Over the last few decades, fiber reinforcements in the matrix organizational structure of composites have yielded excellent performance, making them a preferred choice for high-end applications. To better understand the possibilities of varied composite materials in many areas, composite material classifications as well as the attributes of their constituent constituents have been investigated. Natural and synthetic fibers are the two types of fibers available for the production of fiber-reinforced composites. Synthetic fibers are firmer, while natural fibers are less expensive and compostable, set them in the category of environmentally friendly materials. The success of manufacturing process is influenced by the type of fiber material and matrix configuration or fiber material and volume. So the material dictates the production techniques. Composite constructions have increased in terms of material stiffness and strength while also significantly lowering weight. Composites have also demonstrated unexpected characteristics such as resistance to impact, wear, corrosion, and chemicals; however, these attributes are dependent on the material's structure, fiber type, and production method. Based on the properties required, composite materials are used in a range of sectors. More research will be undertaken in the future to identify fresh composite structures particularly natural hybrids by combining different versions and applying new manufacturing technologies.

The role of composites in aerospace industry is in high demand. For the next-generation aircraft and spacecrafts, natural fiber and bio-composite materials have significant scope in addition to environment friendly because they are biodegradable and have future potential in the aerospace sector, namely in the interior components of aircraft. However, difficulties such as its poor heat and fire resistance, water absorption, degradability, and variability in characteristics, as well as the lack of trustworthy prediction techniques, must be addressed in the future for successful aerospace applications.

References

1. Khan T, Hameed Sultan MTB, Ariffin AH (2018) The challenges of natural fiber in manufacturing, material selection, and technology application: a review. J Reinf Plast Compos 37(11):770–779
2. Khan T, Sultan M, Shah A, Ariffin A, Jawaid M (2019) The effects of stacking sequence on the tensile and flexural properties of kenaf/jute fibre hybrid composites. J Nat Fibers 1–12
3. Jawaid M, Khalil HA, Khanam PN, Bakar AA (2011) Hybrid composites made from oil palm empty fruit bunches/jute fibres: water absorption, thickness swelling and density behaviours. J Polym Environ 19(1):106–109
4. Javadian A (2017) Composite bamboo and its application as reinforcement in structural concrete. ETH Zurich
5. James JD, Manoharan S, Saikrishnan G, Arjun S (2019) Influence of bagasse/sisal fibre stacking sequence on the mechanical characteristics of hybrid-epoxy composites. J Nat Fibers 1–11
6. Yusoff RB, Takagi H, Nakagaito AN (2016) Tensile and flexural properties of polylactic acid-based hybrid green composites reinforced by kenaf, bamboo and coir fibers. Ind Crops Prod 94:562–573
7. Vieira Silva D (2019) New technologies to accelerate additive manufacturing processes on an industrial scale
8. Asim M, Jawaid M, Abdan K, Ishak MR (2017) The effect of silane treated fibre loading on mechanical properties of pineapple leaf/kenaf fibre filler phenolic composites. J Polym Environ 26(4):1520–1527
9. Phanthong P, Reubroycharoen P, Hao X, Xu G, Abudula A, Guan G (2018) Nanocellulose: extraction and application. Carbon Res Convers 1(1):32–43
10. Khan T, Sultan MTH, Shah AUM, Ariffin AH, Jawaid M (2019) The effects of stacking sequence on the tensile and flexural properties of kenaf/jute fibre hybrid composites. J Nat Fibers 1–12
11. Jawaid M, Thariq M (2018) Sustainable composites for aerospace applications. Woodhead Publishing
12. Izzati Zulkifli N, Samat N, Anuar H, Zainuddin N (2015) Mechanical properties and failure modes of recycled polypropylene/microcrystalline cellulose composites. Mater Des 69:114–123

13. Khan MN, Wahab MF (2007) Characterization of chemically modified corncobs and its application in the removal of metal ions from aqueous solution. J Hazard Mater 141(1):237–244
14. Jusoh MSM, Santulli C, Yahya MYM, Hussein N, Ahmad HAI (2016) Effect of stacking sequence on the tensile and flexural properties of glass fibre epoxy composites hybridized with basalt, flax or jute fibres. Mater Sci Eng Adv Res 1(4):19–25
15. Mohammed L, Ansari MN, Pua G, Jawaid M, Islam MS (2015) A review on natural fiber reinforced polymer composite and its applications. Int J Polym Sci 2015
16. Mohamad Hamdan MH, Siregar JP, Thomas S, Jacob MJ, Jaafar J, Tezara C (2019) Mechanical performance of hybrid woven jute–roselle-reinforced polyester composites. Polym Polym Compos 0967391119847552
17. Mishra V (2014) Physical, mechanical and abrasive wear behaviour of jute fiber reinforced polymer composites
18. Mihaela Raluca et al C (2018) Composite material designs for lightweight space packaging structures. Incas Bull 10(1):13–25
19. Malingam SD, Ng LF, Chan KH, Subramaniam K, Selamat MZ, Zakaria KA (2018) The static and dynamic mechanical properties of kenaf/glass fibre reinforced hybrid composites. Mater Res Express 5(9):095304
20. Baley C et al (2019) Specific features of flax fibres used to manufacture composite materials. IntJ Mater Form 12(6):1023–1052
21. Zhuo P, Li S, Ashcroft IA, Jones AI (2021) Material extrusion additive manufacturing of continuous fibre reinforced polymer matrix composites: a review and outlook. Compos B Eng 224:109143
22. Zuo P, Srinivasan DV, Vassilopoulos AP (2021) Review of hybrid composites fatigue. Compos Struct 114358
23. Sreejith M, Rajeev R (2021) Fiber reinforced composites for aerospace and sports applications. In: Fiber reinforced composites. Elsevier, pp 821–859
24. Bazan P, Nosal P, Wierzbicka-Miernik A, Kuciel S (2021) A novel hybrid composites based on biopolyamide 10.10 with basalt/aramid fibers: mechanical and thermal investigation. Compos B Eng 223:109125
25. Yang G, Park M, Park S-J (2019) Recent progresses of fabrication and characterization of fibers-reinforced composites: a review. Compos Commun 14:34–42
26. Sathishkumar T, Naveen JA, Satheeshkumar S (2014) Hybrid fiber reinforced polymer composites—a review. J Reinf Plast Compos 33(5):454–471
27. Yang MFM, Hamid H, Abdullah AM (2018) Potential use of cellulose fibre composites in marine environment—a review. Eng Appl New Mater Technol 25–55
28. Priyanka P, Dixit A, Mali H (2017) High-strength hybrid textile composites with carbon, kevlar, and e-glass fibers for impact-resistant structures. A review. Mech Compos Mater 53(5):685–704
29. Panda JN, Bijwe J, Pandey RK (2019) Optimization of the amount of short glass fibers for superior wear performance of PAEK composites. Compos A Appl Sci Manuf 116:158–168
30. Carvalho GB, Canevarolo SV Jr, Sousa JA (2020) Influence of interfacial interactions on the mechanical behavior of hybrid composites of polypropylene/short glass fibers/hollow glass beads. Polym Test 85:106418
31. Yan L, Chouw N, Huang L, Kasal B (2016) Effect of alkali treatment on microstructure and mechanical properties of coir fibres, coir fibre reinforced-polymer composites and reinforced-cementitious composites. Constr Build Mater 112:168–182
32. Li M et al (2020) Recent advancements of plant-based natural fiber–reinforced composites and their applications. Compos Part B Eng 108254
33. Nurazzi N et al (2021) A review on mechanical performance of hybrid natural fiber polymer composites for structural applications. Polymers 13(13):2170
34. Yohanes G (2020) Experimental investigation on mechanical properties of hybrid composite of polyester matrix reinforced by ceramic particles and jute fiber. ASTU
35. Sanjeevi S et al (2021) Effects of water absorption on the mechanical properties of hybrid natural fibre/phenol formaldehyde composites. Sci Rep 11(1):1–11

36. Algin Z, Gerginci S (2020) Freeze-thaw resistance and water permeability properties of roller compacted concrete produced with macro synthetic fibre. Constr Build Mater 234:117382
37. Ahmed MM, Dhakal H, Zhang Z, Barouni A, Zahari R (2021) Enhancement of impact toughness and damage behaviour of natural fibre reinforced composites and their hybrids through novel improvement techniques: a critical review. Compos Struct 259:113496
38. Sailesh A, Palanikumar K (2021) Mechanical properties of flax-cotton fiber reinforced polymer composites. In: Green composites. Springer, pp 393–411
39. Santulli C (2007) Impact properties of glass/plant fibre hybrid laminates. J Mater Sci 42(11):3699–3707
40. Sinha AK, Narang HK, Bhattacharya S (2017) Mechanical properties of natural fibre polymer composites. J Polym Eng 37(9):879–895
41. Chang BP, Mohanty AK, Misra M (2020) Studies on durability of sustainable biobased composites: a review. RSC Adv 10(31):17955–17999
42. Hernandez Michelena A (2019) Natural fibre reinforced composite materials. University of Plymouth
43. Larsen IL, Thorstensen RT (2020) The influence of steel fibres on compressive and tensile strength of ultra high performance concrete: a review. Constr Build Mater 256:119459
44. Jogur G, Manjunath R, Khatkar V (2020) Efficacy of silk and its proteins in bio-medical applications. In: Advances in functional and protective textiles. Elsevier, pp 359–395
45. Vijay R, Manoharan S, Arjun S, Vinod A, Singaravelu DL (2021) Characterization of silane-treated and untreated natural fibers from stem of Leucas aspera. J Nat Fibers 18(12):1957–1973
46. Ghalia MA, Abdelrasoul A (2019) Compressive and fracture toughness of natural and synthetic fiber-reinforced polymer. In: Mechanical and physical testing of biocomposites, fibre-reinforced composites and hybrid composites. Elsevier, pp 123–140
47. Qadir HH, Faraj RH, Sherwani AFH, Mohammed BH, Younis KH (2020) Mechanical properties and fracture parameters of ultra high performance steel fiber reinforced concrete composites made with extremely low water per binder ratios. SN Appl Sci 2(9):1–12
48. Asma B, Hamdi L, Ali B, Youcef M (2020) Flexural mechanical properties of natural fibre reinforced polymer composites—a statistical investigation. Fibers Polym 21(10):2321–2337
49. Singh VK, Mukhopadhyay S (2020) Hybrid biocomposites. Indian J Fibre Text Res (IJFTR) 45(2):224–246
50. Puttegowda M, Rangappa SM, Jawaid M, Shivanna P, Basavegowda Y, Saba N (2018) Potential of natural/synthetic hybrid composites for aerospace applications. In: Sustainable composites for aerospace applications. Elsevier, pp 315–351
51. Judt PO, Zarges J-C, Feldmann M, Ricoeur A, Heim H-P (2019) Deflecting mode-I cracks in anisotropic materials. Mech Mater 136:103060
52. Vigneshwaran S et al (2020) Recent advancement in the natural fiber polymer composites: a comprehensive review. J Cleaner Prod 124109
53. Zhou J, Guan Z, Cantwell W (2018) The energy-absorbing behaviour of composite tube-reinforced foams. Compos B Eng 139:227–237
54. Bhudolia SK, Perrotey P, Joshi SC (2017) Enhanced vibration damping and dynamic mechanical characteristics of composites with novel pseudo-thermoset matrix system. Compos Struct 179:502–513
55. Jawaid M, Khalil HA (2011) Effect of layering pattern on the dynamic mechanical properties and thermal degradation of oil palm-jute fibers reinforced epoxy hybrid composite. BioResources 6(3):2309–2322
56. Jawaid M, Alothman OY, Saba N, Tahir PM, Khalil HA (2015) Effect of fibers treatment on dynamic mechanical and thermal properties of epoxy hybrid composites. Polym Compos 36(9):1669–1674
57. Mohanty A, Misra MA, Hinrichsen G (2000) Biofibres, biodegradable polymers and biocomposites: an overview. Macromol Mater Eng 276(1):1–24
58. Kumar R, Ul Haq MI, Raina A, Anand A (2019) Industrial applications of natural fibre-reinforced polymer composites–challenges and opportunities. Int J Sustain Eng 12(3):212–220

59. Jothibasu S, Mohanamurugan S, Vijay R, Singaravelu DL, Vinod A, Sanjay M (2018) Investigation on the mechanical behavior of areca sheath fibers/jute fibers/glass fabrics reinforced hybrid composite for light weight applications. J Ind Text 1528083718804207
60. Oliveux G, Dandy LO, Leeke GA (2015) Current status of recycling of fibre reinforced polymers: review of technologies, reuse and resulting properties. Prog Mater Sci 72:61–99
61. Panthapulakkal S, Raghunanan L, Sain M, Birat KC, Tjong J (2017) Natural fiber and hybrid fiber thermoplastic composites: advancements in lightweighting applications. In: Green composites. Elsevier, pp 39–72
62. Alsubari S, Zuhri M, Sapuan S, Ishak M, Ilyas R, Asyraf M (2021) Potential of natural fiber reinforced polymer composites in sandwich structures: a review on its mechanical properties. Polymers 13(3):423
63. Baghaei B, Skrifvars M (2016) Characterisation of polylactic acid biocomposites made from prepregs composed of woven polylactic acid/hemp–lyocell hybrid yarn fabrics. Compos A Appl Sci Manuf 81:139–144
64. Atmakuri A, Palevicius A, Kolli L, Vilkauskas A, Janusas G (2021) Development and analysis of mechanical properties of caryota and sisal natural fibers reinforced epoxy hybrid composites. Polymers 13:864, ed: s Note: MDPI stays neutral with regard to jurisdictional claims in published ..., 2021

Advanced Natural/Synthetic Polymer Hybrid Composites of the Future for the Aerospace Industry

Balbir Singh, Kamarul Arifin Ahmad, M. Manikandan, Raghuvir Pai, Eddie Yin Kwee Ng, and Noorfaizal Yidris

Abstract From the standpoint of economic and ecological compatibility, the term hybrid composites are becoming increasingly important. These are composite materials that are composed two or more separate green and artificial/synthetic fibers that are reinforced with appropriate polymetric matrices to produce a composite material with qualities that are equivalent to those of manufactured composite materials. In addition, to stay up with new technological developments, it has essential to design materials that are less harmful to the environment and to protect our ecosystem for many years. This notion has prompted people to adopt hybrid composite materials in everyday engineering and non-engineering applications. These composites have several beneficial characteristics, cost, recyclability, and biodegradability, which made them an excellent option for composite polymers in a variety of applications. In this respect, a concerted effort has been adopted in this chapter to provide a summary of diverse green and artificial/synthetic fibers, their categorization, and their usage

B. Singh (✉) · K. A. Ahmad · N. Yidris
Department of Aerospace Engineering, Faculty of Engineering, Universiti Putra Malaysia, Selangor Darul Ehsan, 43400 UPM Serdang, Malaysia
e-mail: balbir.s@manipal.edu

K. A. Ahmad
e-mail: aekamarul@upm.edu.my

B. Singh · M. Manikandan
Department of Aeronautical and Automobile Engineering, Manipal Institute of Technology, Manipal Academy of Higher Education, Manipal 576104, Karnataka, India

K. A. Ahmad
Aerospace Malaysia Research Centre, Faculty of Engineering, Universiti Putra Malaysia, Selangor Darul Ehsan, 43400 UPM Serdang, Malaysia

R. Pai
Department of Mechanical and Manufacturing Engineering, Manipal Institute of Technology, Manipal Academy of Higher Education, Manipal, Karnataka 576104, India

E. Y. K. Ng
School of Mechanical and Aerospace Engineering, College of Engineering, Nanyang Technological University, 50, Nanyang Avenue, Singapore 639798, Singapore

T. Khan and M. Jawaid (eds.), *Green Hybrid Composite in Engineering and Non-Engineering Applications*, Composites Science and Technology, https://doi.org/10.1007/978-981-99-1583-5_6

in numerous engineering applications including aerospace; the chapter begins with a discussion of composite materials and their varied features.

1 Introduction

Many researchers are presently being compelled to produce more novel materials that may be used in a variety of applied applications while also fulfilling ecological and financial values due to global technological progress. As a result of this necessity, composite materials have been developed [1]. Composites are made up of two macroscopically distinguishable components that work together to provide a better outcome [2]. Continuous and discontinuous phases are used to describe the two materials, with the reinforcement phase or reinforcing material being harder and stronger than the continuous or matrix phase [3]. The matrix phase contains metallic, polymeric, or ceramic elements. A polymer matrix composite is made of polymers, whereas a reinforced composite is made of particulates or fibers [4]. The growing concern about environmental concerns, as well as the demand for additional green materials, has drawn researchers' attention to the usage of fiber-reinforced polymer composites (FRPCs) [5]. These materials have numerous advantages, which include low weight, high strength, and excellent corrosion and fatigue resistance. FRPCs can be made even with or without filler using natural and synthetic fibers. Figure 1 displays the classification of various fibers [5, 6]. Natural fiber composites are often divided into three categories: Animal fibers, vegetable or cellulosic fibers, and mineral fibers. Cellulosic fibers have been increasingly popular in recent years due to their environmental benefits and capacity to minimize international energy usage and environmental concerns [7]. They are divided into two categories: major utility groups, which include hemp, jute, kenaf, and other plant by-products, and secondary utility groups, which include coir, pineapple, and other plant by-products [8]. Bast fibers shown in Fig. 1. and wood fibres are all examples of these fibres (softwood and hardwood) [9]. They are used in applications such as automotive parts, infrastructure Sports products, and furniture items, pipelines etc. and economical reinforcing material [10].

Animal fiber is also most normal natural fiber used as reinforcement in composites after plant fiber. Wool from sheep, bison, cashmere, alpaca, and other animals, as well as feathers from chickens, silk, and hair from other sources, are used to make these fibers [12]. These are the second most readily accessible natural fiber after plant fiber and, as a result, are more costly [13, 14]. Natural fibers are sustainable and offer various advantages over synthetic fibers in terms of properties and applications [15]. Although natural fibers are more ecologically benign than SFCs, they do have certain limitations when it comes to use them ass raw fibers, for example, moisture, and lthermal stability, making it difficult to produce composites [16]. Natural fibers are fine strings formed by nature, while synthetic fibers are mineral-based artificial fibers, and SFCs are composites constructed from these fibers [17]. These fibers are one of the most significant types especially in industries where material properties

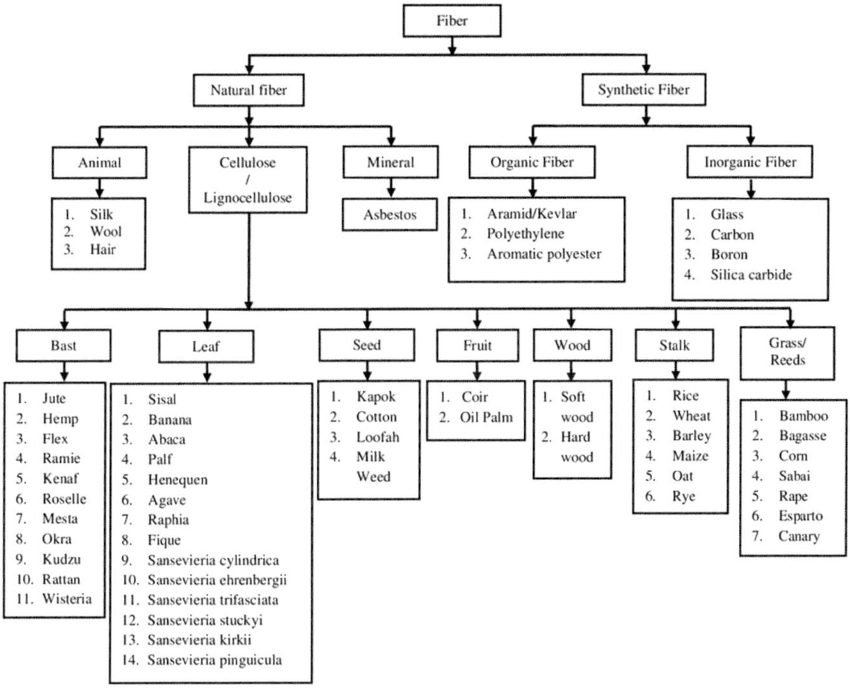

Fig. 1 Natural and synthetic fibers classification [11]

are critical, such as aerospace, automotive, and energy sectors [18]. Glass, carbon, aramid, polyethylene, alumina, and silicon carbide are some of the most common synthetic fibers. Because of its inexpensive cost, glass is more often used fiber in composites [19]. Natural resource use has expanded significantly over time and is now being recognized as a viable alternative to manmade materials. Natural fibers act as reinforcements in composites as a result of this, and they have several benefits over synthetic fibers [20]. The biocompatibility, low density, and processing flexibility of natural fibers are just a few of the benefits that have piqued researchers' curiosity [21]. In reinforcing composites, natural fibers have various economic, technological, and environmental advantages over synthetic fibers [22].

There are three types of matrices, with polymer matrices being the most popular in fiber-reinforced composites [23]. Polymers may be found in practically every component we use in our everyday lives, and they can also be readily processed and produced to a particular size and form while maintaining the needed qualities [24]. When these polymers are suitably reinforced with fibers, their characteristics can be improved. In a polymer matrix, both synthetic and natural fibers can be employed as reinforcements [15, 25]. Because of its inherent features, such as simplicity of manufacture, structural control, productivity, ease of access, reduced physical labor, and cost reduction, polymers have largely supplanted metals [21, 25]. Different thermosetting and thermoplastic polymers have different characteristics.

2 Natural Versus Synthetic Polymer/Hybrid Composites

2.1 *Natural Fibers*

More natural fibers are being used in composite applications as a consequence of the desire for green materials to combat environmental issues [26]. TBecause of their ease of availability and renewability, plant-based natural fibers are the most widely used, whereas animal and mineral fibers are rarely utilized [27]. This section discusses some of the most important plant fibers, and various animal fibers [28].

Kenaf natural fibers

Kenaf is a commercially accessible and inexpensive bast fiber belonging to the Malvaceae family and the plant (Hibiscus cannabinus) [29]. It is primarily cultivated in Asia, Africa, North America, and Europe. A strong, tough, and long-lasting fiber that matures in about less than 100 days. The plant reaches a height of 3 m and a diameter of 3–5 cm [30]. This fiber's low cost, low density, biocompatible, recyclability, relatively high mechanical properties, and insect resistance need for less pesticide use are just a few of its advantages [31]. Figure 2 depicts the kenaf plant with its fibers, and cloth. All bast fibers and kenaf fibers are now us as reinforcement in polymer composites and are used in industries such as rope manufacture, paper production, sporting goods, and the automobile industry [27]. Its mechanical and physical characterizations are shown in Table 1.

Jute natural fibers

Jute is a form of bast fiber that belongs to the Corchoruscapsularis/Corchorusolitorius genus and is mostly grown in south Asia [34]. It is also referred to as "Golden Fiber," and it's gaining popularity owing to its excessive properties, such as high moisture adsorption capacity and low cost in nature [35]. It develops to a scale of 2.5–3.5 m in around 3–6 months, thereafter the fibers may be extracted using a retting technique [36]. Jute fibers are used in a range of polymer-based composites [15] as shown in Fig. 3, with properties Table 1.

Hemp natural fibers

Hemp bast fibers are oldest environmentally acceptable fibers, derived from the plant's stem [27]. It is native to Central Asia and Central Europe and is a member of the Mulberry family and the species Cannabis Sativa. Hemp may grow upto 10 feet and has a rather narrow width [38]. Hemp fibers are stiff, making them an excellent reinforcement for composite materials [20] (Fig. 4 and Table 2).

Coir natural fibers

Coir is a low-cost fruit fiber called Cocos Nucifera botanically. As a consequence of coconut husk extraction, it is cultivated mostly in tropical countries [41]. The coconut fruit has a hard inner shell and a fibrous outer shell, and it is collected manually or using a crushing machine. Coir fibers are available in two colors: brown

Fig. 2 Kenaf **a** stem; **b** leaves; **c** flower; **d** seed [32]

Table 1 Characterizations of kenaf and jute fiber [33]

Properties	Kenaf fiber	Jute fiber
Density (g/cm^3)	1.4	1.3
Tensile strength (MPa)	930	393–773
Young's modulus	20	26.5
Elongation at break (%)	1.6	1.5–1.8
Cellulose content (%)	53–57	58–63
Hemicellulose content (%)	15–19	20
Lignin content (%)	5–11	12–14
Diameter	–	20–200

and white. Coconut fiber is dark in mature coconuts and white in young coconuts [42]. Coir fibers are now employed in a variety of applications, with the manufacture of building boards, roofing materials, building panels, and insulating material boards, as well as fiber-reinforced composites [16] (Fig. 5 and Table 3).

Banana natural fibers

The Musaceae family includes the banana, as one of the world's oldest plants [46]. A tyyical banana plant comprises of 30 large leaves, each around 2 meters lin length and 30-60 cm in width [47]. The fibers are often gathered from trash created during the

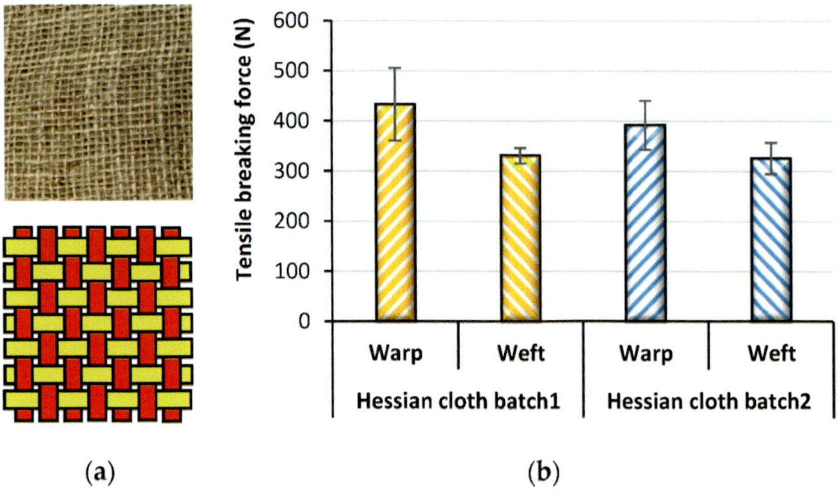

Fig. 3 **a** Hessian Jute fabric with schematic representation of plain weave and **b** properties of jute fabrics in different directions from two different batches [37]

Table 2 Properties of hemp fiber [39]

Properties	Hemp fiber
Density (g/cm^3)	1.4
Tensile strength (MPa)	550–900
Young's modulus	20
Elastic modulus (GPa)	70
Specific strength (s/g)	393–643
Specific modulus ("/g)	50
Elongations at failure (%)	1.6
Moisture absorption (%)	6–12

growth process, its fiber, and the cloth it produces [48]. It is also extensively utilized in the structure, car industries, and equipment components engineering industries. Banana fiber also has a significant amount of cellulose, making it a more helpful strengthening component in composites [49] (Fig. 6 and Table 4).

Bamboo natural fibers

Bamboo is a versatile crop that belongs to the Bambusoideae family and is commonly utilized as durable reinforcing composite material for mechanical purposes [51]. Bamboo plants have a massive assembly and develop swiftly, producing in as little as 4–5 ages after being planted. Within three months, this plant may reach a height of around 200 feet [52]. Bamboo is mostly grown in Asia, Central, and South America [52] (Fig. 7).

Fig. 4 Hemp plant (Cannabis sativa Linn) [40]

Fig. 5 Coconut husk and fiber [43]

Table 3 Properties of coir fiber [44]

Properties	Coir fiber
Density (g/cm^3)	1.2
Tensile strength (MPa)	473
Young's modulus (MPa)	19,500
Elongation at break (%)	1.17
Cellulose content (%)	65
Hemicellulose content (%)	20
Lignin content (%)	12
Diameter	115

Table 4 Properties of banana fiber [45]

Properties	Banana fiber
Density (g/cm^3)	1350 kg/m^3
Tensile strength (MPa)	56
Young's modulus (MPa)	3.5
Elongation at break (%)	2.60
Cellulose content (%)	65
Hemicellulose content (%)	19
Lignin content (%)	5
Diameter	80–250

Fig. 6 Banana Musa sp. plantation [50]

Fig. 7 Bamboo plantation in fields [53]

Since it has ecologically favorable characteristics, such as increased growth rate, less mass, good strength, and biodegradability, as well as the fact that its roots and leaves hold the soil together and shield it from the sun, it is a good candidate for polymeric composite reinforcement [54]. Bamboo fiber composites are used in flooring., ceilings, and buses, as well as furniture, ornamental items, and sporting goods [16] (Table 5).

Animal fibers

These fibers are used widely besides plant fibers. The bulk of them are proteins, and they can function as reinforcement in composites [56]. Some of the most significant forms of animal fibers are addressed in this section. Angora fiber is a multimillion-dollar textile fabric derived from a variety of animals including sheep, goats, camels,

Table 5 Properties of bamboo fiber [55]

Properties	Bamboo fiber
Density (g/cm^3)	0.6–1.1
Tensile strength (MPa)	140–1150
Young's modulus (MPa)	11–17
Elongation at break (%)	2.5.3.7
Cellulose content (%)	26–43
Hemicellulose content (%)	19
Lignin content (%)	21–31
Diameter	25–40

rabbits, and additional creatures [57]. Cashmere fiber is a rich, soft-wool fiber made from the cashmere goat, whereas qiviut wool is a more expensive smooth fabric originating from the musk ox [58]. Wool fibers are mostly produced in South East Asia, and China, with annual sheep wool output exceeding five hundred thousand metric tons. Because pure wool is costly, waste wool is used as reinforcement/filler in composites [59]. Silk, an important natural protein fiber used in textile industry and comes from a range of sources. The fibers are derived from a variety of insects, with the bulk of silk originating from butterfly larvae, Mulberry silk (Bombyx mori) from silkworms, and dragline silk (Nephila) from spiders [60].

2.2 *Artificial/Synthetic Fibers*

These are now the most common form of material utilized in composite products as reinforcement. Glass, aramid, carbon, Kevlar, and other forms of synthetic fibers are addressed in this section.

Carbon fibers

These are a type of new material that contains 90% carbon and is commonly seen as a reinforcement in polymer materials [61]. Carbon fibers are made from polyacrylonitrile fiber, petroleum pitch, and rayon fiber [62] (Fig. 8).

Carbon fibers are used in applications that need high strength, difficulty, less weight, chemical inertness, high oscillation, and excellent fatigue properties [64]. They come in a variety of shapes, including constant filament tows, sprang fibers, and mats. Carbon fibers have good brittleness qualities when associated with glass or aramid fibers [65] (Table 6).

Glass fibers

Originally, heated rods were drawn into filaments to make this fiber. These fibers are available as roving, chopped strand, textiles, matting, and yarns, among other forms [67]. Glass fibers have varied qualities that make them appropriate for use in polymer

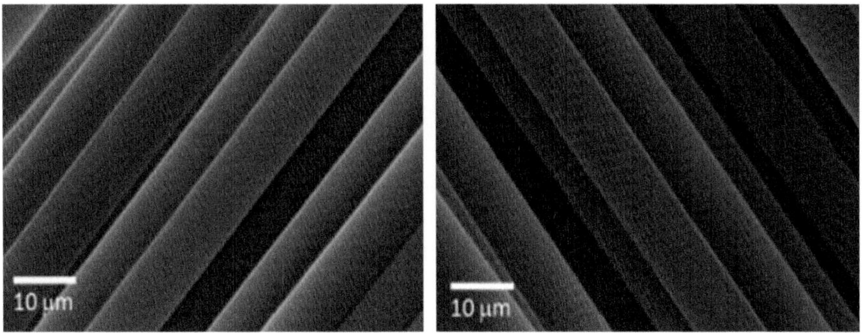

Fig. 8 SEM images of carbon fiber surfaces: original fiber and with ethanol spray on the right [63]

Table 6 Properties of carbon fiber [66]

Properties	Carbon fiber
Density (g/cm^3)	1.8
Tensile strength (MPa)	1900–3700
Young's modulus (MPa)	350–550
Elongation at break (%)	0.4–0.7
Carbon content (%)	93

composites, including high difficulty, strength, elasticity, strong chemical-resistance, and great insulating capabilities [68]. Glass fibers are available in a variety of forms for use as reinforcement in composites, including E-glass, AR-glass, R-glass, and S-glass. E-glass is extensively used fiber among them [69].

Aramid fibers

Aramid is a popular natural fiber of aromatic polyamide that is commonly utilized in composite reinforcement [70]. Aroma diacid derivatives and aromatic diamines are frequently combined to make this fiber. Markets the widely viable aramid fibers under the Kevlar trademark [71]. Aramid fibers have several desirable characteristics, including low density, toughness, and shock resistance [72]. The tensile strength is comparable to Glass fiber. They have several disadvantages, including as poor compression qualities and susceptibility to moisture and heat [73].

2.3 *Types of Natural/Synthetic Fibers Composite Materials*

Awareness of environmental issues has led to greater use of NFCs as polymer matrix composite reinforcement, despite the fact that composites composed of the same reinforcing composite materials do not function well since they are subjected to varying loading conditions over their service lifetimes [74]. To address this issue, hybrid composites are already being employed in the producing mny varities of items that were previously made employing traditional materials [75]. The fundamental benefit is their affordability and strength, that are comparable to traditional materials [76]. Polymers can also be utilized in home furniture items [77]. They're used in the maritime industry to make boat hulls, fishing poles, and other things. They're used to make tennis rackets, bicycles, sports helmets, and other sports equipment, as well as storage items including post boxes, grain storage silos, and biogas tanks [77]. As these hybrid composites are now successfully employed in a variety of industrial and engineering fields.

3 Natural/Synthetic Fibers Composite Materials Application in the Aerospace Industries

The aviation industry is another key use for hybrid composites [78]. The aerospace industry's unquenchable need to enhance marketable and defense aviation has caused the introduction of strong composite materials structurally that are currently being used in the construction of aircraft components [79]. The fact that they have a higher stiffness-to-density ratio, as well as improved physical qualities, is another reason for their appeal [80].

The production of acceptable materials for constructing modern aircraft components has also profited from the advent of stronger reinforcements such as glass and carbon fibers, as well as developments in polymers [81]. Such composite materials are being employed in military fighter planes, large and small commercial cargo planes, helicopters, satellites, and rockets, among other applications [82]. A filament winding procedure is used to make rocket engine casings for launch vehicles and missiles. To make randoms, resin injection molding is employed [83].

4 Technical Requirements in the Aerospace Industry

The choice of aerospace NFC is critical for the design of aviation products and parts [84]. It has an impact on many elements of aircraft performance, including ultimate strength, flying effectiveness, payload, power consumption, safety and dependability, lifespan cost, recyclability, and disposability from design through disposal [85]. Aside from meeting the minimum requirements, increasing structural proficiency in aviation design is attractive and increasingly significant because lightweight structures improve aircraft utilization, such as improved energy efficiency, excitation behaviors, load, flight-endurance, and inferior life cycle cost and greenhouse emissions [86]. Earlier investigations have found that lowering density (rather than improving stiffness or strength) is the utmost effective method to enhance structural efficiency [87]. Aluminum, titanium alloys, high-strength steel, and composite materials make up more than 90% of the weight of airplane structures [88]. Before 2000, lightweight aluminum alloys have been the most common aerospace structural materials (accounting for 70–80% of the weight of most commercial airplane airframes), and they continue to play a significant role [89]. Because of the production of high-performance composite materials, the percentage of composites used in aerospace components has grown since the middle of the 1960s and 1970s. Figure 9 shows material distributions for a variety of Boeing products [90].

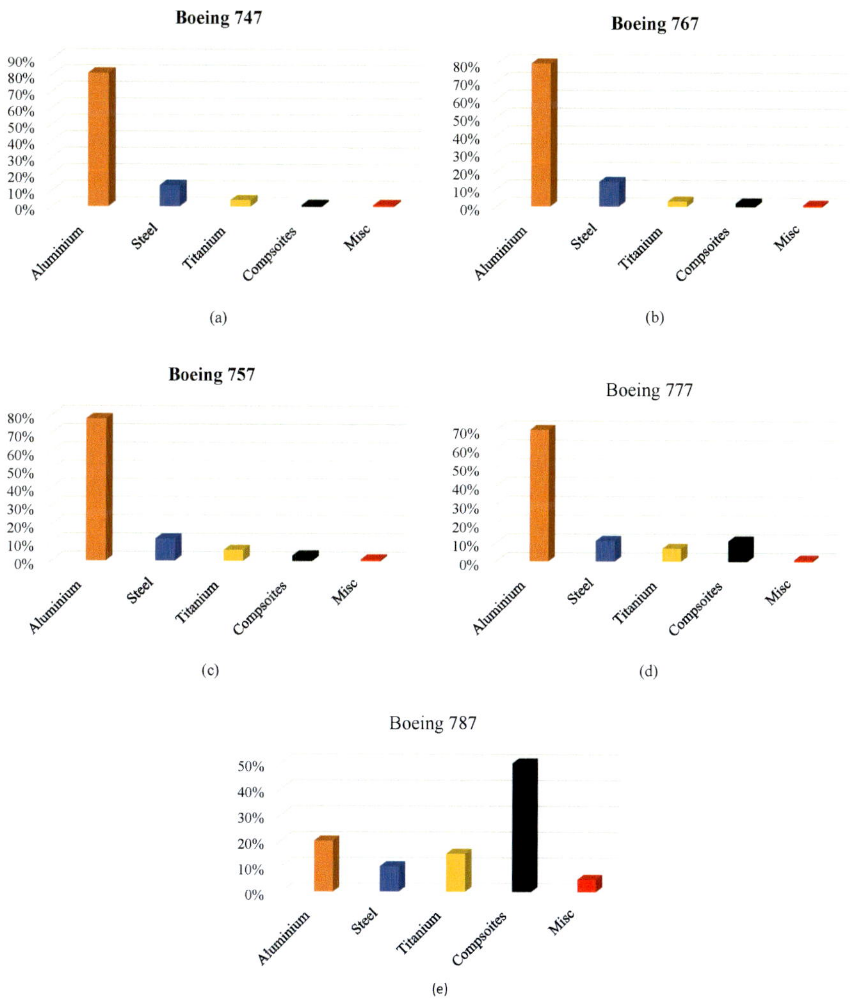

Fig. 9 Selection and use of material and their distributions for different Boeing aircrafts

5 Advantages and Disadvantages

The most extensively utilized resin structures are polyesters [91]. Epoxies are ranked second because of their higher cost, although they have stronger adhesion and shrinking than polyesters [92]. Composite materials have their own set of benefits and drawbacks. When using composites in a design, trade-offs should be addressed based on their intended use [93]. Composite material uses and applications incorporates everything from everyday things to building materials [94]. Window frames, bathtubs, and doors are just a few examples of composite-based products. Carbon nanotubes and CFRPs have also been used in sports products such as rackets and bicycles [95]. Composites have been used in the building of various bridges and utility poles in infrastructural development. Carbon composites and glass composites are among the composites used [96]. Composites are in great demand in the aerospace sector because they have excellent strength-to-weight ratio. In aerospace, weight is a crucial factor. It has an impact on an aircraft's ability to fly [97]. Previously, airplane structures were entirely made of metal, resulting in a bulky overall construction. Composites were shown to be capable of drastically reducing the weight of the aircraft while delivering equivalent or even higher structural integrity than metals once they were introduced [98]. Composites are now being used in the design of more and more aircrafts in order to make them lighter. A lightweight airplane consumes less fuel and emits lower environmental impact as a result [99]. The Airbus A380 and Boeing 787 Dreamliner are two commercial airplanes that use composites in their design [100]. The A380 is made up of around 25–30 tons of composites, with CFRP accounting for 85 percent of the total. The B787 Dreamliner, but at the other hand, is made up of materials to the tune of 50%, with the remaining 20% aluminum and 30% titanium [101]. For Hubble Space Telescope, the stiffness and low thermal expansion coefficient of its high gain antenna was achieved using graphite fibers dispersed in aluminum matrix. Fiber-reinforced SiC is used in airplane brakes to bear the enormous temperatures greater than 1000 °C. For thermal safety of space crafts, ablative synthetic polymer containing zirconia fibers are used due to its high temperature resistance.

6 Future Perspective and Applications of Composites Materials

Composite materials are employed in several industries, including automobile and aviation, wind energy, electrical, sports, home purposes, building engineering, and pharmaceutical and chemical industries [102]. Polymer composites have tremendous potential for structures that are primarily subjected to compression stress [103]. Because of their high compressive strength, adaptability in manufacturing composite structure shells, lightweight, low density, and corrosion resistance, hybrid composites are desirable. Composite materials are used to make a variety of parts for automobiles and aerospace industries because of their excellent qualities [78].

7 Conclusions

This chapter presents an overview of the significance of various natural and synthetic fibers, their classification, production and manufacturing, as well as applications in day-to-day life. The chapter also provides the importance of hybrid type composites, as well as their varied characteristics and applications in industrial and engineering fields particular in aerospace. Furthermore, a few of the benefits of composite materials over it so standard material properties, including high fatigue and microstructural features, structural applications ratio, and, most crucially, elevated properties, have made them a suitable materials for aerospace production industry and a range of other industry applications where they are extensively used and also play a key role.

References

1. Khan T, Hameed Sultan MTB, Ariffin AH (2018) The challenges of natural fiber in manufacturing, material selection, and technology application: a review. J Reinf Plast Compos 37(11):770–779. https://doi.org/10.1177/0731684418756762
2. Tee YL, Peng C, Pille P, Leary M, Tran P (2020) PolyJet 3D printing of composite materials: experimental and modelling approach. Jom 72(3):1105–1117
3. Selvaraj D, Silva F, Campilho R, Baptista A, Pinto G (2019) Influence of the natural additive on natural fiber reinforced thermoplastic composite. Procedia Manuf 38:1121–1129
4. Ramu P, Kumar CJ, Palanikumar K (2019) Mechanical characteristics and terminological behavior study on natural fiber nano reinforced polymer composite–a review. Mater Today: Proceed 16:1287–1296
5. Khan MZ, Srivastava SK, Gupta M (2018) Tensile and flexural properties of natural fiber reinforced polymer composites: a review. J Reinf Plast Compos 37(24):1435–1455
6. Rohit K, Dixit S (2016) A review-future aspect of natural fiber reinforced composite. Polym Renew Resour 7(2):43–59
7. Węcławski BT (2015) The potential of bast natural fibres as reinforcement for polymeric composite materials in building applications. Brunel University London
8. Aaliya B, Sunooj KV, Lackner M (2021) Biopolymer composites: a review. Int J Biobased Plast 3(1):40–84
9. Devireddy SBR (2017) Mechanical, thermal and physical properties of hybrid banana-jute fibers reinforced epoxy and polyester composites: modeling and experiments
10. Mohammed L, Ansari MN, Pua G, Jawaid M, Islam MS (2015) A review on natural fiber reinforced polymer composite and its applications. Int J Polym Sci 2015
11. Sathishkumar T, Naveen J, Satheeshkumar S (2014) Hybrid fiber reinforced polymer composites—a review. J Reinf Plast Compos 33(5):454–471
12. Ramamoorthy SK, Skrifvars M, Persson A (2015) A review of natural fibers used in biocomposites: plant, animal and regenerated cellulose fibers. Polym Rev 55(1):107–162
13. Pickering KL, Efendy MA, Le TM (2016) A review of recent developments in natural fibre composites and their mechanical performance. Compos A Appl Sci Manuf 83:98–112

14. Attar H, Ehtemam-Haghighi S, Soro N, Kent D, Dargusch MS (2020) Additive manufacturing of low-cost porous titanium-based composites for biomedical applications: advantages, challenges and opinion for future development. J Alloy Compd 827:154263
15. Puttegowda M, Rangappa SM, Jawaid M, Shivanna P, Basavegowda Y, Saba NN (2018) Potential of natural/synthetic hybrid composites for aerospace applications. In: Sustainable composites for aerospace applications. Elsevier, pp 315–351
16. Keya KN, Kona NA, Koly FA, Maraz KM, Islam MN, Khan RA (2019) Natural fiber reinforced polymer composites: history, types, advantages and applications. Mater Eng Res 1(2):69–85
17. Mohajerani A et al (2019) Amazing types, properties, and applications of fibres in construction materials. Materials 12(16):2513
18. Mahmud S, Hasan K, Jahid M, Mohiuddin K, Zhang R, Zhu J (2021) Comprehensive review on plant fiber-reinforced polymeric biocomposites. J Mater Sci 56(12):7231–7264
19. Yuhazri M, Zulfikar A, Ginting A (2020) Fiber reinforced polymer composite as a strengthening of concrete structures: a review. In: IOP conference series: materials science and engineering, vol 1003, no 1. IOP Publishing, p 012135
20. Zaghloul MYM, Zaghloul MMY, Zaghloul MMY (2021) Developments in polyester composite materials—an in-depth review on natural fibres and nano fillers. Compos Struct 278:114698
21. Mali P, Sherje AP (2022) Cellulose nanocrystals: fundamentals and biomedical applications. Carbohyd Polym 275:118668
22. Kumar R, Ul Haq MI, Raina A, Anand A (2019) Industrial applications of natural fibre-reinforced polymer composites–challenges and opportunities. Int J Sustain Eng 12(3):212–220
23. Ilyas R, Sapuan S (2019) The preparation methods and processing of natural fibre bio-polymer composites. Curr Org Synth 16(8):1068–1070
24. Gu Y, Zhao J, Johnson JA (2020) Polymer networks: from plastics and gels to porous frameworks. Angew Chem Int Ed 59(13):5022–5049
25. Benin S, Kannan S, Bright RJ, Moses AJ (2020) A review on mechanical characterization of polymer matrix composites & its effects reinforced with various natural fibres. Mater Today Proceed 33:798–805
26. Kuranchie C, Yaya A, Bensah YD (2021) The effect of natural fibre reinforcement on polyurethane composite foams—a review. Sci Afr 11:e00722
27. Sadrmanesh V, Chen Y (2019) Bast fibres: structure, processing, properties, and applications. Int Mater Rev 64(7):381–406
28. Kumar PSS, Allamraju KV (2019) A review of natural fiber composites [Jute, Sisal, Kenaf]. Mater Today Proceed 18:2556–2562
29. Afzal MZ et al (2020) Kenaf (Hibiscus Cannabinus L.) breeding. J Nat Fibers 1–19
30. Xu J, Tao A, Qi J, Wang Y (2020) Bast fibres: kenaf. In: Handbook of natural fibres. Elsevier, pp 71–92
31. Mohamed S, Zainudin E, Sapuan S, Azaman M, Arifin A (2018) Introduction to natural fiber reinforced vinyl ester and vinyl polymer composites. In: Natural fibre reinforced vinyl ester and vinyl polymer composites. Elsevier, pp 1–25
32. Abbas AGN, Aziz FNAA, Abdan K, Mohd Nasir NA, Norizan MN (2022) Kenaf fibre reinforced cementitious composites. Fibers 10(1):3. https://doi.org/10.3390/fib10010003
33. T. Khan, M. T. H. Sultan, A. U. M. Shah, A. H. Ariffin, and M. Jawaid, "The Effects of Stacking Sequence on the Tensile and Flexural Properties of Kenaf/Jute Fibre Hybrid Composites," *Journal of Natural Fibers*, pp. 1–12.https://doi.org/10.1080/15440478.2019.1629148, 2019.
34. Royy S, Lutfar LB (2020) Bast fibres: jute 2. In: Handbook of natural fibres: volume 1: types, properties and factors affecting breeding and cultivation, p 39
35. Peu NI (2019) The market prospect or export potential of diversified jute products in Bangladesh
36. Islam MM (2019) Advanced production technology and processing of jute. In: Agronomic crops: Springer, pp 387–440

37. Shahinur S, Sayeed MMA, Hasan M, Sayem ASM, Haider J, Ura S (2022) Current development and future perspective on natural jute fibers and their biocomposites. Polymers 14(7):1445. https://doi.org/10.3390/polym14071445
38. Wang S (2021) Diagnosing hemp and cannabis crop diseases. CABI
39. Lu N, Swan Jr RH, Ferguson I (2012) Composition, structure, and mechanical properties of hemp fiber reinforced composite with recycled high-density polyethylene matrix. J Compos Mater 46(16):1915–1924. https://doi.org/10.1177/0021998311427778
40. Promhuad K, Srisa A, San H, Laorenza Y, Wongphan P, Sodsai J, Tansin K, Phromphen P, Chartvivatpornchai N, Ngoenchai P, Harnkarnsujarit N (2022) Applications of hemp polymers and extracts in food, textile and packaging: a review. Polymers 14(20):4274. https://doi.org/10.3390/polym14204274
41. Binoj J, Raj RE, Hassan SA, Mariatti M, Siengchin S, Sanjay M (2020) Characterization of discarded fruit waste as substitute for harmful synthetic fiber-reinforced polymer composites. J Mater Sci 55(20):8513–8525
42. Manikantan M, Pandiselvam R, Beegum S, Hebbar K (2018) Value addition through product diversification in coconut. Train Manual 109
43. Khan T, Singh B, Ahmad KA, Pai R (2022) Extraction of coir fibers by different methods. In: Jawaid M (ed) Coir fiber and its composites. Woodhead Publishing, pp 19–42. https://doi.org/10.1016/B978-0-443-15186-6.00022-9
44. Saw SK, Akhtar K, Yadav N, Singh AK (2014) Hybrid composites made from jute/coir fibers: water absorption, thickness swelling, density, morphology, and mechanical properties. J Nat Fibers 11(1):39-53. https://doi.org/10.1080/15440478.2013.825067
45. Venkateshwaran N, ElayaPerumal A (2012) Mechanical and water absorption properties of woven jute/banana hybrid composites. Fibers Polym 13:907–914. https://doi.org/10.1007/s12221-012-0907-0
46. Šimoníková D, Čížková J, Zoulová V, Christelová P, Hřibová E (2022) Advances in the molecular cytogenetics of bananas, family Musaceae. Plants 11(4):482
47. Nascimento RE, Monte J, Cadima M, Alves VD, Neves LA (2021) Rendering banana plant residues into a potentially commercial byproduct by doping cellulose films with phenolic compounds. Polymers 13(5):843
48. Rangappa SM, Siengchin S, Parameswaranpillai J, Jawaid M, Ozbakkaloglu T (2022) Lignocellulosic fiber reinforced composites: Progress, performance, properties, applications, and future perspectives. Polym Compos 43(2):645–691
49. Praveena B, Buradi A, Santhosh N, Vasu VK, Hatgundi J, Huliya D (2021) Study on characterization of mechanical, thermal properties, machinability and biodegradability of natural fiber reinforced polymer composites and its applications, recent developments and future potentials: a comprehensive review. Mater Today Proceed
50. Jumaidin R, Diah NA, Ilyas RA, Alamjuri RH, Yusof FAM (2021) Processing and characterisation of banana leaf fibre reinforced thermoplastic cassava starch composites. Polymers 13(9):1420. https://doi.org/10.3390/polym13091420
51. Rua JJ, Buchely MF, Monteiro SN, Colorado HA (2021) Structure-property relations of a natural composite: bamboo guadua angustifolia, under impact and flexural loads. J Compos Mater 55(21):2953–2966
52. Abhiram Y, Das A, Sharma KK (2021) Green composites for structural and non-structural applications: a review. Mater Today Proceed 44:2658–2664
53. Radzi AM, Zaki SA, Hassan MZ, Ilyas RA, Jamaludin KR, Daud MYM, Aziz SA (2022) Bamboo-fiber-reinforced thermoset and thermoplastic polymer composites: a review of properties, fabrication, and potential applications. Polymers 14(7):1387. https://doi.org/10.3390/polym14071387
54. Manickavasagam V, Vijaya Ramnath B, Elanchezhian C (2015) A review on natural and bamboo fiber composites. Int J Adv Res Sci Eng Technol 2:11
55. Li Z, Liu CP, Yu T (2002) Laminate of reformed bamboo and extruded fiber-reinforced cementitious plate. J Mater Civil Eng 14(5):359–365. https://doi.org/10.1061/(ASCE)0899-1561(2002)14:5(359)

56. Gurunathan T, Mohanty S, Nayak SK (2015) A review of the recent developments in biocomposites based on natural fibres and their application perspectives. Compos A Appl Sci Manuf 77:1–25
57. Baskin Y (2013) A plague of rats and rubbervines: the growing threat of species invasions. Island Press
58. Babu K (2015) Natural textile fibres: animal and silk fibres. In: Textiles and fashion. Elsevier, pp 57–78
59. Chong TY, Law MC, Chan YS (2021) The potentials of corn waste lignocellulosic fibre as an improved reinforced bioplastic composites. J Polym Environ 29(2):363–381
60. Reddy N (2019) Silk: materials, processes, and applications. Woodhead Publishing
61. Das TK, Ghosh P, Das NC (2019) Preparation, development, outcomes, and application versatility of carbon fiber-based polymer composites: a review. Adv Compos Hybrid Mater 2(2):214–233
62. Hassan MF, Sabri MA, Fazal H, Hafeez A, Shezad N, Hussain M (2020) Recent trends in activated carbon fibers production from various precursors and applications—a comparative review. J Anal Appl Pyrol 145:104715
63. Vázquez-Moreno JM, Sánchez-Hidalgo R, Sanz-Horcajo E, Viña J, Verdejo R, López-Manchado MA (2019) Preparation and mechanical properties of graphene/carbon fiber-reinforced hierarchical polymer composites. J Compos Sci 3(1):30. https://doi.org/10.3390/jcs3010030
64. Wang B, Gao H (2021) Fibre reinforced polymer composites. In: Advances in machining of composite materials. Springer, pp 15–43
65. Handwerker M, Wellnitz J, Marzbani H (2021) Review of mechanical properties of and optimisation methods for continuous fibre-reinforced thermoplastic parts manufactured by fused deposition modelling. Prog Addit Manuf 1–15
66. Singh SB, Sivasubramanian MVR, Likhith Reddy A, Chandra Prakash K, Nagarjuna K, Hima Kiran A (2012) Performance of NSM-FRP RC beams in flexure and shear using locally developed CFRP rebars. Int J Sustain Mater Struct Syst 1(1):42–67. https://doi.org/10.1504/IJSMSS.2012.050453
67. Chen Y (2021) Glass fiber-reinforced polymer composites for power equipment. Polym Compos Electr Eng 377–417
68. Srivastava JP, Kumar P (2022) Introduction to glass fiber-based composites and structures. Nat Synth Fiber Reinf Compos Synth Prop Appl 1–16
69. Beckman IP, Lozano C, Freeman E, Riveros G (2021) Fiber selection for reinforced additive manufacturing. Polymers 13(14):2231
70. Zhang B, Wang W, Tian M, Ning N, Zhang L (2020) Preparation of aramid nanofiber and its application in polymer reinforcement: a review. Eur Polymer J 139:109996
71. Mighani H (2020) Schiff Base polymers: synthesis and characterization. J Polym Res 27(6):1–18
72. Nasser J, Lin J, Steinke K, Sodano HA (2019) Enhanced interfacial strength of aramid fiber reinforced composites through adsorbed aramid nanofiber coatings. Compos Sci Technol 174:125–133
73. Moghadam MA, Izadifard RA (2020) Effects of steel and glass fibers on mechanical and durability properties of concrete exposed to high temperatures. Fire Saf J 113:102978
74. Andrew JJ, Srinivasan SM, Arockiarajan A, Dhakal HN (2019) Parameters influencing the impact response of fiber-reinforced polymer matrix composite materials: a critical review. Compos Struct 224:111007
75. Saba N, Jawaid M, Sultan M (2019) An overview of mechanical and physical testing of composite materials. Mech Phys Test Biocompos, Fibre-Reinf Compos Hybrid Compos 1–12
76. Siddika A, Al Mamun MA, Alyousef R, Amran TM (2019) Strengthening of reinforced concrete beams by using fiber-reinforced polymer composites: a review. J Build Eng 25:100798
77. Mayandi K et al (2020) An overview of endurance and ageing performance under various environmental conditions of hybrid polymer composites. J Market Res 9(6):15962–15988

78. Sharma AK, Bhandari R, Aherwar A, Rimašauskienė R, Pinca-Bretotean C (2020) A study of advancement in application opportunities of aluminum metal matrix composites. Mater Today: Proceed 26:2419–2424
79. Hung P-Y, Lau K-T, Cheng L-K, Leng J, Hui D (2018) Impact response of hybrid carbon/glass fibre reinforced polymer composites designed for engineering applications. Compos B Eng 133:86–90
80. Tayeb AH, Amini E, Ghasemi S, Tajvidi M (2018) Cellulose nanomaterials—binding properties and applications: a review. Molecules 23(10):2684
81. Patadiya J, Gawande A, Joshi G, Kandasubramanian B (2021) Additive manufacturing of shape memory polymer composites for futuristic technology. Ind Eng Chem Res 60(44):15885–15912
82. Hussain I, Immanuel R (2022) Composite materials and its advancements for a cleaner engine of future. In: Advances in engine tribology. Springer, pp 169–191
83. Rajak DK, Wagh PH, Linul E (2021) Manufacturing technologies of carbon/glass fiber-reinforced polymer composites and their properties: a review. Polymers 13(21):3721
84. Jayakrishna K, Kar VR, Sultan MT, Rajesh M (2018) Materials selection for aerospace components. In: Sustainable composites for aerospace applications. Elsevier, pp 1–18
85. Chadha C, Crowe KA, Carmen CL, Patterson AE (2018) Exploring an AM-enabled combination-of-functions approach for modular product design. Designs 2(4):37
86. Burston M, Ranasinghe K, Gardi A, Parezanovic V, Ajaj R, Sabatini R (2021) Design principles and digital control of advanced distributed propulsion systems. Energy 122788
87. Tomei V, Imbimbo M, Mele E (2018) Optimization of structural patterns for tall buildings: the case of diagrid. Eng Struct 171:280–297
88. Qiu Y, Yang H, Tong L, Wang L (2021) Research progress of cryogenic materials for storage and transportation of liquid hydrogen. Metals 11(7):1101
89. Aamir M, Giasin K, Tolouei-Rad M, Vafadar A (2020) A review: Drilling performance and hole quality of aluminium alloys for aerospace applications. J Market Res 9(6):12484–12500
90. Gholizadeh S, Gholizadeh S (2022) Impact behaviours and Non-Destructive Testing (NDT) methods in carbon fiber composites in aerospace industry: a review. Authorea Preprints
91. Xu S et al (2018) Renewable atom-efficient polyesters and thermosetting resins derived from high oleic soybean oil. Green Chem 20(5):1106–1113
92. Hoque MS, Hossain MJ, Imtiaz MA, Das S, Rashid M (2018) Scope of dry wood & wood composite alternate to stone in case of acid wash on denim fabric. Int J Curr Eng Technol 8(02):382–388
93. Hamidi F, Aslani F (2019) Additive manufacturing of cementitious composites: materials, methods, potentials, and challenges. Constr Build Mater 218:582–609
94. Xiao T, Liu D (2019) The most advanced synthesis and a wide range of applications of MOF-74 and its derivatives. Microporous Mesoporous Mater 283:88–103
95. Ngo TD (2020) Introduction to composite materials. Compos Nanocompos Mater—From Knowl Ind Appl
96. Okafor CE (2021) Review of natural fiber composite design for sustainable infrastructural development. University-Led Knowledge and Innovation for Sustainable Development, p 64
97. Mansor M, Nurfaizey A, Tamaldin N, Nordin M (2019) Natural fiber polymer composites: utilization in aerospace engineering. In: Biomass, biopolymer-based materials, and bioenergy. Elsevier, pp 203–224
98. Surjadi JU et al (2019) Mechanical metamaterials and their engineering applications. Adv Eng Mater 21(3):1800864
99. Ranasinghe K, Guan K, Gardi A, Sabatini R (2019) Review of advanced low-emission technologies for sustainable aviation. Energy 188:115945
100. Vlasova V (2019) Using composite materials in aircraft. In: AIP conference proceedings, vol 2171, no 1. AIP Publishing LLC, p 030020
101. Alam MA et al (2022) Recent advancements in advanced composites for aerospace applications: a review. Adv Compos Aerosp Eng Appl 319–339

102. Chauhan V, Kärki T, Varis J (2019) Review of natural fiber-reinforced engineering plastic composites, their applications in the transportation sector and processing techniques. J Thermoplast Compos Mater 0892705719889095
103. Alsubari S, Zuhri M, Sapuan S, Ishak M, Ilyas R, Asyraf M (2021) Potential of natural fiber reinforced polymer composites in sandwich structures: a review on its mechanical properties. Polymers 13(3):423
104. Sair S, Oushabi A, Kammouni A, Tanane O, Abboud Y, El Bouari A (2018) Mechanical and thermal conductivity properties of hemp fiber reinforced polyurethane composites. Case Stud Constr Mater 8:203–212
105. Renreng I, Bakri B, Chandrabakty S (2021) Evaluation of mechanical and physical properties of pressed coir fiber/epoxy composite with NaOH and microwave treatment of fiber. J Renew Mater 9(2):325
106. Sathishkumar G et al (2022) Influence of lignite fly ash on the structural and mechanical properties of banana fiber containing epoxy polymer matrix composite. Polym Bull 79(1):285–306
107. Lokesh P, Kumari TS, Gopi R, Loganathan GB (2020) A study on mechanical properties of bamboo fiber reinforced polymer composite. Mater Today Proceed 22:897–903
108. Zhang W, Deng X, Sui G, Yang X (2019) Improving interfacial and mechanical properties of carbon nanotube-sized carbon fiber/epoxy composites. Carbon 145:629–639

Natural/Synthetic Polymer Hybrid Composites in Automotive Applications

K. M. Faridul Hasan, Shuai Chen, György Török, Liu Xiaoyi, Péter György Horváth, and Tibor Alpár

Abstract The composite materials have become popular in automotive sector due to the weight reduction capabilities and cost minimization features. Various synthetic/natural fiber materials are used for producing the automotive composite panels. However, the hybrid composites are showing immense potential due to their extreme feasibility for adequate materials preference, strength to weight ratio, lower cost, feasible materials production cost, and functional properties enhancements in terms of thermal, UV-protection, and so on beside the superior mechanical and dimensional stability. Additionally, synergistic effect is also obtained when reinforcing natural and synthetic fiber with polymeric matrix in composite system which may not be achieved when only natural fibers is used. In this chapter, different composites developed from natural/synthetic fiber reinforcement, their manufacturing and characterization protocol, perceived characteristics, and applications would be addressed in detail. Moreover, their economic aspects would also be demonstrated further.

Keywords Fiber · Hybrid composites · Thermo-mechanical performances · Fabrication · Automotives

K. M. F. Hasan (✉) · G. Török · P. G. Horváth · T. Alpár
Faculty of Wood Engineering and Creative Industry, University of Sopron, Sopron 9400, Hungary
e-mail: hasan.kmfaridul@uni-sopron.hu

T. Alpár
e-mail: alpar.tibor@uni-sopron.hu

S. Chen
Department of Biochemistry and Molecular Biology, College of Basic Medical Sciences, Guizhou Medical University, Guiyang 550025, China

L. Xiaoyi
School of Public Health, the key, Laboratory of Environmental Pollution Monitoring and Disease Control, Ministry of Education, Guizhou Medical University, Guiyang 550025, China

T. Khan and M. Jawaid (eds.), *Green Hybrid Composite in Engineering and Non-Engineering Applications*, Composites Science and Technology,
https://doi.org/10.1007/978-981-99-1583-5_7

1 Introduction

Technological advancements are motivating the researchers/scientists throughout the globe with immense driving force for innovating more novel products which could be feasible toward the practical applications by the manufacturers. The composite materials could be defined as the structural material comprised of two different materials which could be identifiable macroscopically. The composites are becoming a widespread application material for structural products where the different thermal and mechanical properties are getting immense potentiality [1–5]. Different sophisticated fields like automotives, aeronautical, marine, electronics, defence, furniture, packaging, and so on are getting attentions for hybrid composites material [6, 7]. Moreover, the composite markets are booming continuously with the expansion of science and technology around the globe. The value of the composite market was nearly $76 billion in 2016 which is expected to be grown up nearly 8.9% (compound annual growth) within the period 2017 to 2025 [8]. Additionally, depending on the types of reinforcement materials used, there could be 15 to 40% lightweight features attainable from the composites [8]. Interestingly, a significant portion of global composite marked is dominated by polymer composite materials which was nearly around 95% or more until 2016 [8]. The trend is going to reach the peaks day by day. Generally, different thermoplastic/thermosetting polymers are used for composites fabrications especially in the automotive applications.

Engineered composites developed from fibers, mortars, cements, and polymer fabrications are used for many structural applications [3, 9–15]. Hybrid composites are typically used for various industrial applications like as defence, medical, automotive, and so on. However, the arrangement of fibers in the laminated composite system plays an important role for the improved mechanical and structural performances. There are two or more solid materials in the composite system used to produce hybrid composites [16, 17]. Two of the materials could be added in continuous/discontinuous phases. However, the discontinuous phases are comparatively stronger than the continuous phases. Although use of synthetic fibers is seen since 1960s, however the utilization of lignocellulosic materials is a recent concept for composites developments [18]. However, significant results are getting displayed by the developed hybrid composites. Like as Dhakal et al. [19] reported about the unidirectional and cross-ply flax fiber and unidirectional carbon fiber to reinforce with epoxy resin for hybrid composites development through applying compression molding. They claimed that, both mechanical and physical properties improved when the carbon fiber was added in the composite system [19]. Their experiment further suggested that, flax fibers helped for improving the toughening of composites against crack propagations whereas carbon fibers helped for thermal stability enhancements/dimensional stability/stiffness of the hybrid products [19].

Increased pressure created by the environmental specialists throughout the globe for preserving natural resources too is also getting increased gradually [20, 21]. Therefore, a hybridization of both natural/synthetic fibers could facilitate to reduce the pressures from biobased materials too. This is the high time to think about both the utilizations of natural resources and to control the threats of deforestation as well. This chapter will review the possibility of hybrid composites development for automotive applications and efficient uses of resources.

2 Natural/Synthetic Fiber Reinforced Composites

The naturally derived cellulosic fibers are getting used since last 3000 years as the reinforcement materials with different polymeric matrix. The journey of cellulosic materials with phenolic adhesive shown the application since 1908, which was later on extended by melamine and urea derivatives of polymers. The reinforcement of cotton with polymers to produce composites was seen in the case of radar aircraft for military [22, 23]. However, the composites used for automotive applications found in 1950, where the East German Trabant Car Co., used a frame made of cotton fiber reinforced polyester resin. With the span of time, more diverse routes of cellulosic fibers were found to have use for composites fabrication starting from seeded fibers to bast/leaf/fruit/stalk/or even grass or routes fibers. However, the most popularly used fibers (Fig. 1) are flax, hemp, sisal, cotton, sugarcane bagasse, carbon, glass, silk, aramid, Kevlar, rice straw, agave, and so on [24–32]. The images of some natural fibers are shown in Fig. 2.

In addition to natural fibers, synthetic fibers were used since long times to produce composite products too. Generally, natural fibers are hybridized with better strength/functional properties providing materials like glass/carbon/aramid/Kevlar. The hybridization of natural fibers with synthetic fibers provide better mechanical

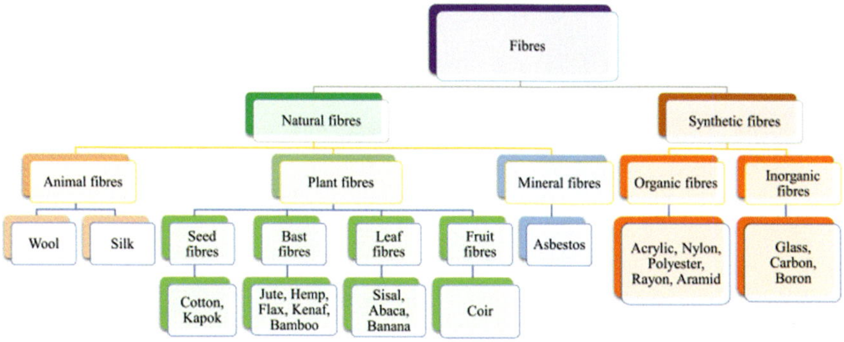

Fig. 1 Classification of natural and synthetic fibers [33]. Published under the Creative Commons CC-BY-NC-ND license agreement

performances, thermal stability, and resistance to moisture [35]. There are versatile research conducted on natural/synthetic fiber reinforced composites [36–38].

Glass fibers are most commonly used synthetic fibers used for composites fabrication which was developed initially around 1600 BC. Glass fibers can be found in different shapes and forms like yarns, rovings, strands, fabrics, and mats. Glass fibers has higher stiffness, strength, chemical/moisture resistance, and flexibility which make them excellent candidate for polymer reinforced composites production. Moreover, glass fibers can be found as S-glass, R-glass, AR-glass, and E-glass, whereas E-glass is the most common types, which is used extensively. Interestingly, market of glass fiber is booming due to their tremendous use in our daily lives especially for the composite products [39–41]. Carbon is another type of most important

Fig. 2 Fibers and fabrics of different naturally derived plants: **a** Sisal, **b** Jute, **c** Kenaf, **d** Hemp, **e** Coir, **f** Banana, and **g** Bamboo. Reprinted with the permission from Elsevier [34]. Copyright, Elsevier 2018

Fig. 2 (continued)

fiber for cementitious and plastic-based composites fabrication. This fiber contains nearly 90% carbon materials and the carbon fibers are used for producing various composite products like as plastics, cementitious materials, and so on [42–44].

3 Categories of Hybrid Composites

The Greek-Latin word "hybrid" demonstrates the mixing of something origin/composition [35]. Hybrid composites exist extreme demand on transportation, structural, and automotive sector. The precise classification of hybrid composite is a bit complex, however according to the type of constituents, hybrid composites could be of following categories [45, 46]:

(I) Intraply hybrid: Two or more reinforcement fibers are mixed together where the layers remain same.
(II) Interplay hybrid: The layers (two/more) having same reinforcements which do not interface with the hybrids.
(III) Selective hybrids: Additional strength is needed to place the reinforcements over the base reinforced laminating layer

(IV) Super hybrid composites: Containing metal composite pile/metal foils which are stacked in specific orientations/stackings.

(V) Intermingled hybrid: The reinforcement fibers are mixed randomly as if no concentrations are present in the material (either type).

4 Possibility of Developing Hybrid Composites Through Reinforcing Natural and Synthetic Fibers

Still now carbon and glass fibers got most attention for the hybridizations with natural fibers due to their outstanding performance characteristics. Besides, the natural fibers synthetic fiber also attained significant attentions for composites production. However, as the synthetic fibers are not eco-friendly, hence their reinforcement with natural fibers could enhance the sustainability besides providing feasible thermo-mechanical performances. Hybrid composites are prepared through applying multiple methods including hand lap-up, hydraulic press, cold press, compression molding, and so on. A typical representation of stacking sequence for flax/glass fiber in the composite system is shown Fig. 3. Chauhan et al. [47] reported about the carbon and Kevlar with jute fibers for producing the composites of automotives spring leaf applications. They found that developed hybrid composites facilitated to improve the flexural properties by 44 and 173% for 3 and 4 mm thickness of the products, respectively [47]. The products also shown a decline in weight by 66 and 57% [47]. Rana et al. [48], hybridized natural sisal with glass fiber in the presence of epoxy resin where they fixed the glass fibers with varying percentages of sisal (0, 2, 4, 6%). They found that [48], upto 4% increase in the sisal fibers facilitated to enhance the tensile properties by 48 MPa which was the highest obtained value and then starts to decline again (in case of 6% sisal fiber loading). The report on sisal/carbon fiber reinforced unsaturated polyester resin composites was also reported [49]. This study also followed the similar trend in increased mechanical properties with the increase in carbon fiber loading [49]. Moreover, alkaline treatment could also help to further increase the mechanical properties. The different combinations of fibers sisal/carbon (100/0, 75/25, 50/50, 25/75, 0/100) provides different mechanical performances, whereas untreated 100% sisal fiber/polyester composites provided 24.16 MPa tensile strength and 100% carbon fiber/polyester composite provided 122.11 MPa strength. However, other composites displayed the strength between these two ranges although the mechanical properties start to increase with the increased values of carbon fiber loading.

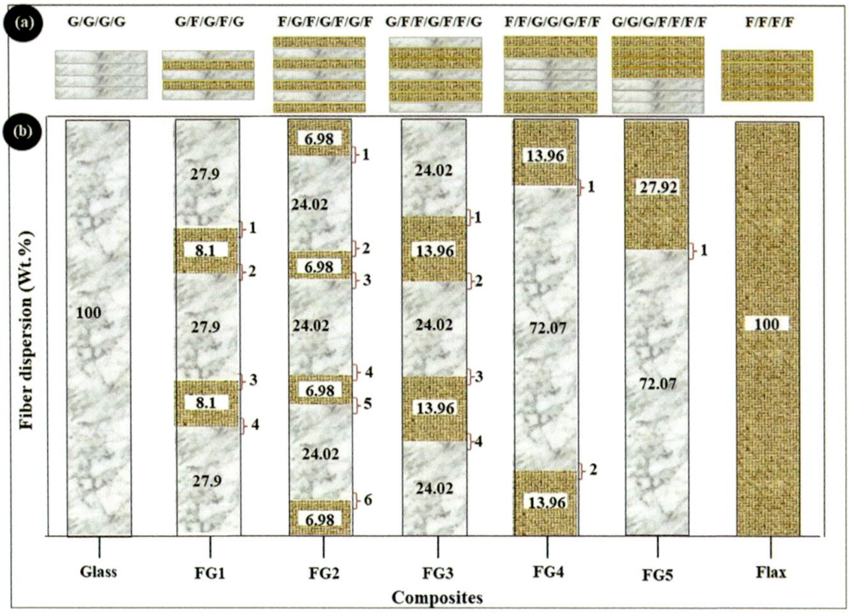

Fig. 3 Different stacking sequence of a typical flax/glass fiber reinforced composites (G indicates glass and F indicates flax fibers): **a** stacking sequence of fibers and **b** wt.% of the fibers.[50]. Reprinted with the permissions from Elsevier. Copyright, Elsevier 2019

5 Characteristics of Hybrid Composites Through Reinforcing Natural and Synthetic Fibers

There are numerous matrix materials used for composites manufacturing especially for their versatility. Typically, various thermosetting and thermoplastic polymers are used for composites fabrications. However, they possess diversified physicochemical properties which determines the performances of the ultimate composites. Properties of some widely used resins are shown in Table 1. On the other hand, synthetic fibers are stronger than that of natural fibers, hence the increased loading of synthetic fibers in the composite system facilitate to enhance the mechanical properties of the composites. In our previous study for glass and flax fiber reinforced MDI (methylene diphenyl diisocyanate) resin bonded hybrid laminated composites, the similar trend was also noticed [16]. Highest tensile strength was displayed by 100% glass reinforcements by 78.61 (8.2) MPa where 100% flax fiber shown the value by 21.19 (1.59) MPa and their hybrid reinforcements shown 49.44 (2.05) MPa [46].

Table 1 Mechanical properties of some adhesives [13, 34]

Different adhesives	Tensile strength (MPa)	Tensile modulus (GPa)	Glass transition temperature	Flexural strength (MPa)	Flexural modulus (GPa)
Epoxy	55–130	2.7–4.1	170–300	110–150	3–4
Phenolic	50–60	4–7	175	80–135	2–4
Vinyl ester	73–81	3–3.5	–	130–140	3
Polyester	34–105	2.1–3.5	130–160	70–110	2–4

6 Performance Characteristics of Hybrid Composites

Typically, hybrid composites are also tested by tensile properties (both strength and modulus), flexural properties (both strength and modulus), thermal conductivity, thermogravimetric analysis, and so on. Some of the mechanical properties of various natural/synthetic fiber reinforced composite products are highlighted in Table 2. It is seen that when synthetic fibers were loaded in the composite system, it helps to enhance the mechanical properties of the natural fiber-based hybrid composites too.

Nowadays, laminated composites are also designed in terms of numerical analysis before going to bulk productions to predict the mechanical properties or other functionalities in advance to minimize the process loss. However, still some deviations in actual values and predicted results are seen (Fig. 4).

The microstructural analysis in terms of SEM (scanning electron microscopy) also provides an explicit picture regarding the fiber to matrix interfaces. The adhesion between the reinforcement and matrix could further be enhanced using nanoparticles, hence the performances of the products also gets increased [58]. Therefore, Mohan et al. [59], conducted a study on multiwalled carbon nanotube (MWCNT) loaded glass/flax/epoxy where they found incorporation of nanoparticles helped to increase adhesion among the reinforcements and epoxy. However, they could not see the presence of nanoparticles cluster in morphological views (Fig. 5), it maybe that due to sonification, MWCNT nanoparticles got incorporated with epoxy resin in the composite system [59].

The cellulosic part of the hybrid composites produced from natural/synthetic fiber reinforcements absorb moisture from the surrounding atmosphere. Additionally, the higher the lignin present in the natural fiber, the lower is the water absorption too [60]. Conversely, synthetic fibers don't absorb water like as the natural fibers as they are free from water absorption molecules. Consequently, natural/synthetic fiber reinforced composites absorb less water. However, the water absorption is initiated at the outer layer of the composites which is proceeded slowly at bulk of the polymer in composite system [61].

Table 2 Mechanical properties and applications of some hybrid composites

Constituent materials	Tensile strength (MPa)	Tensile modulus (GPa)	Flexural strength (MPa)	Flexural modulus (GPa)	Thermal conductivity (W/(m K))	Uses	References
Sugarcane bagasse/bamboo charcoal/epoxy	3.4–15.4	–	72–76		0.13–0.086	Automotive interior	[51]
Hemp/glass/polypropylene	~45–58	~3.8–4.8	~92–102	~4.6–5.6	–	Automotive interior	[52]
Aloe vera- ceramic fiber wool- glass fiber/epoxy	101.86	–	110.95	–	–	Automobile parts	[53]
Sisal fiber- ceramic fiber wool- glass fiber/epoxy	77.98	–	89.35	–	–	Automobile parts	[53]
Sugar palm/glass fiber/polyester	–	1.84	–	–	–	Different parts of boats	[54]
Palm/glass/epoxy	120	–	–	–	–	–	[55]
Date palm/Kevlar/epoxy	221.3–237.9	3.1–3.6	–	–	–	Automotive bumper	[56]
Carbon/flax/epoxy	179.55 ± 3.42–193.72 ± 3.27	76.52 ± 2.94–141.85 ± 10.50	64.8–121.1	50.6–107.1	–	–	[57]

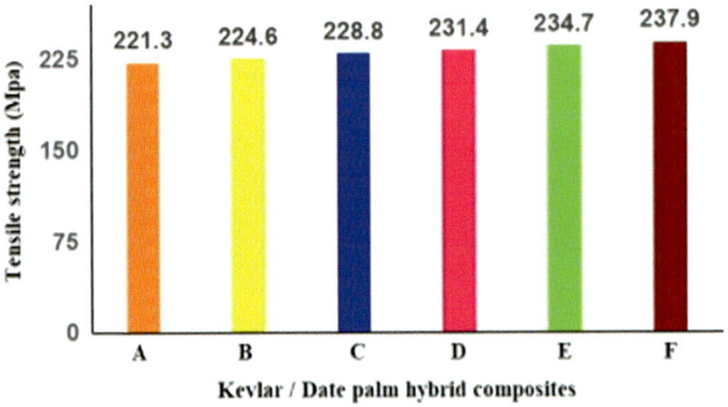

Fig. 4 Tensile strengths of hybrid composites (actual/predicted values by ANSYS). Reprinted with permissions from Elsevier [56]. Copyright, Elsevier 2021

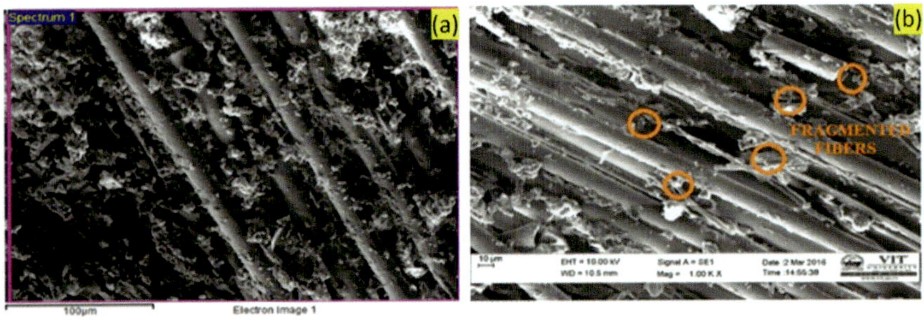

Fig. 5 SEM profile of glass/flax/epoxy nanocomposites (multiwalled carbon nanotube): **a** fiber to matrix bonding and **b** fractured surfaces of pulled composites [59]. Adapted with permission from Elsevier. Copyright, Elsevier 2018

7 Benefits of Hybrid Composites Over Traditional Composites

Hybrid composites exist numerous benefits over conventional composites as mentioned bellow:

- Provides environmental sustainability when both natural/synthetic fibers used as the reinforcement materials.
- Thermosetting polymer reinforced hybrid composites also shown improved mechanical performances.
- The loading of nanoparticles also displaying better thermomechanical performances

- The incorporation of natural fibers with synthetic materials also minimizes cost as they are abundantly available throughout the world.

8 Potential Application for Hybrid Composites Through Reinforcing Natural and Synthetic Fibers

Numerous research addressed about the possibility of hybrid composites applications in the field of automotives [62–64]. Nowadays nearly 50% of the materials used for cars production are polymer-based materials. Globally, the average plastic production is 105 kg, which is around 120 kg for developed nations accumulating 10–12% of total cars production [65]. Another interesting statistical data showing that fiber reinforced composites minimize the cost by 20% and weight reduction by 30% from automotives [25]. Additionally, different processing technologies are used for different applications of composite products. Hybrid composites are getting popularity for versatile applications like as headliners, door panels, roofing sheets, furniture panels, pipelines, storage tanks, pallets, spare tyre covers, seat backs, dash boards, spare-wheel panels, interior panels, and so on for construction and building sectors since long times [4, 66, 67]. However, the automotive companies are also using hybrid composites for their different parts. Hybrid composites are also seen to be used for households like as roofs, suitcases, bath units, tables, chairs, lampshades, etc. The marine sector also uses hybrid composites as but hulls. The applications also found in sports sector for helmets, tennis rackets, and bicycles. Nowadays they are also utilized in the biogas container, storage silos, and post boxes too. However their widespread applications also noticed in aerospace including military aircraft to satellites, helicopters, small and big civil transport aircraft, missiles, and so on [34]. Specifically, different automotive companies are also using fiber reinforced composites for their vehicle parts like as Mercedes-Benz is using jute-based door panels for it's a-class cars since more than two decades. Interestingly, most automotive companies in Germany like as Volkswagen, Mercedes-Benz, Audi, Opel, Ford, Damiel Chrysler, and BMW are using fiber reinforced composites/hybrid composites for their vehicles [68]. However, the European automotive companies are also trying to be increasingly sustainable through using as much as possible eco-friendly products [69]. The images of different composite products for diversified application is shown in (Figs. 6 and 7).

9 Conclusion

Nowadays, the hybrid composites are booming due to their superior thermos-mechanical performances and outstanding fabrication feasibilities through replacing

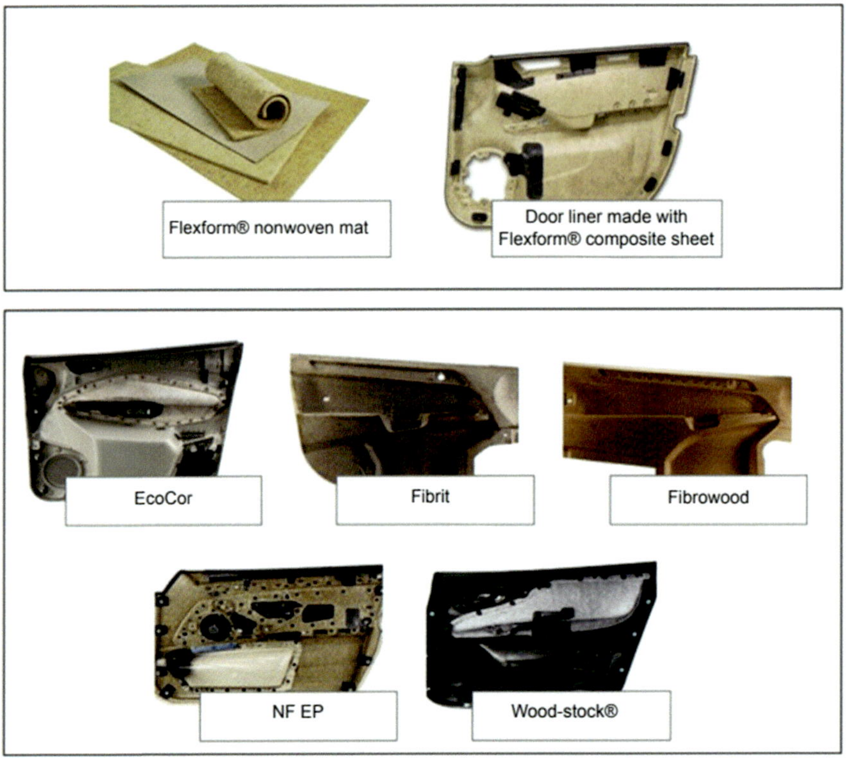

Fig. 6 Application of different sustainable composites for automotive applications. Reprinted with the permission from Elsevier [70]. Copyright, Elsevier 2018

the conventional composite products throughout the globes. Although hybrid composites have multifaced application potentials however, this chapter entails their fabrication, properties and application in automotive sector. The different research and review articles published in this area has been reviewed and critically analysed here. Scientists have reported about different possibilities such as using various natural fibers and biobased nano/micro crystalline cellulose and synthetic fibers like carbon, aramid, and glass with different thermoplastic and thermosetting polymers. To enhance the performances of ultimate products various treatment of materials and incorporation of nanofillers in the composite system also addressed. Additionally, modelling of the composites also carried out nowadays to explore the optimized development of hybrid composites before the fabrication which is also discussed further in this study.

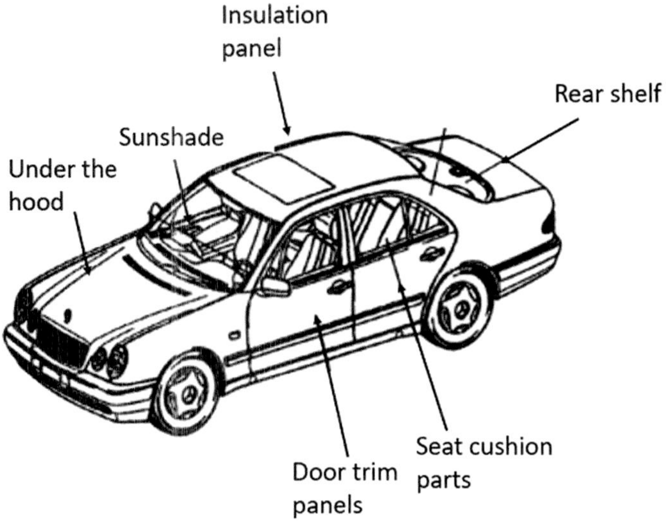

Fig. 7 Different parts of a car made of fiber reinforced polymeric composite. Adapted with the permission from Elsevier [18]. Copyright, Elsevier 2011

References

1. Mahmud S, Hasan KF, Jahid MA et al (2021) Comprehensive review on plant fiber-reinforced polymeric biocomposites. J Maters Sci 56:7231–7264
2. Hasan KF, Horváth PG, Bak M et al (2021) A state-of-the-art review on coir fiber-reinforced biocomposites. RSC Adv 11(18):10548–10571
3. Hasan KMF, Horváth PG, Alpár T (2021) Lignocellulosic fiber cement compatibility: a state of the art review. J Natur Fibers 1–26
4. Hasan K, Horváth PG, Alpár T (2021) Potential fabric-reinforced composites: a comprehensive review. J Mater Sci 56:14381–14415
5. Hasan KF, Csilla C, Mucsi ZM et al (2022) Effects of sisal/cotton interwoven fabric and jute fibers loading on polylactide reinforced biocomposites. Fibers Polym
6. Hasan KF, Horváth PG, Zsolt K et al (2021) Design and fabrication technology in biocomposites manufacturing. Value-added biocomposites: technology, innovation, and opportunity. CRC Press, Boca Raton, USA, pp 158–183
7. Alpár T, Hasan KF, Horváth PG (2021) Introduction to biomass and biocomposites. Value-added biocomposites: technology, innovation, and opportunity. CRC Press, Boca Raton, pp 1–33
8. Adesina O, Jamiru T, Sadiku E et al (2019) Mechanical evaluation of hybrid natural fibre–reinforced polymeric composites for automotive bumper beam: a review. J Adv Manuf Technol 103(5–8):1781–1797
9. Hasan K, Fatima Zohra B, Bak M et al (2020) Effects of cement on lignocellulosic fibres. In: 9th hardwood proceedings. University of Sopron Press, Sopron, Hungary
10. Hasan KF, Horváth PG, Alpár T (2021) Development of lignocellulosic fiber reinforced cement composite panels using semi-dry technology. Cellulose 28:3631–3645
11. Hasan K, Horváth PG, Kóczán Z et al (2021) Novel insulation panels development from multi-layered coir short and long fiber reinforced phenol formaldehyde polymeric biocomposites. J Polym Res 28(12):1–16

12. Hasan KF, Horváth PG, Kóczán Z et al (2021) Semi-dry technology-mediated coir fiber and Scots pine particle-reinforced sustainable cementitious composite panels. Constr Build Mater 305:124816
13. Hasan KF, Horváth PG, Bak M et al (2021) Rice straw and energy reed fibers reinforced phenol formaldehyde resin polymeric biocomposites. Cellulose 28:7859–7875
14. Hasan KF, Horváth PG, Baş S et al (2021) Industrial flame retardants for polyurethanes. Materials and chemistry of flame-retardant polyurethanes volume 1: a fundamental approach. ACS Publications, Washington, DC, USA, pp 239–264
15. Hasan KMF, Horváth PG, Alpár T (2020) Effects of alkaline treatments on coconut fiber reinforced biocomposites in 9th Interdisciplinary doctoral conference. Doctoral Student Association of the University of Pécs, Pecs, Hungary
16. Hasan KF, Horváth PTGR, Alpár T (2021) Thermomechanical behavior of methylene diphenyl diisocyanate-bonded flax/glass woven fabric reinforced laminated composites. ACS Omega 6(9):6124–6133
17. Hasan KF, Horváth PG, Miklos B et al (2021) Hemp/glass woven fabric reinforced laminated nanocomposites via in-situ synthesized silver nanoparticles from Tilia cordata leaf extract. Compos Interfaces
18. Jawaid M, Khalil HA (2011) Cellulosic/synthetic fibre reinforced polymer hybrid composites: a review. Carbohydr Polym 86(1):1–18
19. Salman SD, Leman Z, Sultan MT et al (2015) Kenaf/synthetic and Kevlar®/cellulosic fiber-reinforced hybrid composites: a review. BioResources 10(4):8580–8603
20. Das PP, Chaudhary V (2021) Moving towards the era of bio fibre based polymer composites. Clean Eng Technol 4:100182
21. Diyana Z, Jumaidin R, Selamat MZ et al (2021) Physical properties of thermoplastic starch derived from natural resources and its blends: a review. Polymers 13(9):1396
22. Piggott M (2002) Load bearing fibre composites. Kluwer Academic Publishers, New York, USA
23. Lubin G (2013) Handbook of composites. Van Nostrand Reinhold Company, New York, USA
24. Al-Oqla FM, Sapuan S (2014) Natural fiber reinforced polymer composites in industrial applications: feasibility of date palm fibers for sustainable automotive industry. J Clean Prod 66:347–354
25. Ahmad F, Choi HS, Park MK (2015) A review: natural fiber composites selection in view of mechanical, light weight, and economic properties. Macromol Mater Eng 300(1):10–24
26. Hasan K, Horváth PG, Kóczán Z et al (2021) Colorful and facile in situ nanosilver coating on sisal/cotton interwoven fabrics mediated from European larch heartwood. Sci Rep 11(1):1–13
27. Hasan KF, Wang H, Mahmud S et al (2022) Enhancing mechanical and antibacterial performances of organic cotton materials with greenly synthesized colored silver nanoparticles. Int J Cloth Sci Technol
28. Hasan K, Pervez M, Talukder M et al (2021) Community entrepreneurship and environmental sustainability of the handloom sector. Handloom Sustain Cult. Springer, Singapore, pp 21–47
29. Hasan KF, Liu X, Kóczán Z et al (2021) Nanosilver coating on hemp/cotton blended woven fabrics mediated from mammoth pine bark with improved coloration and mechanical properties. J Text Inst 1–10
30. Hasan KMF, Horváth PG, Alpár T (2021) Silk protein and its nanocomposites. Biopolymeric nanomaterials: fundamentals and applications. Elsevier, pp 309–323
31. Hasan KF, Horváth PG, Kóczán Z et al (2021) Coloration of flax woven fabric using *Taxus baccata* heartwood extract mediated nanosilver. Colour Technol 00:1–11
32. Pervez M, Talukder M, Shafiq F et al (2018) Effect of heat-setting on UV protection and antibacterial properties of cotton/spandex fabric. In: IOP Conference series: materials science and engineering. IOP Publishing
33. Awais H, Nawab Y, Amjad A et al (2021) Environmental benign natural fibre reinforced thermoplastic composites: a review. Compos Part C: Open Access 4:100082
34. Puttegowda M, Rangappa SM, Jawaid M et al (2018) Potential of natural/synthetic hybrid composites for aerospace applications. Sustainable composites for aerospace applications. Elsevier, Duxford, United Kingdom, pp 315–351

35. Begum S, Fawzia S, Hashmi M (2020) Polymer matrix composite with natural and synthetic fibres. Adv Mater Process Technol 6(3):547–564
36. Salman SD (2021) Partial replacement of synthetic fibres by natural fibres in hybrid composites and its effect on monotonic properties. J Indus Text 51(2):258–276
37. Naidu VNP, Reddy GR, Kumar MA et al (2011) Compressive & impact properties of sisal/glass fiber reinforced hybrid composites. Int J Fibre Text Res 1(1):11–14
38. Aslan M, Tufan M, Küçükömeroğlu T (2018) Tribological and mechanical performance of sisal-filled waste carbon and glass fibre hybrid composites. Compos Part B: Eng 140:241–249
39. Monticeli FM, Cioffi MOH, Voorwald HJC (2022) Mode II delamination of carbon-glass fiber/epoxy hybrid composite under fatigue loading. Int J Fatigue 154:106574
40. Ng LF, Yahya MY, Muthukumar C (2022) Mechanical characterization and water absorption behaviors of pineapple leaf/glass fiber-reinforced polypropylene hybrid composites. Polym Compos 43(1):203–214
41. MJ, R. and G. Goud, (2022) Development of Calotropis procera-glass fibers reinforced epoxy hybrid composites: dynamic mechanical properties. J Nat Fibers 19(2):395–402
42. Massarwa E, Tabrizi IE, Yildiz M (2021) Mechanical behavior and failure of glass/carbon fiber hybrid composites: multiscale computational predictions validated by experiments. Compos Struct 260:113499
43. TG, Y.G., V. Kushvaha, S. MR, et al (2021) A new study on flax-basalt-carbon fiber reinforced epoxy/bioepoxy hybrid composites. Polym Compos 42(4):1891–1900
44. Baihaqi NN, Khalina A, Nurazzi NM et al (2021) Effect of fiber content and their hybridization on bending and torsional strength of hybrid epoxy composites reinforced with carbon and sugar palm fibers. Polimery 66(1):36–43
45. Hwang J-H, Seo D-W, Park K-T et al (2014) Experimental study on the mechanical properties of FRP bars by hybridizing with steel wires. Engineering
46. Ravishankar B, Nayak SK, Kader MA (2019) Hybrid composites for automotive applications–a review. J Reinf Plast Compos 38(18):835–845
47. Chauhan A, Kumar M, Kumar S (2022) Fabrication of polymer hybrid composites for automobile leaf spring application. Mater Today Proc 48:1371–1377
48. Rana R, Rana S, Purohit R (2017) Characterization of properties of epoxy sisal/glass fiber reinforced hybrid composite. Mater Today Proc 4(4):5445–5451
49. Noorunnisa Khanam P, Abdul Khalil H, Jawaid M et al (2010) Sisal/carbon fibre reinforced hybrid composites: tensile, flexural and chemical resistance properties. J Polym Env 18(4):727–733
50. Kumar CN, Prabhakar M, Song J-I (2019) Effect of interface in hybrid reinforcement of flax/glass on mechanical properties of vinyl ester composites. Polym Test 73:404–411
51. Abedom F, Sakthivel S, Asfaw D et al (2021) Development of natural fiber hybrid composites using sugarcane bagasse and bamboo charcoal for automotive thermal insulation materials. Adv Mater Sci Eng 2021:1–10
52. Sanjay M, Yogesha B (2017) Studies on natural/glass fiber reinforced polymer hybrid composites: an evolution. Mater Today Proc 4(2):2739–2747
53. Bharat K, Abhishek S, Palanikumar K (2017) Mechanical property analysis on sandwich structured hybrid composite made from natural fibre, glass fibre and ceramic fibre wool reinforced with epoxy resin. In: IOP conference series: materials science and engineering. IOP Publishing
54. Misri S, Leman Z, Sapuan S et al (2010) Mechanical properties and fabrication of small boat using woven glass/sugar palm fibres reinforced unsaturated polyester hybrid composite. In: IOP conference series: materials science and engineering. IOP Publishing
55. Raju K, Balakrishnan M (2020) Evaluation of mechanical properties of palm fiber/glass fiber and epoxy combined hybrid composite laminates. Mater Today Proc 21:52–55
56. Muthalagu R, Murugesan J, Kumar SS et al (2021) Tensile attributes and material analysis of kevlar and date palm fibers reinforced epoxy composites for automotive bumper applications. Mater Today Proc 46:433–438
57. Cheng M, Zhong Y, Kureemun U et al (2020) Environmental durability of carbon/flax fiber hybrid composites. Compos Struct 234:111719

58. Hasan KF, Horváth PTGR, Alpár T (2022) Nanotechnology for waste wood recycling. In: Nanotechnology in paper and wood engineering. Woodhead Publishing, Duxford, United Kingdom, pp 61–80
59. Mohan K, Rajmohan T (2018) Effects of MWCNT on mechanical properties of glass-flax fiber reinforced nano composites. Mater Today Proc 5(5):11628–11635
60. Gupta M, Srivastava R (2016) Mechanical properties of hybrid fibers-reinforced polymer composite: a review. Polym Plast Technol Eng 55(6):626–642
61. Mohammed L, Ansari MN, Pua G et al (2015) A review on natural fiber reinforced polymer composite and its applications. Int J Polym Sci 2015:1–15
62. Fiore V, Valenza A, Di Bella G (2012) Mechanical behavior of carbon/flax hybrid composites for structural applications. J Compos Mater 46(17):2089–2096
63. Singh J, Kumar M, Kumar S et al (2017) Properties of glass-fiber hybrid composites: a review. Polym Plast Technol Eng 56(5):455–469
64. Sanjay M, Arpitha G, Yogesha B (2015) Study on mechanical properties of natural-glass fibre reinforced polymer hybrid composites: a review. Mater Today Proc 2(4–5):2959–2967
65. Annandarajah C, Langhorst A, Kiziltas A et al (2019) Hybrid cellulose-glass fiber composites for automotive applications. Materials 12(19):3189
66. Vijay R, Singaravelu DL, Vinod A et al (2019) Characterization of novel natural fiber from saccharum bengalense grass (Sarkanda). J Nat Fibers 17(12):1739–1747
67. Arpitha G, Sanjay M, Senthamaraikannan P et al (2017) Hybridization effect of sisal/glass/epoxy/filler based woven fabric reinforced composites. Exp Tech 41(6):577–584
68. Suddell BC, Evans WJ (2005) Natural fiber composites in automotive applications. Natural fibers, biopolymers, and biocomposites. CRC Press, Boca Raton, United States, pp 253–282
69. Peijs T (2003) Composites for recyclability. Mater Today 4(6):30–35
70. Mayandi K, Rajini N, Manojprabhakar M et al (2019) Recent studies on durability of naturalsynthetic fiber reinforced hybrid polymer composites. Durability and life prediction in biocomposites, fibre-reinforced composites and hybrid composites. Elsevier, Duxford, United Kingdom, pp 1–13

Application of Hybrid Reinforced Cellulose-Glass Fiber Based Composites in Automotive Industries

I. O. Oladele, L. N. Onuh, G. S. Ogunwande, and S. G. Borisade

Abstract The application of light weight composite materials has been a fast growing trend in the fabrication of intricate and body parts in the automotive industry. Since the automotive industry covers all the modes of land transportation, covering the development of components such as engines and body parts (excluding tires, batteries and fuel), it has focused on developing novel high-safety materials that possess excellent properties. Amongst all transportation mechanisms in the automotive industry, automobile cars are the most affordable and easiest mode of transportation globally and there are several automobile industries across the globe with series of models annually. However, with government regulations on environmental issues, automotive industries have introduced the development and fabrication high biodegradable materials for different automotive parts, hence, the need for hybrid cellulose-glass fiber reinforced materials to compensate for strength, light weight and degradation after use. This review presents the properties and usefulness of hybrid cellulose-glass fiber reinforced materials and its need in the automotive industry.

Keywords Cellulosic fiber · Glass fiber · Automotive industry · Automobile · Environment · Composites

I. O. Oladele (✉) · L. N. Onuh · G. S. Ogunwande · S. G. Borisade
Department of Metallurgical and Materials Engineering, Federal University of Technology, PMB, 704, Akure, Ondo State, Nigeria
e-mail: iooladele@futa.edu.ng

I. O. Oladele
Centre for Nanomechanics and Tribocorrosion, School of Metallurgy, Chemical and Mining Engineering, University of Johannesburg, Johannesburg, South Africa

S. G. Borisade
Department of Materials and Metallurgical Engineering, Federal University Oye-Ekiti, Oye-Ekiti, Ekiti State, Nigeria

T. Khan and M. Jawaid (eds.), *Green Hybrid Composite in Engineering and Non-Engineering Applications*, Composites Science and Technology,
https://doi.org/10.1007/978-981-99-1583-5_8

1 Introduction

Prior to the present material age, various material ages has evolved over centuries from the Stone age to Aluminium age. The present material age has been discovered to be plastic age having transcended different eras to what it is today [1]. All these changes in materials ages has been greatly influenced by many research efforts for recycling existing scientific ideas to promote improved concepts for the development of new materials. All technological innovations emanating from material development have been linked with the ancient philosophies that are being adapted progressively. Hence, composite material development remains one of the most appropriate means for improving materials properties to suit the present and future dynamics needs and applications of man [2]. After the discovery of composite materials, the era of metal forgery, foundry and steel production began to fade away gradually, thus, industrial-scale manufacture of the first synthetic composites symbolized another milestone in the evolution of green-house industrialization. Ever since the innovation of the first set of synthetic composite materials, it has greatly encountered an exponential development. Due to the fast-rising developments of novel technologies in the automotive industries, it is safe to say that composite materials have been integrated into all production routes of the automotive industry. The occurrence of this unique type of engineering material was primarily based on economic considerations, durability and mechanical performance [2].

Composite materials are unique engineering materials formed by the merging of reinforcement materials (particle/fiber/special forms) and base materials (known as matrix). The matrix aids in binding of the reinforcements to form strong bonds and transfers functional loads in the composite. It also protects the reinforcements from encompassing disturbances and external factors when they are in a continuous phase [3]. Hybrid composites on the other hand are composite materials that are developed by mixing two or more distinct types of reinforcement materials (particle/fiber, particle/particle or fiber/fiber) in a single matrix.

Hybridization is a technique used for the incorporation of two or more reinforcement materials (particle/fiber) into a predetermined matrix material. The introduction of hybridization in composite production has been proved to yield improvements in the mechanical and physical properties of composite materials [2, 4]. Interlaminate and intralaminates hybridization are the two major types of hybridization. Here, the interlaminate (or sometimes known as simple laminate is made up of layers of distinct fibers deposited on top of each other, whereas intralaminates have two fibers bound together within the same layer [5].

Several cellulose based bio-fiber/particle reinforced composites exist in different areas of applications. Currently, cellulose fibers are majorly used as substitutes for synthetic fibers due to their 'green' image [6, 7]. Cellulose fibers are renewable and can be incinerated after use without causing air pollution. Interestingly too, the amount of CO_2 released during the incineration process is very minute when compared to the amount that the plant takes up throughout its lifetime [2]. In today's world, cellulose based hybrid composites are becoming an efficient option in the manufacturing sector and aids mitigating negative environmental effects. Cellulose based fibers are derived from plant fibers. They may also be derived from a variety of sources, including trees and seaweed [7]. Hemp, kenaf, flax, and jute are examples of natural fibers derived from the wood of plants. The leaves of the seeds are used to extract sisal, abaca, and banana fibers [8]. In the automobile industry, the hybrid composite material substitutes other polymer materials since it has a high carbon property and is non-corrosive. Natural fibers like sisal and roselle and others can be utilized in making hybrid composites to manufacture automotive components [8, 9].

Hybrid composites are displacing other basic engineering materials by meeting the needs of a variety of industries specifically in the automotive manufacturing industries. In comparison to using a single kind of a traditional standard material such as steel, plastics and aluminum, the use of a cellulose-based/glass fiber hybrid based composites has various advantages [10]. The most compelling reason is that they provide better properties with less weight and higher economic value, they produce more output with less consumption and a longer lifespan, making them a cost-effective option for a variety of applications in the automotive industry.

Various studies have shown that polymer-based composites have emerged as the leading group of composites that are fast displacing all other materials in several applications due to their inherent properties. Polymer-based composites can be entirely synthetic, completely natural, or a mixture of synthetic and natural-based. However, a recent desire for eco-friendly materials has shifted attention from complete synthetic-based materials to natural fibers, whether in a partial or total replacement [2]. Thus, this chapter provides an overview of research trends for cellulose-glass fiber based polymer composites.

2 Why Cellulose-Glass Fiber Hybrid Composites?

To begin with, the hybridization of a natural fiber with a synthetic fiber provides better wear and mechanical properties when compared to the hybridization of a natural fiber composite with another natural fiber [11]. The hybridization of two distinct fibers have shown to be a technique of high efficiency to develop high profile materials that meet a variety of needs. In order to utilize the greatest features of both natural and synthetic fibers, they can be blended in the same matrix to create hybrid composites that completely exploit the best properties and strengthen the weaker properties of each fiber component. The addition of fibers (artificial or natural) to any material used as a base matrix (can be metal, polymer or ceramics) is known to modify the mechanical characteristics of such material yielding a composite-based significant material [2, 12].

Due to possible dangers that could emanated from technological advancements in the automotive industries, the growing concern about global warming and its impact on environmental degradation has shifted researchers and scientists' interest to the use of more cellulose-glass fiber-based hybrid composite in the production of environmentally friendly products in order to reduce non-decomposable wastes that pose harm to the environment [11, 13]. This shift in focus is geared toward the development of environmentally friendly products, particularly in automotive parts. The area of natural fiber-reinforced composites (NFRCs) have seen a high rise and have also gained much ground in research and development since last decade. This is because NFRCs including hemp, jute, silk, kenaf, and cotton-based composites possess various significant advantages, including low thickness, low cost, accessibility, corrosion resistance, biodegradability, toughness, and morphological flexibility [14]. These advantageous properties give them the prospect to be used in the production of on a wide range of automotive parts, including car assembly and development as presented in Table 1.

In nonstructural applications, a few attempts have been made to harmonize natural fibers with artificial fibers. As a result, numerous vehicle parts that were formerly fabricated using just artificial or natural fiber reinforced composites are currently enhanced by hybridization with environmentally friendly fiber-based reinforced composites [15]. Hybridizing natural fiber reinforced composites with high strength

Table 1 Automakers and parts produced from polymer matrix composites [1]

Automakers	Models	Parts
Volkswagen	Bora, Golf, Passat Variant	Seatback, door panel, boot liner, boot-lid finish panel
BMW	3, 5, 7 series	Door lining panel, seatback, boot lining
Opel	Vectra, Zafira, Astra	Door panel, Headliner panel, Instrument panel
Fiat	Marea, Brava, Punto	Boot liner, door panel
Audi	A2, A3, A4, A6, A8	Spare tire lining, door panel, hat rack, boot lining, seatback
Mercedes Benz	A-class, E-class, S-class	Seatback, door panel, cover panel, dashboard, windshield, engine cover, roof cover, instrument panel, bumper
Toyota	ES3, Brevis, Celsior, Raum	Spare tire lining, seatback, floor mat, door panel
Volvo	V70, C70	Natural foam, seat padding
Mitsubishi	Outlander, and Eclipse cross	The instrument panel, cargo area floor, door panel
Daimler Chrysler	A, C, E, and S class	Door panels, car windshield, dashboard, and pillar cover panel
General Motors	Cadilac De Ville, Chevrolet Trial Blazer	Seatbacks, cargo area floor mat
Peugeot	406	Front and rear door panels, seatbacks, and parcel shelves
Ford	Mondeo CD 162, Focus	Door panels, boot-liner, door inserts, and floor trays
Renault	Twingo and Clio	Rear parcel shelf

synthetic fibers can improve their qualities. Hybrid composites are majorly weaved or inter-twined together with at least two distinct fibers in a grid pattern.

In various literatures, cellulose-glass fiber hybrid composite has been investigated for auto applications using different cellulose fibers like hemp, banana, kenaf and sisal [3, 10, 14]. In comparison to non-renewable derived components in these composites, cellulose-glass fiber hybrid composite has proved to be a low-cost choice composite. One of the main constituents of cellulose-glass fiber hybrid composite are the presence of natural fibers, since they are sustainable, may be able to reduce the natural weight of the fiber portion of the composite material. When compared to other composite materials with equivalent mechanical characteristics, the densities of cellulose fiber composites are low and fiber structures influence the composite materials properties [16]. The surface morphologies as revealed from the SEM images of fractured surfaces of pawpaw fiber-glass fiber hybrid reinforced epoxy composites were shown in Figs. 1 and 2.

Fig. 1 SEM images of hybrid 6 wt% glass fiber/treated network and linear pawpaw fiber reinforced epoxy composites [16]

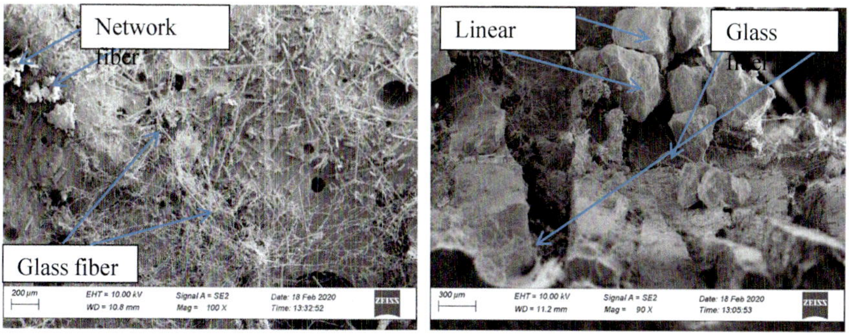

Fig. 2 SEM images of hybrid 15 wt% glass fiber/treated network and linear pawpaw fiber reinforced epoxy composites [16]

The application of standard fiber composite materials in the automotive industry can result to the reduction of vehicle weight without affecting the utilitarian performance [17].

In the development of cellulose-glass fiber based composites, cellulose is the weaker of the two components, for this fact, they are connected and inter-twined to form a much more grounded core. However, while there are different types of cellulose based fibers, it is only a few of them that is exploited for automotive applications. Cellulose-based fibers are gathered on a constant basis, whereas kenaf, jute, and hemp fibers are collected three to four times each year. These fibers have a higher yield content than the other regular filaments [18]. Kenaf, jute, and hemp are agronomic partners that are resistant to climate variations and require less water to grow.

The application of cellulose-based fibers in the auto industry has become an interesting trend and is gaining many grounds due to their easy processing technique. By combining regular strands into an item, the eco-friendliness and biodegradability of cellulose fibers are enhanced. Banana, Coconut, Sisal, and Flax are some of the other often used regular filaments used to enhance the biodegradability of some cellulose fibers. The hybridization of at least two strands in a single matrix results to a significant increase in mechanical and physical properties of cellulose based fiber composites [19].

Due to the factors of environmental impact, polymer based matrix reinforced with cellulose fibers is another viable choice and also a practical solution for automotive and aviation applications [20]. Cellulose-glass fiber hybrid composite offers certain advantages over traditional materials, but they also have some disadvantages which researchers are putting into consideration. To mitigate these disadvantages, researchers are on the quest to determine the best matrix material that will encompass the hybrid fiber reinforcements to give optimum properties. There is an additional obligation to learn and adapt while working with components made up of natural fibers. Novel techniques are also being developed by zealous researchers to improve the chances of obtaining the optimum filament networks and best water absorption properties of cellulose and glass fibers [21]. Thus, hybridizing glass fibers and natural fibers in base matrix materials yields composites with potentials to increase the amount of sustainable and biodegradable materials for fabrication of automotive parts. It has been used as a process for fulfilling the Corporate Average Fuel Economy (CAFE) Standards until 2025, and its use has the potential to reduce a vehicle's overall life cycle weights (NHTSA 2012).

Weight reduction, environmental friendliness, and energy conservation are vital factors that are put into considerations in the fabrication of automobile parts. Each of the three factors are linked, so if one of them is dealt with, the others will surely be resolved. Since fiber-based hybrid composite materials are formed from the interlocking and intertwining of the fiber network, the resulting properties of the developed hybrid material will differ, yielding unique and distinctive physio-chemical and mechanical properties (which includes weight reduction). The alteration of the properties of engineering materials has in time past been an existing trend in the automotive industry, and it is the best solution that an automotive engineer has in dealing with the aforementioned issues. The weight of an automobile can be reduced by the application of better materials and a lot of plan improvement, which also increases the eco-friendliness of IC motor-powered automobiles while conserving energy in electric and hybrid vehicles [22].

In conclusion, Glass fibers as synthetic fiber show better properties when applied in the fabrication used in hybrid composites when compared to carbon fibers because they are economic efficient and possess higher stiffness properties, as well as good strength, toughness, and impact resistance. For these reasons stated above, cellulose-glass fiber hybrid composite is a unique material that need to be exploited in the automotive industry [23].

3 Mechanical Properties of Hybrid Cellulose-Glass Fiber Composites

Cellulose-based fibers, a class of natural fibers are grown and harvested naturally from plants. They fibers are spun into threads and yarns before being sewn into natural textiles. They contain hemicellulose, which regulates moisture absorption, heat deterioration, and biodegradation. They are equally lightweight, renewable, and resource-rich [24]. Glass fibers on the other hand are also widely used in the production of thermoplastics and thermosets polymer-based composites. They are rigid, have a lot of compound blocking, have a lot of layered stability and they are protected [25].

Natural/synthetic fiber-based hybrid composites when developed show unique intermediate properties between natural and synthetic fibers. The individual property of a fiber (usually the natural fiber) in the hybrid composite may outweigh the property of the synthetic fiber, though, this depend on the predetermined proportion [26]. To reduce dependency on synthetic filaments and promote the usage of environmentally friendly materials, cellulose-based fibers are being employed in composite materials. In terms of low thickness, low cost, inexhaustibility, carbon-impartiality, highly significant mechanical characteristics, and biodegradability, natural fibers surpass synthetic filaments [27]. Regardless of the benefits provided by cellulose-based fibers, limiting factors such as high dampness responsiveness, low mechanical strength, ineffective interfacial attachment between fiber/matrix, and insignificant thermal stability have hampered and limited the use of cellulose-based fibers as reinforcements and impede their applications [28]. Subramaniam et al. [29] confirmed this when they investigated the effects of stacking arrangement on the tensile and semi-static intrusion of kenaf/glass hybrid fiber metal overlays. When compared to non-hybrid kenaf and glass supported fiber metal overlays, the composite with glass demonstrated a positive hybrid impact in terms of tensile and intrusion resistance [29]. Feng et al. [30] studied the fatigue and tensile strength features of hybrid kenaf/glass fiber metal laminates. In their work, the insertion of glass fiber enhanced the tensile strength of the developed composite material. Thus, when kenaf is hybridized with glass fiber, the desired fatigue life is achieved, showing optimum enhancements [30]. Veettil et al. [31] in their study developed composites materials from alkali (NaOH) treated eucalyptus and lotus natural fibers reinforced in epoxy-hardener by the hand lay-up molding technique followed by soft compression in criss-cross, biaxial, and uniaxial orientation. They discovered that the flexural strength of composites was enhanced by the addition of E-glass fibers [31]. Oladele et al. [16] studied the influence of structural physiognomies of pawpaw fiber–glass fiber hybrid–based green composites on mechanical properties and biodegradation potential of epoxy composites and revealed that the mechanical properties were mostly enhanced by treated linear structure pawpaw fiber while biodegradation was highly promoted by treated network structure pawpaw fiber.

It is worth noting that, while cellulose-based fibers are not only environmentally friendly but also have high elasticity, they lack some specific desirable properties such as excellent mechanical properties, and in order to fill this gap, the synergy effect of hybridization with E-glass fiber makes them ideal for use in the automobile industry [32]. And as a result, excellent outputs of mechanical properties such as bending are obtained, indicating that the hybridization of natural fibers with E-glass fibers improves the resultant composite's usefulness, particularly in practical applications [33].

In the same vein, Abd El-Baky and his research team investigated the effects of stacking pattern and relative fiber amounts on the mechanical characteristics of economically efficient innovative hybrid fiber metal laminates based on aluminum alloy and jute/glass fiber reinforced epoxy composites. The hand lay-up technique was used in the production process, followed by compression molding. They discovered that reinforcement hybridization can potentially enhance the mechanical characteristics of jute/glass hybrid-based composites, making them suitable for the fabrication of automotive parts. Incorporating high strength fibers into the composite base matrix, on the other hand, results in higher tensile properties but lower flexural resistance [34]. The fatigue and tensile strength features of hybrid kenaf/glass fiber metal laminates were studied by Feng et al. [30]. According to their findings, the insertion of glass enhanced the material's tensile strength. Therefore, when kenaf fibers are reinforced alongside glass fibers, an enhanced fatigue life is achieved [30].

Another study combined date palm wood flour with glass fiber. The samples were manufactured with an injection molding machine, and the mixing was done in an extruder. The mechanical properties of the composites, notably tensile strength, young's modulus and hardness, were discovered to be increased by 18%. The limited zone generated by the absence of agglomeration in the composites' interstice was found to be responsible for the improved mechanical characteristics and hardness level. This made them appropriate for developing parts for automobiles that need reasonably mechanical qualities as well as a significant level of hardness [35].

Several researchers have also concentrated on different features of natural/synthetic fiber-based hybrid polymer composites. According to Thew and Liao [36], the mechanical properties of bamboo/glass fiber supported hybrid composites were affected by fiber length, fiber weight proportion, and matrix-fiber adhesion properties. Goud and Rao [37] observed that increasing the glass fiber stacking enhances the flexibility, flexural, effect, and hardness properties of Roystonearegia/glass fiber hybrid composites. Thomas et al. [38] also studied banana-glass hybrid composites and observed that the layering design or composite computation has a significant influence on the composites' unique behavior. In another research output carried out by Cui and Tao [39], wood-glass fiber reinforced hybrid composites were prepared by compounding reused high-thickness polyethylene from metropolitan strong waste, garbage wood fibers, and slashed glass fibers. Mechanical tests of the developed composites with varying compositions of glass fibers content were performed. The results revealed that by using type L chopped Glass Fibers, the tensile and impact strength of the composite can be enhanced at the same time, but would be reduced when type S sliced glass filaments were used [39].

Jute-E-glass hybrid fibers was reinforced in a PVC matrix and the corresponding composite was fabricated via compression molding. The ultimate tensile strength and modulus, flexural strength and modulus, and impact strength of the fabricated hybrid composites were all examined and compared. It was discovered that enhancements occurred in the tensile strength and modulus, bending strength and modulus, and impact strength by 44, 80, 47, 92, and 37.5%, respectively [40].

Glass-kenaf-waste tea leaf fiber-reinforced hybrid composites were fabricated by Prabhu et al. [41]. The mechanical properties including tensile, flexural, and impact strength of the fabricated fiber-reinforced composite were investigated. The interfacial characteristics, internal fissures, and intricate structures of the fractured composite surfaces were also studied using scanning electron microscopy (SEM). The results of the testing demonstrated that a combination of 10% glass fiber, 20% jute fiber, and 10% tea leaf fiber had the greatest mechanical properties and may be utilized as a replacement for pure glass fiber reinforced polymer composites especially in the automotive applications [41]. One major factor driving the automobile industry is that a lot of researches are being carried out on regular basis for the application of natural-synthetic hybrid fibers to replace some of the vehicle's parts or structures (Fig. 3).

Different surface modifications, for example, basic acetylation, silane treatment, benzoyl chloride, and potassium permanganate arrangement have been applied on fibers used for composite fabrication and has been confirmed to show optimum enhancements in the mechanical properties of both natural and synthetic fibers by increasing mechanical interfacial holding capabilities [7, 8]. It is undeniably true that there is a vivid measure of increase in the tensile modulus and strength in relation to the expansion in fiber weight to a specific sum. Natural fibers are economically feasible and lighter, but they have inferior mechanical qualities when compared to synthetic fibers. As a result, rather than relying solely on cheap natural fibers, the hybrid arrangement is also feasible and cost-effective. To find a solution to this problem, hybrid fiber-reinforced composites can be fabricated and tested by applying them to bended lines. This trend was illustrated by Ramesh et al. [42] when the mechanical properties of jute-sisal-glass fiber supported crossover thermosetting polyester composites was investigated. The composite was found to have the most enhanced tensile, flexural, and flexural strength [42].

Fig. 3 Vehicle with several parts produced from polymer matric based composites (PMCs) [1]

The mechanical properties of jute and sisal fiber reinforced with glass fiber composites were studied by Ramesh et al. [43], they discovered a significant improvement in the composite properties, which include flexural, tensile, and impact strength, as well as the percentage of moisture absorption behavior. When compared to a single reinforced natural fiber, neither jute nor sisal fiber reinforced composite possessed better mechanical properties with respect to glass fiber. This finding further proves that adding glass fiber to the top and bottom parts of composite coverings improves the strength of the material. In the investigation of hybrid composite materials consisting of glass and sisal fibers with varying epoxy compositions in the matrix, the mechanical properties of the hybrid composite were investigated. The test values differed amongst the various compositions in each sample. The compression stress in the hybrid composite was concluded to be around 915.5 kg/cm^2 with a low degree of deflection. It also showed a higher hardness, with a hardness value of around 83 HB. This composite is normally appealing with the end goal of vehicle applications due to the outstanding attributes produced from it [44].

Due to the fact that hemp fibers are relatively inexpensive, renewable, and fully biodegradable, they are popularly utilized as the source of cellulose fiber in natural-synthetic hybrid composites. Recently, the hybridization of hemp fibers with glass fibers improved the mechanical properties of the material. The bending strength and bending modulus of 25% hemp fibers and 15% glass fibers were improved [45]. Hybrid reinforced composite-based vehicle parts also possess better results compared to non-hybrid composite-based vehicle parts.

The effects of hemp/glass fiber hybrid composite hybridization on mechanical properties such as tensile property, impact property, and hardness level was tested experimentally in a study carried out by Nachippan et al. [46]. The hybrid composite was created using the hand layup method. SEM micrographs were also utilized to assess the composite's microstructural characteristics and was also used to conduct a failure analysis. According to the findings, the hemp/glass fiber hybrid composite showed good tensile qualities, impact strength, and hardness level. As a result, they're recommended for usage in applications like automobiles that require great impact strength and hardness [46].

Fiber hybridization has been a critical aspect of study for many years to improve the strength and other essential properties of fiber-based composites. In the case of obtaining a high fracture stress and toughness, brittle fibers are usually substituted with ductile fibers. But in some cases, hybridization with both ductile and brittle fibers in the composite yields optimum mechanical and high fracture toughness properties. Pseudo ductility is a major property obtained from the hybridization of fibers in composite materials. Pseudo ductility also show proper enhancements in the plastic deformation of the base matrix and this refers to the crystallographic effects of fiber reorientation and matrix formation [47]. The broad application of cellulose-based glass fiber composites has given rise to many novel innovations in the automobile sector.

Recently, the mechanical properties of epoxy-based composite reinforced with hybridized glass and sisal fibers treated with molybdic acids was discovered to be high compared to unreinforced sample [48]. Mechanical properties such as tensile strength and hardness were analyzed. It was determined that due to their biodegradable properties, they possessed enhanced properties with an improved surface strength and light weight; these particular hybridized composite-based materials are more intriguing to be worked on. With the aid of a scanning electron microscope (SEM), the internal morphology of aggregates and fissured details of the composite was examined. Distinctions in the fiber and matrix content were also used to test the properties of the hybrid composite material, the effects showed that 40% fiber with 60% matrix exhibited excellent properties. With the results obtained, the use of hybridized cellulose-based/synthetic fibers have in no doubt become feasible to meet the structural endurance requirements of automobile interiors and exterior parts, cases, and other required parts [48]. Comparable results were also obtained by Mishra et al. [49] in a research work which involved the development of composite with palm and glass fiber hybrid reinforcements in a polypropylene matrix mixed in a ratio of 40:60 and at different proportions by weight. A positive influence was indicated on the elastic modulus from the hybridized reinforcements.

Harish et al. [50] also investigated the effects of palm and glass fiber reinforcements in a polypropylene matrix. The results obtained demonstrated that adding palm and glass fibers to the composites enhanced the tensile strength, heat conductivity, water absorption capacity and impact resistance. Efforts have been targeted towards understanding the mechanical performance of cellulose-based fiber with glass fiber hybrid reinforced polymer composites. The hybridization of cellulose-based fiber with glass fiber offers a way to enhance the mechanical qualities above natural fiber alone.

Finally, even with the obvious enhancements in the mechanical properties of composites developed from cellulose-based/glass fiber composite, and their desirable properties, it has been observed that hybrid composites comprising of two different cellulose-based fibers with glass fiber have been discovered to be less common than a single cellulose-based/glass fiber composites, although their applications are growing.

4 Fire/Heat Resistivity/Thermal Stability of Cellulose-Glass Fiber Hybrid Composite

When considering cellulose-glass fibers hybrid composite materials in the fabrication of automobile parts, engine parts and parts needed for automotive applications, thermal stability is a non-negotiable property that must not be compromised. As these parts are typically exposed to high temperatures and harmful synthetics, exposure circumstances may impact their prolonged application. For these reasons, researchers have been working relentlessly to characterize cellulose-glass fiber hybrid composite materials in harsh environments in order to assess their resistance to thermal disturbances, or thermal stability.

The thermal stability of natural/synthetic fibers was demonstrated by Zhang et al. [51] when they added glass fibers to wood flour and, as a result, improved the resistance of the resulting hybrid composite to decomposition temperature. This is not surprising given that glass fiber possess a high resistance to thermal degradation. Their findings conclude that glass fiber reinforced composites are potential reinforcement materials for the fabrication of composite-based automobile parts where there is a high risk of continuous heat exposure, thereby extending their life span [51].

Jute-E-glass hybrid-based composites fabricated by reinforcing hybridized jute and E-glass fibers in a PVC matrix was developed via compression molding [40]. Thermal characteristics of the Jute-E-glass hybrid-based composites were investigated, and it was discovered that the E-glass fiber in the PVC framework instigated a greater thermal stability [40]. In addition, in a separate research, it was discovered that by incorporating glass fiber into hemp and PP matrix, substantial improvements in the thermal stability were achieved [52].

5 Sound Absorption of Cellulose-Glass Fiber Hybrid Composite

Aesthetics are crucial both before and after a car's design, and one of the most important aspects of aesthetics is the ability of an automobile to produce little or no sound when in motion, which is directly related to its ability to absorb sound or noise from external or internal sources. for example, In the design underbody shield part of an automobile which is meant to reduce the amount of noise that enters the automobile from the outside. It is advantageous to hybridize glass fiber with cellulose-based fibers since kenaf fibers are a sound-absorbing material that is both renewable and biodegradable. To prove this, Parikh et al. [53] investigated the sound absorption coefficient of an automobile floor system made of needle perforated kenaf fiber, polyethylene terephthalate (PET), and woven polypropylene (PP) fabric in a frequency range of 700–3200 Hz. The absorption coefficient for 3200 Hz was 0.54 and 0.71 for the 8 and 10 mm specimens, respectively, which increased to a value close to 1 when a polyurethane underlay was added to the flooring system.

Lim et al. [54] also discovered a sound absorption coefficient greater than 1.5 kHz of 0.8 for 40 mm thick kenaf fiber samples. They observed that as the sample thickness grew, so did the absorption coefficient. Intuitively, it may be predicted that including cellulose-based fibers alongside glass fibers in hybrid composites will reduce the noise or sound disturbances coming out of such composites. Nandanwar et al. [55] further stated that in the frequency range of 125–4000 Hz, the sound absorption coefficient of kenaf and compression molded glass fiber hybrid composites reinforced composites increased with decreasing bulk density. The composite's higher porosity contributes to the material's ability to absorb sound. Because the underbody shield is located around car components that generate a lot of heat, such as the exhaust pipe, the produced product must be able to withstand harsh weather conditions such as heat, moderate fire assault, and elevated temperatures while preserving mechanical integrity and dimensional stability while using it outside.

6 Crashworthiness of Cellulose-Glass Fiber Hybrid Composite

Crashworthiness is defined as the assessment of a composite product's or part's resistance to a quantifiable reactive force against it or a rapid impact, and it acts as a method of prevention from significant damage or death in the automobile sector. It has been stated that a vehicle's crashworthiness is directly related to its energy absorption level [56].

According to recent studies, a well-designed cellulose-glass fiber hybrid composite can absorb more energy than antique materials used in the automotive industry. Literatures have been reviewed on the crashworthiness of cellulose-based-glass fiber composites. In one of the reviewed literatures, Alkbir et al. [57] made an emphasis on the massive crashworthiness improvements of cellulose-glass fiber hybrid composites in the past two decades. Mahdi et al. [58] also investigated the crash resistance of circular tube and circular cone-like cylinders composed of glass/epoxy, carbon/epoxy, and oil palm frond fiber/epoxy composites. In their research, it was subsequently focused on the semi-static axial crash resistance of hybrid and non-hybrid fiber/polyester composites made from oil palm and coir fibers [58, 59].

They made conclusions which indicated that hybridization has a large effect on the increased amount of resistance to crash obtained in the composites. In another study, fiber-based hybrid composites were fabricated using the wet-wrapping process technique and tested under uniaxial quasi-static compression loading up to complete crushing to determine the level of resistance of Jute-Glass composites to crash. The results showed that hybrid composites have the potential to be used where energy absorbing parts are required in automobile parts because they have a relatively high load carrying ability and energy absorption efficiency [60].

Currently, a lot of study is being done on the possibility of cellulose-based-glass fiber composite to be used for the development of parts in the auto sector. In the automotive industry, regular silk/epoxy rectangular composite cylinders can be used as an energy retaining component.

7 Water Absorption of Cellulose-Glass Fiber Hybrid Composite

Hybridization as a technique for incorporating one or more reinforced fibers into a matrix material has been recorded to enhance the water absorption properties of composite materials for auto applications [61]. In other words, in automotive applications, one of the most effective technique applied to minimize the rate of water absorption in natural fiber-based composites (i.e., enhancing the water absorption properties) is by hybridization with synthetic fibers, as is evident in the case of cellulose-based glass fiber composite. Excessive water absorption has shown to pose a detrimental impact of causing components to expand, compromising their mechanical and dimensional integrity [62]. Swelling, material friction, crack initiation and localized deformation/deterioration of the matrix and fiber-matrix interface are all problems that occur when cellulose-based fibers absorb excessive amount of water, resulting in reduction of properties. Moisture is carried by diffusive or capillary mechanisms in composite materials. In polymer composites, the direct diffusion of water molecules into the spaces between the polymer chains is the first technique of water absorption. Capillary action carries additional moisture to the material through defects at the fiber-matrix interface, voids in the matrix formed during processing, or pathways formed due to inflammation and induced micro-cracking. localized matrix damage.

In many cases, it has been concluded that cellulose-based fibers and their related composites are abundantly available sustainable materials but still have a few drawbacks, such as poor wettability, incongruence for particular polymeric grids, and excessive dampness absorption by the filaments. In the quest to mitigate these drawbacks for effective auto productions, researchers have developed considerable efforts towards understanding the water absorption behavior of cellulose-based/glass fiber hybrid composite targeted towards automobile applications.

A recent study has been targeted at accessing the physical, mechanical, and water absorption characteristics of coir/glass fiber hybrid epoxy-composites [25]. The effects of fiber stacking and length on composite mechanical parameters such as tensile strength, flexural strength, and hardness were studied. According to the research observations, the composite with 10% fiber stacking at 15 mm length has the best strength qualities. The composites with 10% fiber stacking had a maximum flexural strength of 63 MPa at 15 mm fiber length. Similarly, at 20 mm fiber length, composites with 10% fiber stacking had the maximum extreme hardness value of 21.3 Hv [25]. Thus, natural/synthetic fiber hybridization reduced water flow while increasing thermal stability. The hybridization of natural with synthetic fiber (such as fiberglass, carbon, or aramid) can also improve the tensile modulus, strength, and moisture resistance of composites. The hybridization of natural fibers with synthetic fibers, such as glass or carbon fibers, can also increase the composite's stiffness, resistance, and humidity resistance characteristics. The resilience of polypropylene reinforced with bamboo fibers can be increased by combining it with modest quantities of glass fiber [63].

In a comprehensive study carried out by Tamrakar et al. [64], the researchers studied the water absorption behavior of three different types of compression molded composites: all glass fiber reinforced composite (25% by weight), all kenaf fiber reinforced composite (25% by weight), and hybridized kenaf/glass fiber reinforced composite (12.5% glass and 12.5% kenaf by weight) with a matrix of polyethylene terephthalate and an Acrodur binder at ambient (23 °C). The composite material fabricated was to be applied in the development of automotive components that is environmentally friendly by replacing glass fibers with natural fibers for underbody protection applications. Hygrothermal treatments included freeze–thaw cycles between 23 and −29 °C, extended freezing at −29 °C, and re-drying of saturated at room temperature specimens. Water absorption has a considerable influence on mechanical qualities, according to routine flex testing. In dry conditions, hybrid and all-kenaf composites outperform all-glass composites in terms of flex modulus and strength. Despite the significant effect of water absorption, the mechanical characteristics of all-kenaf and hybrid composites after saturation are comparable to those of saturated all-glass composites. The mechanical characteristics of the saturated composites at 70 °C were substantially degraded, as predicted, with the modulus reducing by approximately 55% for the hybrid and all-kenaf composites. Freeze–thaw cycles and extended freezing, on the other hand, have no influence on the mechanical characteristics of saturated composites. Surprisingly, the flexural strength of all composites improved when the tempered specimens were dried again. The enhanced cross-linking in the Acrodur matrix is responsible for this. Furthermore, sound absorption and muffler experiments have shown that glass fibers may be substituted with kenaf fibers without sacrificing performance [64].

Based on the facts that the void content of composites varies substantially depending on the production method, the manufacturing process has a significant impact on the water absorption behavior of composites. Studies with hybrid kenaf composites by Ghani et al. [65] and Mazuki et al. [66] revealed that the water source has no effect on the time to saturation (rain, sea or tap water). Salleh et al. [67] further tested water absorption on polyester matrix hybrid composites made by compression molding with long kenaf fibers (20 wt%) and woven glass fibers (16 wt%). After 28 days of immersion, the specimens were saturated with a total moisture content of 4–5.5%. Water absorption at saturation was also discovered to be comparable in the hybrid composite to that of the all-kenaf composite. A roughly estimated 33% drop in modulus and strength characteristics was also reported.

Panthapulakkal and Sani [52] in their study discovered that the incorporation of glass fiber into hemp and PP matrix substantially enhance the water absorption rates of composites. Similarly, when the water absorption properties of jute and sisal fiber hybridized with glass fiber composites were studied, it was discovered that there was significant improvement in the moisture absorption property when compared to a single natural fiber reinforcement [42].

In a study carried out by Pai et al. [68], the physio-chemical and mechanical properties of hybrid fiber reinforced composites (E-glass fiber and banana fibers) in various weight percentages with epoxy. The composite demonstrated qualities that are crucial for usage as reinforcements in composite materials. The experiment results were scaled up to a 100 kN. The corresponding ideal composite result was analyzed using a Servo-Hydraulic (UTM) Universal Testing Machine (50% E-glass, 100% banana fiber, and hence 400 epoxy resin). Water absorption was observed using C-1 (20% E-glass, four hundredths of banana fiber, and hence four hundredths of epoxy resin). Finally, it was concluded that the amount of banana fiber used in the experiment determined how much water was absorbed [68]. Currently, cellulose-based/glass fiber hybrid composite is used for underbody protection applications and other important parts of the automobile.

8 High Strength-to-Weight and Stiffness-to-Weight Ratio Applications

In an ongoing investigation of general composites for automotive and aerospace applications, essential features such as high strength-to-weight and stiffness-to-weight ratio were discovered in composites reinforced with long lignocellulosic fibers such as hemp, flax, jute, and cotton. The primary goal of cellulose-based/glass fiber hybridization is to increase composite strength, rigidity, and ductility [69]. In another research work carried out by Raju and Balakrishnan [70], a polymer matrix composite reinforced with glass and palm fiber palm was developed and intended to reduce weight, and increase the hardness and resistance of vehicle parts. From the attained conclusions, it was reported that by increasing the phase ratio of the fiber particles, the tensile strength of the composite is increased. After flexural characterization was carried out on the hybridized composites, it was reported that the hybridized fiber reinforced composites showed better properties than the neat composite. The hardness characterization also indicated that the composite are potential materials for high strength and light weight applications, especially in the automobile sector [70].

One of the primary reasons cellulose-based/glass fiber hybrid composite is widely considered and applied in the automobile industry is to replace heavier traditional materials such as magnesium, copper, and steel, thereby giving way for lighter and more efficient automobile vehicles. Automakers intend to choose the most efficient source of cellulose-based/glass fiber hybrid composite in order to improve the feasibility and reduce environmental effect. Due to this fact, the emission of CO_2 gas has encountered a drastic decreased, and energy efficiency has simultaneously increased. Furthermore, it is expected that several million automobiles reach the end of their useful life each year. Even though certain pieces may be recyclable, around 25–40% of the vehicle's weight (fibers and plastics) ends up in landfills as garbage.

Hybridization with high-strength synthetic fibers such as glass fibers can improve the qualities of composites driven by natural fibers. Hybrid composites are made by intercalating two or more kinds of fibers in a common matrix. One of the key advantages of composites is the notion of hybridization, which allows the design engineer to change the material properties according to the needs.

9 Environmental Impact of Cellulose-Based/Glass Fiber Hybrid Composite

Due to the high varying levels of contamination, most automotive components are built of non-perishable composite materials such as polystyrene, polypropylene (PP), and polyethylene (PE), as well as glass fiber. These high level occurrences of contamination if not mitigated will result in releasing dangerous emissions, putting the ecology at risk. In the quest to mitigate these contamination, applying bio-based components in automotive fabrication operations have been proved by researchers to be a potential technique to minimize the negative environmental effect of the production process of automotive parts. Bio-based resources also possess a number of potential advantages in terms of controlling gas emissions and other environmental consequences that pose detrimental effects to human lives [71].

As the global environmental concerns take a high rise, there have been an increasing desire in adding ecologically friendly and sustainable components like cellulose-based fibers when fabricating composite materials. Cellulose-based fibers are potential alternative to traditional fillers like carbon fiber because of their low carbon content, low cost, biodegradability, non-toxicity, minimal health risk, low abrasion on processing equipment, and acceptable specialized properties. Unfortunately, the susceptibility to moisture and humidity of cellulose-based fibers, as well as the subsequent poor mechanical properties makes it difficult to be used as a single reinforcement component in composite fabrication for external applications in the automotive industry. However, these disadvantages in the application of cellulose-based fibers as a single reinforcement component in composite fabrication can be mitigated by combining the cellulose-based fillers with synthetic fibers to form hybrid composites where the best properties of one type of fiber match the shortcomings of another. Accordingly, rigorous design studies for hybrid-based composites will strike the perfect balance between performance, sensitivity to environmental degradation and cost/weight reductions.

10 Applications of Cellulose-Based/Glass Fiber Hybrid Composite

There are four objectives to be attained for the effective and long-term application of cellulose-based/glass fiber hybrid composite in the automobile industry. Firstly, it must enhance design flexibility for easier assembly and disassembly of automotive parts. Secondly, it must be able to minimize carbon dioxide emissions to the lowest minimum from automobiles. Thirdly, it should be able to reduce the net weight of the automobile for reasonable fuel consumption. Finally, it should reduce the reliance on non-renewable resources such as petroleum-based plastics [40]. To completely comprehend the application regions, materials engineers must first realize that cellulose-based/glass fiber hybrid composite components for automobile applications are classified into four categories: structural elements and fuel systems, interior trim, underhood applications and external parts.

Sealants, paint, varnish, adhesives, varnishes, and electronic and electrical components are all examples of "hidden" automotive elements that support the possibility of utilizing cellulose-based/glass fiber hybrid composite. Seat backs, dashboards, parcel trays, trunk liners, door panels, and other interior trim elements were originally produced from polymer-based composites derived from natural fibers in Europe and then in North America [46]. In the 1960s, coir fibers were initially employed in vehicle seats. Cellulose-based/glass fiber hybrid composite has been increasingly popular in recent years, and they are still utilized in headrests, back cushions, and headquarters funds [17]. Interestingly, cellulose-based/glass fiber hybrid composite in automotive application is less expensive than traditional materials, despite the fact that the latter's is still widely used.

As indicated in Table 2, the innovative applications of cellulose-based/glass fiber hybrid composite composites are expanding fast across all industries with a particular emphasis on the automotive engineering industry. According to prior studies by Yang et al. [45], a 25% reduction in automobile weight results in a reduction of 250 million barrels of crude oil per year. Furthermore, American, European and Asian manufacturers such as General Motors, Ford, Honda, Toyota and Volkswagen are already employing cellulose-based/glass fiber hybrid composites (e.g., jute, hemp, kenaf, glass etc.) as bio-composites for their seats, roofs, dashboard coverings, and trunk covers [46]. Table 2 present the country, manufacturing industries and car models for parts developed from cellulose-glass fiber based composites.

Table 2 Country, manufacturer and areas of applications of cellulose-based and synthetic fibers in automotive industry [72]

Country	Manufacturers	Automobile model	Applications in automobile
France	Peugeot	406	Seat backs, parcel shelf, front and rear door panels
France	Citroen	C3 Picasso, C5	Parcel shelves, boot linings, door panels, interior door paneling, and mud guards
Germany	Opel	Astra, Zafira, Vectra	Pillar cover panel, door panels, head-liner panel, and instrumental panel
Germany	Volkswagen	Passat Variant, Golf, A4, Bora	Seat back, door panel, boot-lid finish panel, and boot-liner
Germany	Audi	A2, A3, A4, A4Avant, A6, A8, Roadstar, Coupe	Boot-liner, side and back door panel, spare tire-lining, seat back, and hat rack
Germany	Daimler Chrysler	A, C, E, and S class, EvoBus (exterior)	Pillar cover panel, door panels, car windshield/car dashboard, and business table
Germany	BMW	3, 5 and 7 series and other Pilot	Noise insulation panels, seat back, headliner panel, boot-lining, door panels, and molded foot well linings
Germany	Mercedes Benz	C, S, E, and A classes	Door panels (flax/wood/sisal fibers with epoxy resin/UP matrix), instrument panel support, insulation (cotton fiber), glove box (cotton fibers/wood molded, flax/sisal), molding rod/apertures, seat backrest panel (cotton fiber), seat surface/backrest (coconut fiber/natural rubber) and trunk panel (cotton with PP/PET fibers)
Italy	Fiat	Alfa Romeo 146, 156, 159, Punto, Brava, Marea	Door panel
Japan	Toyota	ES3	Pillar garnish and other interior parts

(continued)

Table 2 (continued)

Country	Manufacturers	Automobile model	Applications in automobile
Japan	Mitsubishi	Fiat SpA	Floor mats, indoor cladding, seat back linings, cargo area floor, door panels, instrumental panel, and floor panels
United Kingdom	Rover	2000 and others	Rear storage shelf/panel, and insulations
United Kingdom	Lotus	Eco Elise (July 2008)	Body panels, spoiler, seats, and interior carpets
United Kingdom	Vauxhall	Corsa, Astra, Vectra, Zafira	Headliner panel, interior door panels, pillar cover panel, and instrument panel
United States	General Motors	Cadillac De Ville, Chevrolet Trail	Blazer seat backs, cargo area floor mat
United States	Volvo	V70, C70	Seat padding, natural foams, cargo floor tray, dash boards and ceilings
United States	Ford	Mondeo CD 162, Focus	Floor trays, door inserts, door panels, B pillar, and boot-liner

11 Conclusion

The review presented in this work showed that cellulose-glass fiber based composites are dependable materials for automobile applications. The development of this hybrid composites offer solution to environmental issues, shortage of low cost, light-weight and strong materials needed for weight reduction in automobiles and advanced the development of more natural fiber based composite materials for various applications. With the obvious areas of application of polymer based composites from this hybrid composite in automobile, more efforts are still needed for further improvement and applications. Current and future material needs call for more research and development in this field of study.

References

1. Oladele IO, Omotosho TF, Adediran AA (2020) Polymer-based composites: an indispensable material for present and future applications. Int J Polym Sci (Hindawi) 1–12
2. Oladele IO, Omotosho TF, Ogunwande GS, Owa FA (2021) A review on the philosophies for the advancement of polymer-based composites: past, present and future perspective. Appl Sci Eng Prog 2021:1–27
3. Sanjay MR, Arpitha GR, Yogesha B (2015) Study on mechanical properties of natural-glass fiber reinforced polymer hybrid composites: a review. Mater Today Proc 2:2959–2967

4. Maslinda AB, Abdul Majid MS, Ridzuan MJM, Afendi M, Gibson AG (2017) Effects of water absorption on the mechanical properties of hybrid interwoven cellulose-cellulosic fiber reinforced epoxy composites. J Compos Struct Compos Struct 167:227–237
5. Muddappa PPY, Rajanna T, Giridhara G (2021) Effects of different interlaminar hybridization and localized edge loads on the vibration and buckling behaviour of fiber metal composite laminates. Compos Part C Open Access 4:100084
6. Xin C, Zhang T, Tsai SB, Zhai YM, Wang J (2020) An empirical study on greenhouse gas emission calculations under different municipal solid waste management strategies. Appl Sci 10(5):10–33
7. Oladele IO, Omokafe MS, Olusegun JS (2016) Influence of chemical treatment on the constituents and tensile properties of selected agro-fibres. West Indian J Eng 38(2):4–12
8. Oladele IO, Omotoyinbo JA, Adewuyi BO, Kavishe FPL (2013) The effects of production processes on the mechanical properties of sisal fiber reinforced polypropylene composites. Philip J Sci 142(2):189–198
9. Njoku CE, Alaneme KK, Omotoyinbo JA, Daramola MO (2019) Natural fibers as viable sources for the development of structural, semi-structural, and technological materials—a review. Adv Mater Lett 10(10):682–694
10. Manikanta PS, Chowdary MS, Niranjan-Kumar MSR (2020) Evaluation of mechanical properties of banana and s-glass fiber reinforced hybrid composite. Adv Appl Mech Eng 775–789
11. Oladele IO, Akinwekomi AD, Ibrahim OI, Adegun MH, Talabi SI (2021) Assessment of impact energy, wear behavior, thermal resistance and water absorption properties of hybrid bagasse fiber/$CaCO_3$ reinforced polypropylene composites. Int Polym Proc 36(2):205–212
12. Ramasamy M, Daniel AA, Nithya M, Kumar SS, Pugazhenthi R (2021) Characterization of natural-synthetic fiber reinforced epoxy-based composite-hybridization of kenaf fiber and kevlar fiber. Mater Today Proc 37:1699–1705
13. Bhardwaj D, Gupta A, Chaudhary V, Gupta S (2021) Hybridization of natural fibers to develop the polymeric composite materials: a review. Adv Eng Mater 355–363
14. Saroj S, Nayak RK (2021) Improvement of mechanical and wear resistance of natural fiber reinforced polymer composites (NFRPC) through synthetic fiber (glass/carbon) hybridization. Trans Indian Inst Met 74(11):2651–2658
15. Chandgude S, Salunkhe S (2021) In-state-of-art: mechanical behavior of natural fiber-based hybrid polymeric composites for application of automobile components. Polym Compos 42(6):2678–2703
16. Oladele IO, Makinde-Isola BA, Adediran AA, Ayanleye OT, Taiwo SA (2021) Influence of structural physiognomies of pawpaw fiber-glass fiber hybrid-based green composites on mechanical properties and biodegradation potential of epoxy composites. J Reinf Plast Compos 40(23–24):867–879
17. Boland CS, De Kleine R, Keoleian GA, Lee EC, Kim HC, Wallington TJ (2016) Life cycle impacts of natural fiber composites for automotive applications effects of renewable energy content and lightweighting. J Ind Ecol 20(1):179–189
18. Neto J, Queiroz H, Agular R, Lima R, Cavalcanti D, Banea MD (2022) A Review of recent advances in hybrid natural fiber reinforced polymer composites. J Renew Mater 10(3):561
19. Wang B, Huang S, Yan L (2021) Natural/synthetic fiber-reinforced bioepoxy composites. In: Bio-based epoxy polymers, blends and composites: synthesis, properties, characterization and applications, pp 73–116
20. Jagadeesh P, Puttegowda M, Rangappa SM, Siengchin S (2021) A Review on extraction, chemical treatment, characterization of natural fibers and its composites for potential applications. Polym Compos 42(12):6239–6264
21. Ho M, Wang H, Lee JH, Ho C, Lau K, Leng J, Hui D (2012) Critical factors on manufacturing processes of natural fiber composites. Compos B Eng 43(8):3529–3562
22. Suriani MJ, Ilyas RA, Zuhri MYM, Khalina A, Sultan MTH, Sapuan SM, Ruzaidi CM, Nik-Wan F, Zulkifli F, Harussani MM, Azman MA, Radzi FSM, Sharma S (2021) Critical review of natural fiber reinforced hybrid composites: processing, properties, applications and cost. Polymers 13(20):3514

23. Alves C, Ferrao PMC, Silva AJ, Reis LG, Freitas M, Rodrigues LB, Alves DE (2010) Ecodesign of automotive components making use of natural jute fiber composites. J Clean Prod 18(4):313–327
24. Gaurav H, Gohal V, Kumar MN (2020) Study of natural fiber composite material and its hybridization techniques. Mater Today Proc 02277
25. Bhagat VK, Biswas S, Dehury J (2014) Physical, mechanical, and water absorption behavior of coir/glass fiber reinforced epoxy based hybrid composites. Polym Compos 35(5):925–930
26. Dhar MS, Jumaat FA, Ng LF, Subramaniam K, Ab-Ghani AF (2018) Tensile and impact properties of cost-effective hybrid fiber metal laminate sandwich structures. Adv Polym Technol 37(7):2385–2393
27. Shahroze RM, Ishak MR, Salit MS, Leman Z, Asim M, Chandrasekar M (2018) Effect of organo-modified nanoclay on the mechanical properties of sugar palm fiber-reinforced polyester composites. BioResources 13(4):7430–7444
28. Abd El-baky MA, Attia MA (2019) Water absorption effect on the in-plane shear properties of jute–glass–carbon-reinforced composites using Iosipescu test. J Compos Mater 53(21):3033–3045
29. Subramaniam K, Malingam SD, Feng NL, Bapokutty O (2019) The effects of stacking configuration on the response of tensile and quasi-static penetration to woven kenaf/glass hybrid composite metal laminate. Polym Compos 40(2):568–577
30. Feng NL, Malingam SD, Zakaria KA, Selamat MZ (2019) Investigation on the fatigue life characteristic of kenaf/glass woven-ply reinforced metal sandwich materials. J Sandw Struct Mater 21(7):2440–2455
31. Veettil AA, Thayyil AD, Thamba NB, Mangalaraja RV (2020) Study on flexural properties of lotus, eucalyptus composites reinforced with epoxy and e-glass fibers in different orientations. J Nat Fibers 1838991
32. Ramesh M, Palanikumar K, Reddy KH (2013) Mechanical property evaluation of sisal–jute–glass fiber reinforced polyester composites. Compos B Eng 48(12):1–9
33. Sanjay MR, Arpitha GR, Yogesha B (2015) Study on mechanical properties of natural-glass fibre reinforced polymer hybrid composites: a review. Mater Today Proc 2(4–5):2959–2967
34. Abd El-Baky MA, Alshorbagy AE, Alsaeedy AM, Megahed M (2020) Fabrication of cost effective fiber metal laminates based on jute and glass fabrics for enhanced mechanical properties. J Nat Fibers 1739594
35. Bichang'a DO, Aramide FO, Oladele IO, Alabi OO (2022) A review on the parameters affecting the mechanical, physical and thermal properties of natural/synthetic fiber hybrid reinforced polymer composites. Adv Mater Sci Eng 16:1–15
36. Thwe MM, Liao K (2003) Environmental effects on bamboo-glass/polyoropylene hybrid composites. J Mater Sci 38(2):363–376
37. Goud G, Rao RN (2011) Effect of fiber content and alkali treatment on mechanical properties of Roystonea regia-reinforced epoxy partially biodegradable composites. Bull Mater Sci 34(7):1575–1581
38. Thomas S, Pothan LA, Cherian BM (2009) Advances in natural fiber reinforced polymer composites: marco to nanoscales. Int J Mater Prod Technol 36(1–4):317–333
39. Cui Y, Tao J (2009) Fabrication and mechanical properties of glass fiber-reinforced wood plastic hybrid composites. J Appl Polym Sci 112(3):1250–1257
40. Khan RA, Sharmin N, Khan MA, Das AK, Dey K, Saha S, Islam T, Islam R, Nigar F, Sarker B, Debnath KK, Saha M (2010) Comparative studies of mechanical and interfacial properties between jute fiber/PVC and e-glass fiber/PVC composites. J Reinf Plast Compos 29(7):1078–1088
41. Prabhu L, Krishnaraj V, Sathish S, GokulKumar S, Karthi N (2022) Mechanical, chemical and sound absorption properties of glass-kenaf-waste tea leaf fiber-reinforced hybrid epoxy composites. J Ind Text 51(10):1674–1700
42. Ramesh M, Palanikumar K, Reddy KH (2013) Reinforced, mechanical property evaluation of sisal-jute-glass fiber. Compos B 48(2013):1–9

43. Ramesh M, Palanikumar K, Reddy KH (2016) Evaluation of mechanical and interfacial properties of sisal-jute-glass fiber reinforced polymer composites. Trans Indian Inst Metals 69:1851–1859
44. Kumar MG, Vasanthanathan A (2021) Investigation of mechanical and interfacial characterization of hybrid sisal/glass fiber reinforced composites. Mater Today Proc 47:7041–7044
45. Yang Z, Peng H, Wang W, Liu T (2010) Crystallization behavior of poly (ecaprolactone)/layered double hydroxide nanocomposites. J Appl Polym Sci 116(5):2658–2667
46. Nachippan NM, Alphonse M, Raja VKB, Shasidhar S, Teja GV, Reddy RH (2021) Experimental investigation of hemp fiber hybrid composite material for automotive application. Mater Today Proc 44:3666–3672
47. Swolfs Y, Gorbatikh L, Verpoest I (2014) Fiber hybridization in polymer composites: a review. Compos Part A 67:181–200
48. Kumar M, Uthranarayan C (2021) Scheming of mechanical properties of molybdic acid added sisal glass fiber reinforced epoxy composite. Mater Today Proc 47:7054–7059
49. Mishra S, Mohanty AK, Drzal LT, Misra M, Parija S, Nayak SK, Tripathy SS (2003) Studies on mechanical performance of biofiber/hybrid reinforced polyester hybrid composites. Compos Sci Technol 63:1377–1385
50. Harish S, Peter MD, Bensely A, Mohan LD, Rajadurai A (2009) Mechanical property evaluation of natural fiber palm composite. Mater Charact 60:44–49
51. Zhang X, Yang H, Lin Z, Tan S (2013) Polypropylene hybrid composites filled by wood flour and short glass fiber: effect of compatibilizer on structure and properties. J Thermoplast Compos Mater 26(1):16–29
52. Panthapulakkal S, Sain M (2007) Injection-molded short hemp fiber/glass fiber-reinforced polypropylene hybrid composites-mechanical, water absorption and thermal properties. J Appl Polym Sci 103:2432–2441
53. Parikh DV, Chen Y, Sun L (2006) Reducing automotive interior noise with natural fiber nonwoven floor covering systems. Text Res J 76(11):813
54. Lim ZY, Putra A, Nor MJM, Yaakob MY (2018) Sound absorption performance of natural kenaf fibres. Appl Acoust 130:107–114
55. Nandanwar A, Kiran MC, Varadarajulu KC (2017) Influence of density on sound absorption coefficient of fibre board. Open J Acoust 7(01):1
56. Bisagni C (2009) Experimental investigation of the collapse modes and energy absorption characteristics of composite tubes. Int J Crashworthiness 14:365–378
57. Alkbir MFM, Sapuan SM, Nuraini AA, Ishak MR (2016) Fiber priorities and crashworthiness parameters of natural fiber-reinforced composite structure: a literature review. Compos Struct 148:59–73
58. Mahdi E, Hamouda A, Sahari B, Khalid Y (2002) Effect of material and geometry on crushing behavior of laminated conical composite shells. Appl Compos Mater 9:265–290
59. Mahdi E, Hamouda A, Sen A (2004) Quasi-static crushing behavior of hybrid and non-hybrid natural fiber composite solid cones. Compos Struct 66(1–4):647–663
60. Mohamed AA, Abd El-Baky MA, Hassan MA, Sebaey TA, Mahdi E (2018) Crashworthiness characteristics of carbon–jute–glass reinforced epoxy composite circular tubes. Polym Compos 39(S4):E2245–E2261
61. Maslinda AB, Abdul Majid MS, Ridzuan MJM, Afendi M, Gibson AG (2017) Effects of water absorption on the mechanical properties of hybrid interwoven cellulosic-cellulosic fiber reinforced epoxy composites. J Compos Struct Compos Struct 167:227–237
62. Alamri H, Low IM (2012) Mechanical properties and water absorption behavior of recycled cellulose fiber reinforced epoxy composites. Polym Test 31(5):620–628
63. Thwe MM, Liao K (2002) Effects of environmental aging on the mechanical properties of bamboo-glass fiber reinforced polymer matrix hybrid composites. Compos Part A 33(1):43–52
64. Tamrakar S, Kiziltas A, Mielewski D, Zander R (2021) Characterization of kenaf and glass fiber reinforced hybrid composites for underbody shield applications. Compos B Eng 216:108805
65. Ghani MAA, Salleh Z, Hyie KM, Berhan MN, Taib YMD, Bakri MAI (2012) Mechanical properties of kenaf/fiberglass polyester hybrid composite. Procedia Eng 41:1654–1659

66. Mazuki AAM, Akil HM, Safiee S, Ishak ZAM, Bakar AA (2011) Degradation of dynamic mechanical properties of pultruded kenaf fiber reinforced composites after immersion in various solutions. Compos B Eng 42(1):71–76
67. Salleh Z, Hyie KM, Berhan MN, Taib YM, Nik Roseley NR, Razib ARM (2014) Effect of water absorption on mechanical properties of kenaf and kenaf hybrid polyester. Appl Mech Mater 465:967–972
68. Pai LR, Bangarappa KS, Shrinivasa LD, Mayya CR, ThomasPinto N (2018) Hybridization effect on water absorption and flexural properties of e-glass/banana fibre/epoxy composites. Polym Compos 43(2):198–209
69. Shah N, Fehrenbach J, Ulven CAA (2019) Hybridization of hemp fiber and recycled-carbon fiber in polypropylene composites. Sustainability 11:3163–3178
70. Raju K, Balakrishnan M (2020) Evaluation of mechanical properties of palm fiber/glass fiber and epoxy combined hybrid composite laminates. Mater Today Proc 21:52–55
71. Nurazzi NM, Khalina A, Sapuan SM, Ilyas RA (2019) Mechanical properties of sugar palm yarn/woven glass fiber reinforced unsaturated polyester composites: effect of fiber loadings and alkaline treatment. Polimery 64(10):12–22
72. Mohammed L, Ansari MNM, Pua G, Jawaid M, Islam MS (2015) A review on natural fiber reinforced polymer composite and its applications. Int J Polym Sci 2015

A Review on Composite Aerostructure Development for UAV Application

Shahrul Malek Faizsal Bin Shahrul Hairi, Siti Juita Mastura Binti Mohd Saleh, Ahmad Hamdan Ariffin, and Zamri Bin Omar

Abstract Composite material is employed in various aerospace, automotive, and civil structure applications. Composite material offers excellent properties such as lightweight, high strength to weight ratios, excellent corrosion resistance and many more, as proved by much research. Since composite is a reliable new material in the aviation and aerospace industry, the unmanned aerial vehicle (UAV) industry is also looking forward to using this material. It is capable of enhancing the performance of the UAV by reducing the weight, thus increasing the flight time or payload. The extended usage of UAVs in the world nowadays increases the demand for UAV production in many fields such as military, transportation, and logistics, leading to the production of C-Drone in Universiti Tun Hussein Onn Malaysia (UTHM), the first giant drone cargo in Malaysia. C-Drone is designed to weigh 400 kg with a payload of 180 kg. The structure of the C-Drone is made of aluminium alloy 6061 T-6. An extensive literature review was conducted to compare and replace the existing material with a lighter material. Therefore, the discussion and material comparison will be reviewed in this manuscript to get a clear direction for the new material of the existing C-Drone. Researchers prove that composite material such as carbon fibre reinforced plastic (CFRP) can substitute current material, aluminium alloy, due to its extensive strength to weight ratio.

Keywords Composite material · Unmanned aerial vehicle · Lightweight performances

---

S. M. F. B. S. Hairi · S. J. M. B. M. Saleh · A. H. Ariffin (✉) · Z. B. Omar
Faculty of Mechanical and Manufacturing Engineering, Universiti Tun Hussein Onn Malaysia (UTHM), 86400 Parit Raja, Batu Pahat, Johor Darul Takzim, Malaysia
e-mail: ahmadh@uthm.edu.my

S. M. F. B. S. Hairi
e-mail: cd180140@siswa.uthm.edu.my; malekfaizsal@gmail.com

S. J. M. B. M. Saleh · A. H. Ariffin
Research Centre for Unmanned Vehicle (ReCUV), Faculty of Mechanical and Manufacturing Engineering, Universiti Tun Hussein Onn Malaysia (UTHM), 86400 Parit Raja, Batu Pahat, Johor Darul Takzim, Malaysia

T. Khan and M. Jawaid (eds.), *Green Hybrid Composite in Engineering and Non-Engineering Applications*, Composites Science and Technology,
https://doi.org/10.1007/978-981-99-1583-5_9

1 Introduction

Technological advancement nowadays in fulfilling the demand of fabrication has forced researchers to look into the innovation of material. It is crucial to make sure the innovative materials can be applied in various practical applications, leading to economical and environmentally friendly.

Innovation of materials could be varying according to the solid materials available in primary classifications such as metals, ceramics, and polymers [1]. By having the variety or mixing the chemical and the atomic structure, this process could produce new types of material such as composites, semiconductors and biomaterials, as shown in Fig. 1.

Today, drone technology has been a game-changer and break various barriers in many fields such as military and commercial industries and even can be an exciting hobby. For example, prime companies such as Amazon and Google are now attempting to deliver packages with drones, and Facebook is using drones to provide Internet connections in remote areas, which keep the number of drone production increase at a 66.8% compound annual growth rate (CAGR) as the article from Insider stated [2, 3]. Malaysia is also looking for this opportunity as

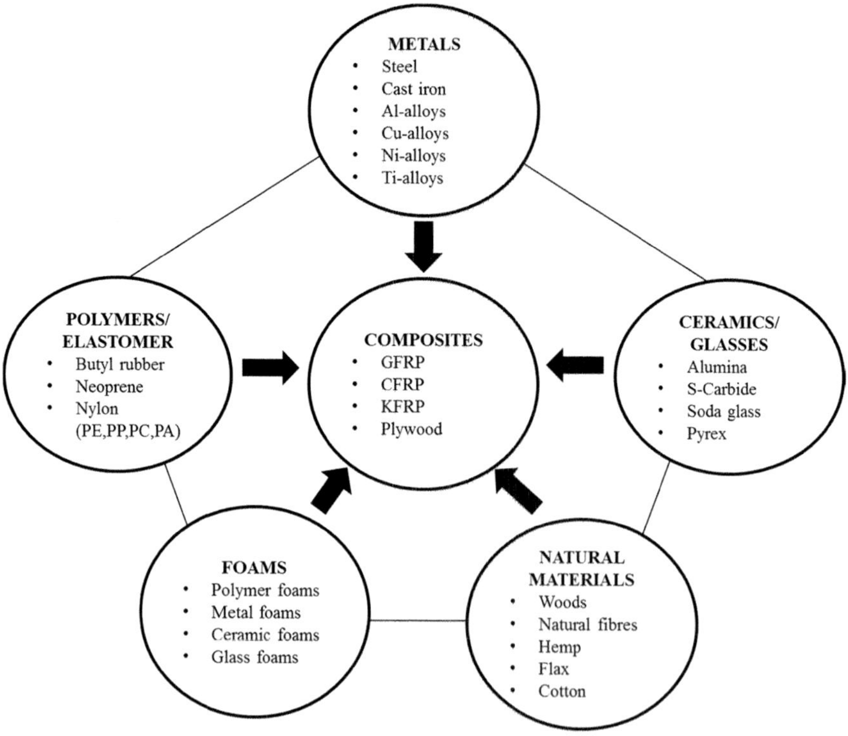

Fig. 1 Classification of materials [1]

Fig. 2 C-Drone, UTHM

the game-changer in various sectors such as military, transportation and logistics, and other services by estimating the revenue of $127 billion in value by 2025 and supporting UAV development [4]. For drone or UAV, also being innovated, consisting of carbon fibre-reinforced composites (CRFCs), thermoplastics such as polyester, nylon, polystyrene, aluminium and many more [5].

C-Drone was built in Universiti Tun Hussein Onn Malaysia (UTHM), as in Fig. 2 is the first giant drone cargo aimed to carry a payload over 180 kg. Based on the problems faced by the group of UTHM, the weight of cargo drones made of aluminium alloy 6061 T-6 is too significant and could affect the flight time and reduce the expected payload of the C-Drone. In this research, the original design and strength of the material should remain, but the cargo drone's weight must be reduced. In the initial stage, the expected value and maximum take-off for the weight of C-Drone is 680 kg. With that, lighter material can be proposed to reduce the weight of the initial design of the C-Drone. Therefore, an extensive review is conducted in this chapter to compare the potential material in replacing the existing material employed in the C-Drone.

2 Composite Material

Composite materials have been applied in aviation and aerospace applications due to their enhanced strength and stiffness-to-density ratios with superior physical properties [6]. This is also supported because the designers keep developing a new way or method to improve the structural material and play a significant role in current and future aviation or aerospace components. For instance, non-critical structural material in an aeroplane such as panelling and aesthetic interiors are now made of

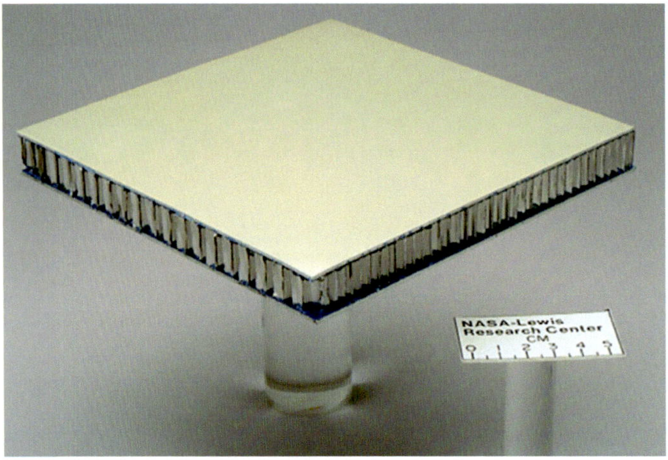

Fig. 3 Example of honeycomb composite sandwich structure use in aerospace or aviation industries [8]

CFRPs and honeycomb materials, as illustrated in Fig. 3, which offer lightweight combining with high strength and bending stiffness to the structure [7].

Composite materials can be defined as a material structure consisting of two macroscopically identifiable materials that work together to accomplish a superior result and consist of relatively strong, stiff fibres in a rigid resin matrix [9, 10]. The two materials can be classified into two categories which are continuous and discontinuous phases. The continuous matrix phase is the reinforcement phase, such as metallic, polymeric, or ceramic materials, while the discontinuous phase is much harder and more robust as particulate or fibre-reinforced composite. In other words, combining two or more materials with different properties will produce unique composite properties [8].

The composite material consists of two categories which are natural such as wood and bone, and synthetic, such as CFRP and glass fibre reinforced plastic (GFRP). For natural composite material, the wood consists of cellulose fibres in a lignin matrix, and bone consists of hydroxyapatite particles in a collagen matrix, while for the synthetic composite material, those materials are stiff and robust in terms of their density but brittle in a polymer matrix, which is tough but neither particularly stiff nor strong [11]. CFRP and GFRP are always being applied in aerospace and other industries by combining the materials with corresponding properties and producing the benefits such as high strength, stiffness, toughness and low density [11].

Besides, the composite material can also be divided into other classes: particulate composites and fibrous composites. CFRP and GFRP are the best examples of a fibrous composite material consisting of lightweight, high modulus fibres embedded in a surrounding material called the matrix and the properties that vary with the direction of interest [12]. For particulate composites, metal matrix composites (MMC) combine non-metallic particles in a metallic matrix such as silicon carbide combined in a metallic matrix aluminium alloy. Composites are good in corrosion resistance compared to metal alloys and show the excellent result in fatigue tests when it meets the tougher resin matrix [7, 8, 11, 13]. Figures 4 and 5 show the properties comparison of several materials.

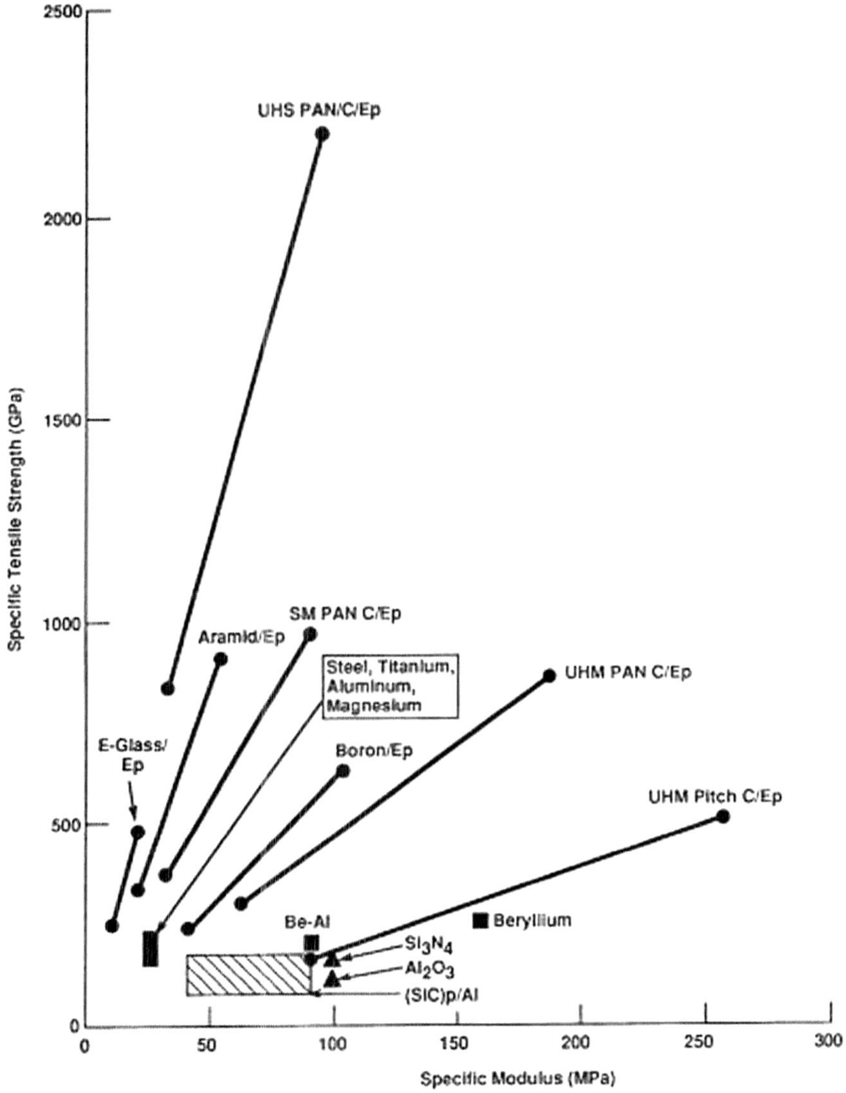

Fig. 4 Specific tensile strength as a function of the specific modulus of composite materials and other materials [13]

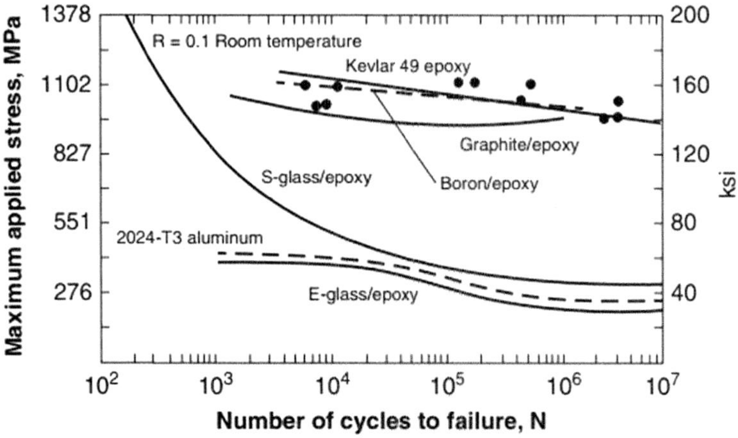

Fig. 5 Number of cycles to failure as a function of maximum stress for aluminium and composite material subjected to tension with a stress ratio R = 0.1 [13]

Figure 4 shows the composites offer significant order-of-magnitude improvements over metals in both specific strength and stiffness. For instance, the composite material, ultra-high-strength polyacrylonitrile (UHS PAN), recorded the highest specific tensile strength, approximately between 1000 and 2000 GPa, compared to other materials such as metal and ceramic. Figure 5 illustrates the number of cycles to failure, N, varies with maximum stress, S, for aluminium and other types of material subjected to tension with maximum stress, R, is 0.1. The composites consist of epoxy matrices reinforced with key fibres such as aramid, boron, SM carbon, high-strength (HS) glass, and E-glass shows the maximum applied stress are higher with the more significant number of cycles to failure compared to other material [13].

Besides, composites material can also be shaped into many forms that produce flexibility in design and are easier than a metal alloy. It could lead to the development of one-piece designs and will continue to reduce the number of components in overall assemblies, reducing the number of parts making up the component, the number of joints and potential failure points at the joint.

In addition, using composite material will cause more components to be manufactured in getting near-net-shape, reducing the amount of machining and waste of material [7–11]. With that, the summary of the composite material compared to metal for UAV construction is as follows:

i. Low weight.
ii. Excellent corrosion resistance.
iii. High fatigue resistance.
iv. Reduced machining.
v. It can be fabricated according to tapered sections and intricate contoured parts.
vi. Reinforcement fibres can be oriented in the direction of maximum stiffness and strength.

vii. Reduced number of assemblies and fasteners in composite manufacturing processes.
viii. Low thermal expansion reduces operational problems in high altitude flights.

Aluminium is not feasible in design options for UAV construction due to its isotropic properties. The strength and stiffness of the material are the same when measured in any direction through the material, thus limiting the design optimisation to the structure [10]. Meanwhile, plenty of studies proves that CFRP could contribute to anisotropic properties, reducing material usage and offering exact strength or higher in different directions through the material following the intended design structure.

Even though composites material comes with many benefits, as discussed, the nature of the material will come with few disadvantages. For composite materials, the raw materials will be cost higher than most metal alloys due to the high cost of fabricating composite components, and their vulnerability to moisture may exist in certain cases [8].

2.1 Mechanical Properties of Composite

The critical parameter for the composite can be determined through their strength, stiffness, lower weight, high chemical inertness, high damping, and outstanding fatigue characteristics in the material applications [9].

The strength of a material can be determined by dividing its density by force per unit area at failure [14]. Material that exhibits strong and lightweight properties would be favourable for strength to weight ratio. Table 1 compares strength at various materials.

Carbon fibre shows promising tensile strength compared to other materials with a range of between 3500 and 5500 MPa. Since the aviation and aerospace industry mostly uses aluminium alloy, the data in Table 1 shows significant differences for the strength of both materials with aluminium alloy tensile strength only in between 248 and 330 MPa. The only competitors for carbon fibre are Kevlar and Spectra fibre which is among the new material developed from the further reinforcement.

Table 1 The strength comparison between different materials

Material	Tensile strength MPa	References
Spectra fibre	2600–3700	[14–16]
Kevlar	2500–3600	[6, 14–17]
Carbon fibre	3500–5500	[6, 14, 16, 18, 19]
Glass fibre	1500–4200	[14, 16, 18, 20]
Steel alloy	500–745	[21–23]
Aluminium alloy 6061-T6	248–330	[14, 24, 25]

Furthermore, carbon fibre's density is 59% lower than aluminium, making it lighter when applied.

The fatigue characteristic is essential in determining the nature of a structure in the cyclically loaded systems [26]. The fatigue crack grows on the intensity and cycle frequency of the applied load. Table 2 shows the fatigue behaviour of various composite. Carbon composite is the best compared to aramid and E-glass fibre. All the materials satisfy the requirement for tensile strength, while for compressive strength, aramid fibre reinforced polymers would fail due to their poor compressive strength. CFRP does satisfy the compressive requirement compared to E-glass. Besides, fibre composite made of carbon shows an excellent indicator for modulus of elasticity compared to other materials that can resist the material towards non-permanent or elastic deformation. The cost must be economically optimal in fabricating parts using specific materials. Table 2 also indicating the price of carbon fibre indicates adequate, which is acceptable with the properties served by the fabric. However, choosing carbon fibre to upgrade the structure will be more economical than using the standard steel which the amounts could be reduced up to 20% of the total cost of the construction [27].

Table 3 shows different reinforcement and matrix resin types in the composite category and their properties [28]. Carbon with polyether-ether-ketone (PEEK) thermoplastic polymer recorded the highest tensile strength with 419 ksi due to the heat that can be applied to the material for treatment. Carbon with epoxy polymer would also be recommended because it also recorded among the highest tensile strength, 400 ksi, and only a few lower than carbon with PEEK. Even though both carbons with different polymers have high strain value among other compositions, all the strain rates for carbon fibre can be considered low compared to other materials outside this category, such as metal.

Table 2 Qualitative comparison between carbon, aramid and E-glass in fibre composite [27]

Criterion	Carbon	Aramid	E-glass
Tensile strength	Very good	Very good	Very good
Compressive strength	Very good	Inadequate	Good
Modulus of elasticity	Very good	Good	Adequate
Long term behaviour	Very good	Good	Adequate
Fatigue behaviour	Excellent	Good	Adequate
Bulk density	Good	Excellent	Adequate
Alkaline resistance	Very good	Good	Inadequate
Price	Adequate	Adequate	Very good

Table 3 The comparison of different fibre material and their properties [28]

Material type	Nomenclature	Tensile strength (ksi)	Modulus (Msi)	Strain (%)
Carbon/epoxy	T300/934	245	20	1.0–1.2
	IM7/8551-7	400	24	1.62
	P75/934	135	44	0.2–0.5
	AS4/3501-6	100	10	1.0
	IM6/3501-6	330	23	1.5
Glass/epoxy	E-glass/934	150–170	6–8	2.75
Kevlar/epoxy	K-49/7934	80–85	4	1.85
Carbon/PEEK	IM7/APC-2	419	24	1.6
Carbon/phenolie	FM5055	15–20	2.6–2.8	1.0–1.2

2.2 *Thermal Properties of Composite*

For thermal characteristics of carbon fibres, ultrahigh-modulus fibres linear expansion coefficients for 30 million-psi modulus fibres is approximately -1.3×106 in./(in.-°F), which differ from almost all other materials. It increases the possibility of the material in structure design with zero or very low linear and planar thermal expansion, suitable for precision instruments. Other properties in thermal, which are transverse coefficients of expansion, are quite different for carbon fibre, typically 15 $\times$ 106 in./(in.-°F) [28].

Tables 4 and 5 show the comparison of the thermal properties of several composites [13]. Carbon fibre shows relatively low plane coefficient thermal expansion, thermal conductivity and through-thickness thermal conductivity. From Table 4, both standard modulus carbon PAN (SM) and intermediate modulus carbon PAN (IM) have a low coefficient of thermal expansion compared to aluminium 7075-T6. For instance, both carbon PAN recorded in-plane thermal conductivity between 2.8 and 6 W/m K compared to aluminium 7075-T6 with 130 W/m K.

With that, the specific thermal conductivity of the material is also being compared as in Table 5, showing both carbon PAN having lower specific thermal conductivity between 1.8 and 4.8 W/m K compared to aluminium 7075-T6 with 46 W/m K. Tables 6 and 7 show a comparison of heat expansion and heat conduction between carbon fibre and other materials as in [29].

Table 6 shows carbon fibre has heat expansion six times less than aluminium and more than three times less than steel. It shows that the carbon fibre is less likely to change shape when exposed to heat during operation than other materials, and it is also crucial to ensure the part can withstand high temperatures.

Table 7 shows carbon fibre contain heat conductivity of 40 times less than aluminium and ten times less than steel. It shows that carbon fibre is an excellent insulator and suitable to be applied as part of a UAV.

Table 4 The comparison of in-plane coefficient thermal expansion, thermal conductivity and through-thickness thermal conductivity [13]

Fibre/metal	Density $\left(\frac{g}{cm^3}\right)$	In-plane CTE 10^6/K	In-plane thermal conductivity (W/m K)	Through-thickness thermal conductivity (W/m K)
7075-T6	2.8	24	130	130
E-glass	2.1	10	0.9	0.6
Aramid	1.38	1.4	0.9	0.1
Boron	2.0	6.5	1.4	0.7
SM carbon (PAN)	1.58	3.1	2.8	0.5
IM carbon (PAN)	1.61	2.3	6	0.5
UHM carbon (PAN)	1.66	0.4	23	0.5
UHM carbon (pitch)	1.80	−0.4	195	10
UHK carbon (pitch)	1.80	−0.4	300	10

Table 5 The comparison of specific thermal conductivity of different material [13]

Fiber/metal	Specific thermal conductivity (W/m K)
7075-T6	46
E-glass	0.4
Aramid	0.7
Boron	0.7
SM carbon (PAN)	1.8
IM carbon (PAN)	4.8
UHM carbon (PAN)	14
UHM carbon (pitch)	110
UHK carbon (pitch)	170

Table 6 The comparison of heat expansion of different material in inch/degree Fahrenheit [29]

Material	Heat expansion (in/°F)
Aluminium	13
Steel	7
Glass fibre-epoxy composite	7–8
Kevlar/aramid-epoxy composite	3
Carbon fibre-epoxy composite	2

Table 7 The comparison of heat conduction of different material in (W/m) [29]

Material	Heat conduction (W/m)
Carbon fibre-epoxy composite	5–7
Steel	50
Aluminium	210

2.3 *Impact Behaviour*

Impact testing is a procedure to determine the impact resistance or toughness of materials which be obtained by calculating the amount of energy absorbed during fracture by the collision between the weight and specimen. The results will show the transfer of energy between them and could determine the fracture mechanics of the material. This test will also be able to obtain the toughness, impact strength, fracture resistance and impact resistance of the material [1, 2].

Three different types of materials, which are aluminium alloy Al 6061-T6 (SS), steel AISI 1045 and CFRP-Sandwich composite, were compared in terms of impact characteristics. Those materials are being tested with a certain amount of forces ranging between 800 and 2500 N through simulation in Solidwork and obtain the maximum displacement between the material [30]. Throughout the experiment, the maximum displacement of CFRP-Sandwich composite was recorded higher than steel and aluminium. For instance, the CFRP Sandwich composite recorded 1600 mm displacement compared to the other material, which recorded zero displacements when 2500 N force was applied [30]. Besides, the researchers also compare the forces and stress for those materials to determine the strength, fracture toughness, tensile strength and others from the graph. From the experiment, indicating the Steel AISI 1045 and CFRP-Sandwich composite recorded the highest value of maximum stress compared to Al 6061-T6 (SS) [30]. The reading of maximum stress between Steel AISI 1045 and CFRP-Sandwich composite is possibly the same when forced at 2000N and above is being applied to result in the stress of $1.7763 \times 10^7 \frac{N}{m^2}$ for both materials. Al 6061-T6 (SS) recorded the lowest reading of maximum stress [30]. With these results, CFRP-Sandwich composite recorded the highest value of maximum displacement compared to Steel AISI 1045 and Al 6061-T6 (SS), indicating the material has high flexibility in material properties during impact, which can prevent the material from deteriorating [30].

2.4 *Application of Composites Material in UAV Components*

Drones consist of many parts, and selection must be applied in selecting materials to fulfil specific tasks. For current cases, aluminium is among the most common metal used in UAVs, but it does not benefit weight reduction and leads manufacturers to look forward to applying new types of material.

One of the materials is the composites which believes can reduce overall UAV weight by 15–45% depending on the extent of composite use, which could be extensive in increasing the flight time or increasing the payload [31]. However, material density in UAV must be considered in reducing the weight and extending the performance, and the detail is tabulated in Table 8.

Composites from carbon fibre could be varied, such as long and continuous or short and fragmented, and the orientation could be directionally or randomly. For instance, short fibres will consume less cost, and this could optimise the fabrication process, but the properties would be a compromise compared to longer or continuous fibres. The critical factors that need to be determined from the carbon composites are their high strength-to-weight and stiffness-to-weight ratios.

The previous subtopic proved that carbon fibre could be a substitute to the current C-Drone structure, which is the aluminium alloy 6061-T6. Carbon fibre has proven to have excellent strength to weight ratio due to the density of carbon fibre being lighter than other materials, especially aluminium alloy but offering higher strength which could benefit the weight of the structure. Besides, carbon fibre is also being suggested due to it has better thermal properties such as lower coefficient thermal expansion and lower thermal conductivity compared to other composite materials. Since the drone is dealing with components that produce heat, such as high rpm motors, electronic board and vibration, choosing carbon fibre as the structure material could reduce the possibilities for the structure to fail or break due to exposure to excess heat. Last but not least, carbon fibre has also proven to be an excellent impact resistance with higher displacement recorded. Since the drone will undergo take-off and landing rapidly, having high flexibility in material properties during impact can prevent the material from deteriorating due to the material being tough.

3 Selection of Fibre Material and the Orientation

Fibre orientation is vital in dealing with composite material such as CFRP laminate which affects its outstanding strength and stiffness to weight ratio such as carbon [35]. Various ways or techniques were imposed to laminate design for composite panels that are angularly oriented to produce various structural properties [35, 36]. The properties could be classified as in Table 9.

Carbon fibres are innovative materials about 5–10 μm in diameter that contain almost 90% carbon and are widely used as reinforcements in composite materials such as CFRP, carbon composites, carbon fibre-reinforced cement and many more. Due to fibres being in micrometres in diameter, the atoms are bonded together in crystals that are aligned parallel to the long axis of the fibre or can be a narrow sheet of honeycomb crystals as illustrated in Fig. 7 [37, 38]. Several advantages that encourage the study of composite material, which is carbon fibre, are high stiffness, high tensile strength, low weight, high chemical resistance, high-temperature tolerance and low thermal expansion. This could contribute to higher strength in lengthwise orientation compared to across the fibre and advantages to the designers which they could specify

Table 8 Parts of drone using composite materials and its explanation

No.	Parts	Function	Explanation
1	Frame	It holds all the subsystems in place and gives a drone its shape	(a) The material commonly used is thermoplastics such as variants of nylon, polyester, and polystyrene due to being inexpensive and easy to make into a complex part (b) Thermoplastics offer good strength and low density, with several varieties producing tensile strengths over 100 Mpa and densities below 2 g/cm^3 (c) The top choice for high-performance drone frames are carbon fibre-reinforced composites due to their lowest-density and highest-strength materials properties, which offer high strength, low density, and high stiffness to make light and rigid drone frames [5] (d) Fibres can be classified into two based on the carbon composition, which is are carbon (93–95% carbon) and graphite (>95% carbon) (e) Using carbon fibre for the fuselage or frame of UAV causes the structure to be 31.5% lighter than the metal structure, the parts can be reduced by 61.5%, the fasteners reduced up to 61.3%, and the fatigue and corrosion resistance is greatly improved [32] (f) Glass fibre is also commonly used for its low cost and is more common in civilians due to the former's less severe operating conditions

(continued)

Table 8 (continued)

No.	Parts	Function	Explanation
2	Landing gear	To manoeuvre during ground operations such as take-off and landing	(a) The landing gear will be facing dynamic loads and shock-induced vibration, which could lead to fatigue damage and dynamic stressing on the UAV structure[33] (b) Suitable landing gears could ensure the UAV can operate safely, especially during take-off and landing and consider the integrity of the structure (c) Lightweight and high-strength materials are needed to prevent accidents during the operation [34]
3	Motors	To spin the propellers of multirotor drones to enable them to fly	(a) Motors can generate massive heat, which requires materials with high thermal conductivity, such as aluminium (b) Thermoplastics motor housing is being proposed, which serve top strength-to-weight ratios for weight reduction
4	Propellers	Provide lift for the aircraft by spinning and creating an airflow	(a) High speeds rotors tend to absorb the most wear-and-tear when a drone flies (b) Most rotor blades can be made of carbon fibre-reinforced composites to maximise energy and minimise weight (c) Thermoplastic blades are common material due to blades being regularly broken and replaced, reducing the cost of changing them (d) Kevlar/epoxy composites are also being applied in propeller production, as it's far lighter than CFRP, which could reduce the inertia of the propeller and reduce the vibration creating UAV more solid during flight [5, 10]

Table 9 The properties of carbon fibre [35, 36]

Num	Properties	Details
1	Isotropic	(a) The strength and stiffness of the material are equal in any direction if the load is placed along its 0°, 45°, and 90° within the material (b) The properties are independent of direction within the material which the modulus of elasticity *I* will be the same in each direction (E0° = E45° = E90°)
2	Quasi-isotropic	(a) The strength and stiffness are equal in all directions but only within the plane, as in Fig. 6 (b) The laminate results-oriented produce an isotropic [A] matrix $A = \begin{bmatrix} A_{11} & AA_{12} & 0 \\ A_{21} & A_{22} & 0 \\ 0 & 0 & A_{66} \end{bmatrix}$ (c) Several rules apply for a quasi-isotropic laminate: • The number of layers must be $n > 3$ • All layers must have identical orthotropic elastic and identical thickness • The orientation of the k th layer of an n-layer laminate is $\theta_k = \frac{\pi(k-1)}{n}$
3	Anisotropic	(a) The strength and stiffness of the material are different in different directions through the material (b) For carbon fibre, anisotropic can be achieved if the fibres are all oriented in one direction but differences in each direction cause the modulus of elasticity to become (E0° ≠ E45° ≠ E90°) (c) This will affect other properties such as ultimate strength, Poisson's ratio, and thermal expansion coefficient

the direction of the fibre to maximise strength and rigidity. With that, carbon fibre-based from PAN has higher strength and also can be added by aligning the direction towards the greatest stress in a specific orientation [14].

Common properties of PAN carbon fibres are influenced by the type of modulus applied to the same density of PAN [9]. In a nutshell, the high modulus (HM) recorded the highest Young's modulus at 450 Gpa and above with strain to failure rate between 0.4 and 0.7%. For intermediate modulus (IM), Young's modulus recorded could be between 200 and 350 Gpa, while for additional information, low modulus high-tensile (HT) recorded modulus at lower than 100 Gpa with tensile strength higher than 3.0 Gpa [9, 19].

3.1 Type of Fibre Material and Selection of Resin Matrix

The landing gear is commonly made of a composite material consisting of glass fibre, carbon fibre and epoxy resin [39]. The quantity of glass or carbon fibre will be determined according to our requirement where the glass fibre assures flexibility, and

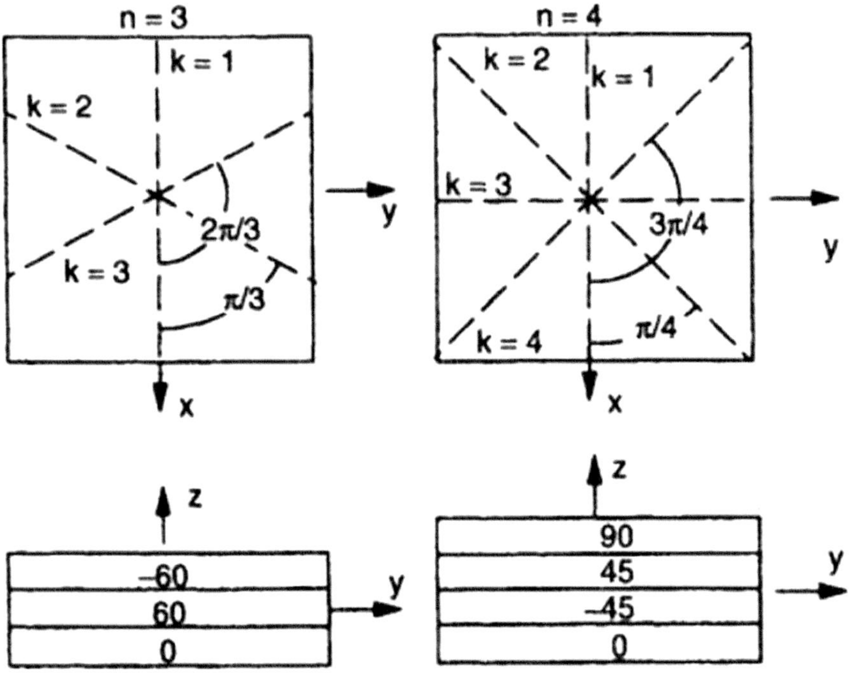

Fig. 6 The example of ply orientation for quasi-isotropic laminates [35]

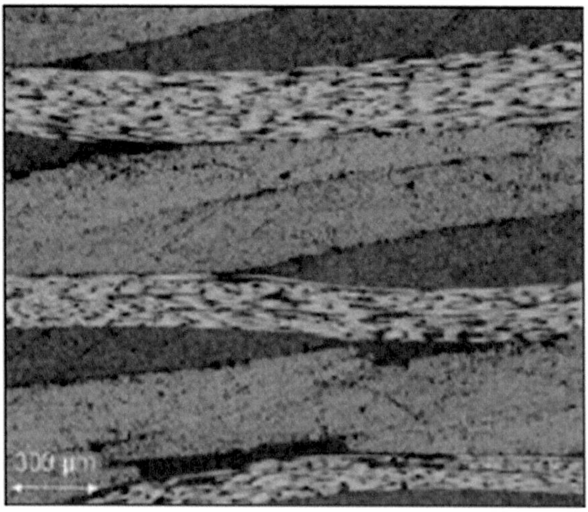

Fig. 7 The microstructure of the carbon fibre epoxy composite [37]

the carbon fibre assures rigidity. During manufacturing processes, epoxy resin will be applied to the landing gear product to ensure the layer does not detach in time. The suggested proportion of the fibre/resin must be approximately 55/45 at 100% [39].

Other researches also suggest various reinforcements such as carbon fibre with different modulus types, glass fibre with different classes and also aramid that can fit with the requirement of the UAV. Based on Table 10, the best material that can be applied for the landing gear are the high modulus aramid from the aramid fibre category with modulus between 160 and 170 Gpa and strength between 2.3 and 2.4 Gpa suitable for the high loaded part application. For the carbon fibre category, high modulus or ultra-high-strength category are suitable to be used for the UAV with modulus between 290 and 450 Gpa and strength between 4.0 and 7.5 Gpa, which is applied in the aerospace industry.

Other than that, the laminated structure of the composites and the fibre-matrix interfaces would prevent delamination or debonding by imposing some practical structural features. For these types of fibre, the designer could apply discontinuous plies in creating different thicknesses and sharp bends according to the stiffening members. This feature is crucial concerning the damage of the landing gear structure due to impact during landing, which could be occurring inside the material and not visible at the structure [40].

The best thermoplastic polymer suggested by the writer for resin matrix systems is PEEK [40]. Resin matrix is vital in dealing with impact damage tolerance and hygrothermal degradation. Besides, this polymer also will increase the failure strain of matrix resin, which help in translating the higher performance of the improved fibre to the composite by achieving better transfer of load from fibre to resin compression strength up to 630 Mpa, tensile modulus up to 50 Mpa and tensile strength up to 750 Mpa as illustrated in Table 11.

3.2 *Fibre Orientations*

According to the article from the carbon fibre and composite manufacturer, which is the Rock West Composite website, fibres can be oriented in any direction between 0° and 180°. For fibre orientation, above 90° will be assigned as a negative angle value. For instance, the fibre angle being oriented at 135° will be assigned as a −45° angle. Carbon fibre for the application of products nowadays can be a combination of two or more of these orientations, which could create different effects of fibre orientation as shown in Table 12:

From all the fibre orientation and their properties explained, a few suggestions have been proposed to obtain the best carbon fibre material according to the requirement of the UAV as follows:

(a) If the parts need to perform in a wide variety of conditions, the bidirectional layup is ideal.

Table 10 The fibres commonly used for aerospace and aeronautical industry [40]

Fibre	Density (g/cc)	Modulus (GPa)	Strength (GPa)	Application area
Glass				
E-glass	2.55	65–75	2.2–2.6	Small passenger aircraft parts, aircraft interiors, secondary parts, radomes, rocket; motor casings
S-glass	2.47	85–95	4.4–4.8	Highly loaded parts in small passenger aircraft
Aramid				
Low modulus	1.44	80–85	2.7–2.8	Fairing; non-load bearing parts
Intermediate modulus	1.44	120–128	2.7–2.8	Radomes, some structural parts; rocket motor casings
High modulus	1.48	160–170	2.3–2.4	Highly loaded parts
Carbon				
Standard modulus (high strength)	1.77–1.80	220–240	3.0–3.5	Widely used for almost all types of parts in aircraft, satellites, antenna dishes, missiles, etc.
Intermediate modulus	1.77–1.81	270–300	5.4–5.7	Primary structural parts in high-performance fighters
High modulus	1.77–1.80	390–450	2.8–3.0 4.0–4.5	Spaces structures, control surface in aircraft
Ultra-high strength	1.80–1.82	290–310	7.0–7.5	Primary structural parts in high performance fighters, spacecraft

Table 11 The properties of PEEK thermoplastic polymer [41]

Property	Value
Compressive strength-longitudinal (MPa)	630
Density $\left(\frac{g}{cm^3}\right)$	1.6
Tensile modulus-parallel to plane (GPa)	50
Tensile strength-Longitudinal (MPa)	750
Volume fraction of fibres (%)	62

Table 12 The fibre orientations and their effects [42]

Fibre orientation	Orientation explanation	Effect of fibre orientation
0° fibre angle	(a) Strongest and stiffest property when fibres are oriented in the direction of the load (b) Bending stiffness strength will be at 0° direction along the length for tubing	(a) Suitable for parts being loaded in one direction (b) Pultruded rod and tubing are parts that only contains 0° fibres but it can be maximised if bending with no twisting will be taken into account
90° fibre angle	(a) Can be applied when bending in both directions is required (b) 90° fibre will be oriented to the circumference for tube which keeps from crushing or buckling	(a) More resistant to buckling and crushing (b) Can be found in tubes or vessels inside the pressure vessel trying to enlarge the tube but being resisted by the 90° orientation layers
±45° fibre angle	(a) This angle will create a quasi-isotropic layup where a positive 45° layer is often paired adjacently to a negative 45° layer (b) 45° layers will provide twisting stiffness and strength for a tube	(a) Laminate can be balanced by pairing +45° with a −45° layer in preventing twisting when loaded (b) When the 45° layers are used together equally with 0° and 90° layers, the plate will become quasi-isotropic (c) For tube, 45° layers will add torsional strength and stiffness

(b) For a tube that needs to perform well in twisting, the layers should be oriented 45° or more.
(c) Woven material is the best choice to increase the thickness quickly.

4 Conclusion

Current researchers are looking into new materials such as carbon fibre or carbon-reinforced plastics, which are also among the material that is widely applied in aerospace industries due to their properties such as lightweight, high strength to weight ratios, excellent corrosion resistance and many more that could be beneficial in structure and weight optimisation.

This analysis presents that CFRP contributes better properties leading to weight and structure optimisation by various methods or techniques imposed in laminate design for composite panels. Understanding the properties of layering such as isotropic, quasi-isotropic, and anisotropic could lead to better angularly oriented in fulfilling the strength requirement. With the current problem faced by C-Drone UTHM, the properties of the CFRP that offer lightweight and better strength through a lot of studies could benefit in the weight reduction in achieving the expected weight of 680 kg value. It could enhance flight time or the cargo drone's payload. This research strengthens the fact that using CFRP is the best material to substitute with the current C-Drone material, aluminium alloy 6061 T-6.

References

1. CES information guide—materials science engineering [Online]. https://depts.washington.edu/matseed/ces_guide/materialclasses.htm. Accessed 05 Apr 2021
2. Future drone technology I future of drones I aircrafts. California University of Pennsylvania [Online]. https://www.calu.edu/academics/undergraduate/associate/drone-technology/future-drone-technology-and-uses.aspx. [Accessed 05 Apr 2021]
3. Insider Intelligence, Drone industry analysis 2021: market trends & growth forecasts. Business Insider, 2021 [Online]. https://www.businessinsider.com/drone-industry-analysis-market-trends-growth-forecasts?r=AU&IR=T. Accessed 05 Apr 2021
4. Kumar PP, Malaysia moves to become a drone hub for Southeast Asia—Nikkei Asia. Nikkei Asia, 25 Oct 2019 [Online]. https://asia.nikkei.com/Economy/Malaysia-moves-to-become-a-drone-hub-for-Southeast-Asia. Accessed 06 Apr 2021
5. Lenning W, What drones are made of?. MatMatch, 23 Oct 2019 [Online]. https://matmatch.com/blog/what-are-drones-made-of/. Accessed 05 Apr 2021
6. Basri EI et al (2019) Performance analysis of composite ply orientation in aeronautical application of unmanned aerial vehicle (UAV) NACA4415 wing. Integr Med Res 8(5):3822–3834. https://doi.org/10.1016/j.jmrt.2019.06.044
7. Standridge M, Aerospace materials—past, present, and future, Aerospace Manufacturing and Design, 13 Aug 2014 [Online]. https://www.aerospacemanufacturinganddesign.com/article/amd0814-materials-aerospace-manufacturing/. Accessed 16 Apr 2021
8. William J, The science and technology of composite materials. Australian Academy of Science, 18 Jun 2015 [Online]. https://www.science.org.au/curious/technology-future/composite-materials. Accessed 16 Apr 2021
9. Jawaid M, Shivanna P, Basavegowda Y (2018) Potential of natural/synthetic hybrid composites for aerospace applications. Elsevier Ltd
10. Gameros A, The use of composite materials in unmanned aerial vehicles (UAVs). AZO Materials, 06 Aug 2015 [Online]. https://www.azom.com/article.aspx?ArticleID=12234. Accessed 06 Apr 2021
11. Quilter A (2004) Composites in aerospace applications. Aviation Pros, 01 Oct 2004 [Online]. https://www.aviationpros.com/engines-components/aircraft-airframe-accessories/article/10386441/composites-in-aerospace-applications. Accessed 16 Apr 2021
12. Herakovich CT (2017) Fibrous composite materials. In: A concise introduction to elastic solids. Springer International Publishing, pp 101–119
13. Zweben C (2015) Composite materials. In: Mechanical engineers' handbook. Wiley, Inc., Hoboken, NJ, USA, pp 1–37
14. Christine, Carbon fiber properties [Online]. https://www.christinedemerchant.com/carboncharacteristics.html [Accessed 10 Apr 2021
15. Nussbaum J, Faderl N, Nuesing R, Boussu F (2012) Development of a new method to determine dynamic tensile strength, vol 01053, pp 1051–1054
16. Bagherpour S (2012) Polyester. In: Polyester, H. E.-D. M. Saleh, Ed. Intech Open
17. Ramadhan AA, Abu Talib AR, Mohd Rafie AS, Zahari R (2013) High velocity impact response of Kevlar-29/epoxy and 6061-T6 aluminum laminated panels. Mater Des 43(January):307–321. https://doi.org/10.1016/j.matdes.2012.06.034
18. Limited WP (2012) Fibre–polymer composites for aerospace structures and engines. Introd Aerosp Mater, no chapter 20, pp 338–393. https://doi.org/10.1533/9780857095152.338
19. Bhatt P, Goe A (2017) Carbon fibres: production, properties and potential use. Mater Sci Res India 14(1):52–57. https://doi.org/10.13005/msri/140109
20. Sathishkumar TP, Satheeshkumar S, Naveen J (2014) Glass fiber-reinforced polymer composites—a review. J Reinf Plast Compos 33(13):1258–1275. https://doi.org/10.1177/0731684414530790
21. Obinna N, Ugochukwu O, Christian O, Imhade O (2017) Evaluation of chatter vibration frequency in CNC turning of 4340 alloy steel material, no February

22. Sherkatghanad E, Nielsen KB (2018) Design of an innovative multi-stage forming process for a complex aeronautical thin-walled part with very small radii Design of an innovative multi-stage forming process for complex aeronautical thin-walled part with very small radii, no January. https://doi.org/10.1016/j.cja.2018.01.019
23. Boztepe MH, Bayramoglu M, Geren N (2016) Adhesion surface bonding capability of different metal sheets for fiber metal laminate (FML) applications adhesion surface bonding capability of different metal sheets for fiber metal laminate (FML) applications, no October
24. Cho YH, Kim TS (2016) Estimation of ultimate strength in single shear bolted connections with aluminum alloys (6061–T6). Thin-Walled Struct 101(April):43–57. https://doi.org/10.1016/j.tws.2015.11.017
25. Hellier AK, Chaphalkar PP, Prusty BG (2017) Fracture toughness measurement for aluminium 6061-T6 using notched round bars. In: 9th Australasian congress on applied mechanics. ACAM 2017, vol 2017-Novem, no November
26. The value of fatigue testing. Test Device by Schenck, 28 Sep 2018 [Online]. https://www.testdevices.com/blog/the-value-of-fatigue-testing. Accessed 10 Apr 2021
27. Meier U (1995) Strengthening of structures using carbon fibre/epoxy composites. Constr Build Mater 9(6):341–351. https://doi.org/10.1016/0950-0618(95)00071-2
28. Mraz S, Basics of aerospace materials: aluminum and composites. Machine Design, 19 Jun 2014 [Online]. https://www.machinedesign.com/materials/article/21831769/basics-of-aerospace-materials-aluminum-and-composites. Accessed 10 Apr 2021
29. Aluminium vs carbon fiber—comparison of materials. DexCraft, 07 Oct 2015. [Online]. http://www.dexcraft.com/articles/carbon-fiber-composites/aluminium-vs-carbon-fiber-comparison-of-materials/. Accessed 18 Apr 2021
30. Mohamed M et al (2018) Comparison between carbon fibre reinforced plastic, aluminium alloy and alloy steel for car hood application for impact test by using finite element analysis. J Fundam Appl Sci 10(2S):464
31. Borchardt JK, Unmanned aerial vehicles spur composites use—Materials Today. Materials Today, 05 Apr 2004 [Online]. https://www.materialstoday.com/composite-applications/features/unmanned-aerial-vehicles-spur-composites-use/. Accessed 06 Apr 2021
32. Application of carbon fiber composite materials for drones. Fly Dragon Drone Tech, 2013 [Online]. http://www.dronefromchina.com/new/carbon-fiber-composite-materials-for-drones.html. Accessed 29 May 2021
33. Son L, Bur M, Rusli M, A new concept for UAV landing gear shock vibration control using pre-straining spring momentum exchange impact damper. https://doi.org/10.1177/1077546316661470
34. Liang YC, Chin PC, Sun YP, Wang MR (2021) Design and manufacture of composite landing gear for a light unmanned aerial vehicle. Appl Sci 11(2):1–12. https://doi.org/10.3390/app11020509
35. Staab GH (1999) Laminate analysis. Sci Direct
36. Carbon fiber 101: what do isotropic, quasi-isotropic, and anisotropic mean? Dragon Plate, 25-Nov-2019 [Online]. https://dragonplate.com/carbon-fiber-101-what-do-isotropic-quasi-isotropic-and-anisotropic-mean Accessed 11 Apr 2021
37. Lukyanov AA (2010) An equation of state of a carbon-fibre epoxy composite under shock loading. Eur Phys J B 74(1):35–45. https://doi.org/10.1140/epjb/e2010-00043-4
38. Varley D, Yousaf S, Youseffi M, Mozafari M, Khurshid Z, Sefat F (2019) Fiber-reinforced composites. In: Advanced dental biomaterials. Elsevier, pp 301–315
39. Ciobanu (2019) Manufacturing of a landing gear using composite materials for an aerial target. J Phys Conf Ser. https://doi.org/10.1088/1742-6596/1297/1/012038
40. Mangalgiri PD (1999) Composite materials for aerospace applications. Indian Acad Sci 22(3):657–664
41. Carbon Fibre/Polyetheretherketone (PEEK) Matrix Composite Sheet. AZO Material, 26 May 2003 [Online]. https://www.azom.com/article.aspx?ArticleID=2002. Accessed 01 Jun 2021
42. How fiber direction influences tube and plate properties. Rock West Composite, 13 Feb 2018 [Online]. https://www.rockwestcomposites.com/blog/how-fiber-direction-influences-tube-and-plate-properties/. Accessed 1 Jun 2021

Natural/Synthetic Polymer Hybrid Composites—Lightweight Materials for Automotive Applications

M. R. M. Asyraf, M. R. Ishak, M. Rafidah, R. A. Ilyas, N. M. Nurazzi, M. N. F. Norrrahim, Mochamad Asrofi, Tabrej Khan, and M. R. Razman

Abstract Hybrid natural/synthetic fibres reinforced composites have become prominent raw material to reduce overall weight of automobile. This happened due to its various in versatile properties such lightweight, low cost, ease in structural development, strength to weight ratio, and high mechanical properties. Due to this factors, many automotive carmakers have been implemented hybrid natural/synthetic

M. R. M. Asyraf (✉)
Engineering Design Research Group (EDRG), Faculty of Mechanical Engineering, Universiti Teknologi Malaysia, 81310 Johor Bahru, Johor, Malaysia
e-mail: muhammadasyraf.mr@utm.my

Centre of Advanced Composite Materials (CACM), Universiti Teknologi Malaysia, 81310 Johor Bahru, Johor, Malaysia

M. R. Ishak
Department of Aerospace Engineering, Faculty of Engineering, Universiti Putra Malaysia, 43400, UPM Serdang, Selangor, Malaysia

M. Rafidah
Department of Civil Engineering, Universiti Putra Malaysia, 43400, UPM Serdang, Selangor, Malaysia

R. A. Ilyas
Faculty of Chemical and Energy Engineering, Universiti Teknologi Malaysia, 81310 Johor Bahru, Johor, Malaysia

N. M. Nurazzi
Bioresource Technology Division, School of Industrial Technology, Universiti Sains Malaysia, 11800 Pulau Pinang, Malaysia

M. N. F. Norrrahim
Research Center for Chemical Defence, Universiti Pertahanan Nasional Malaysia (UPNM), Kem Perdana Sungai Besi, 57000 Kuala Lumpur, Malaysia

M. Asrofi
Department of Mechanical Engineering, University of Jember, Kampus Tegalboto, Jember 68121, East Java, Indonesia

T. Khan
Department of Engineering Management, College of Engineering, Prince Sultan University, Riyadh 11586, Saudi Arabia

T. Khan and M. Jawaid (eds.), *Green Hybrid Composite in Engineering and Non-Engineering Applications*, Composites Science and Technology,
https://doi.org/10.1007/978-981-99-1583-5_10

reinforced polymer composites in many applications including interior and exterior components. These hybrid composites are developed as it allows enhancement especially in functional requirement to replace the current conventional materials such as steel. Moreover, the hybrid composites fabricated from blending of natural and synthetic fibres in a polymer resins exhibit synergistic result on properties which cannot be achieved from normal composites. Thus, in this article, it will be explained on hybrid natural/synthetic polymer composites as lightweight materials for automotive applications. The development of a many type of hybrid composites in brake lever, antiroll bar, impact attenuator, and crush box are also reviewed.

Keywords Natural fibre · Synthetic fibre · Hybrid composites · Lightweight material · And automotive applications

1 Introduction

The emerging demand of natural fibres in composite industry because of low density, biodegradability, cost effectiveness, less toxicity, and abundant availability in nature [1]. Natural fibre composites (NFC) has been applied extensively for many parts of automotive, optical devices, biomaterial for medical sector, spacecraft, flight, tissue repair, cross arm structures and construction [2–8]. The natural fibres originates from various plant sources such as oil palm, sugar palm, kenaf, flax, water hyacinth, pineapple, sisal, hardwood and softwood [9–13]. In general, three major compositions that build natural fibres such as cellulose, hemicellulose and lignin [14]. They are considered composites of hollow cellulose fibrils collected by lignin as a binder in a hemicellulose matrix [15, 16]. The component allows the natural fibre to have good structural integrity is the crystalline cellulose [17, 18]. Currently, most of composite products are shifting to plant cellulosic fibre reinforced thermoplastic biocomposites due to its high specific strength with low density and high stiffness [19, 20]. Moreover, the toughness and strength of manufactured products and goods are improved than their ancestors [21–24].

The biocomposites plays significant roles as compared to other synthetic composites due to its mechanical performance as well as its green technology [25, 26]. Embedding polymeric resins establish the structural integrity for composite structure and allows the pass overs of the shear forces between the fibres which subsequently, protect the composites from extreme forces, temperature, radiation, pressure and humidity [27–31].

M. R. Razman
Research Centre for Sustainability Science and Governance (SGK), Institute for Environment and Development (LESTARI), Universiti Kebangsaan Malaysia, 43600, UKM Bangi, Selangor, Malaysia

2 Demand for Natural Fibre Composites Automotive Industry

NFCs have been one of promising candidates in replacing the conventional materials such as steels and synthetic composites in various automotive applications including antiroll bars, spoilers, lever brakes, organic light emitting diodes (OLED), automobile engine mounting rubber and side door impact beam [32–37]. Table 10.1 displays the recent development and advancement of composite automotive products.

Till this date, various automotive components are fabricated using polymer resin such as polyester and polypropylene-based composites and integrated with cotton, flax, kenaf, kenaf, sisal, jute, or hemp. Moreover, the global demands toward NFCs has been growing exponentially in automotive sector due to their attributes, such as economical cost and marketing, light weight, durability, high mechanical strength, and resistance to corrosion, rather than the technical demands [42]. These biocomposites in car components have been used in various automotive companies such as Proton, Perodua, Peugeot, Misubishi, Citroen, Rover, Lotus, Vauxhall, Toyota, General Motors, Volvo, and Ford. Concerned with being environmentally friendly and developing a sustainable technology, several automotive manufacturers employed polymeric composites for numerous components as shown in Table 10.2 [56].

3 Non-Wood Natural Fibre Segment to Dominate the Market

Natural fibres especially plant cellulosic fibres are categorized into wood and non-wood fibres. In accordance to Pecas et al. [47], non-wood cellulosic fibres are considered as the largest market in the global natural fiber reinforced composite market. These non-wood fibres are included those in straw, leaf, fruit, bast, seed, grass and reed fibres categories [48, 49]. For example, those jute, flax, kenaf, hemp and sisal reinforced polypropylene and polyester polymer composites have been discovered to be an significant material in the automotive sector in order to manufacturing covered components for instrument panels, covered inserts, carriers for covered door panels, seat back panels, door panels, carriers for hard and soft armrests, door bolsters, pillars, center consoles, headliners, side and back walls, rear deck trays, load floors and trunk trim. On top of that, the construction sector also use non-wood fibres in many products such as jute thermoplastic board. The jute thermoplastic board had been seen as potential candidate to produce chequered boards, doors, fence, corrugated sheets, furniture, window frames, boards or sheets, etc. in the construction industry. Bast fiber reinforced polymer composites are applied as the main product in the progress of advanced biocomposite materials since they offer a good prospect for new application area leading to future developments and the growing market share for non-wood fibers.

Table 10.1 Recent developments and advancements of polymeric composite for automotive products

Product	Design illustration	Material	Application	Refs.
Parking brake lever		Kenaf fiber polymer composite	Automotive	[33]
Automobile engine rubber composite	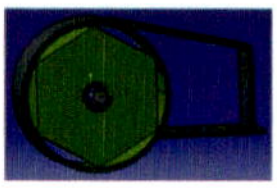	Kenaf fiber polymer composite	Automotive	[34]
Automotive bumper beam	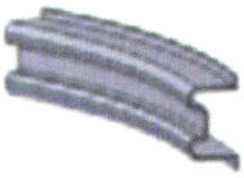	Hybrid natural fiber polymer composite	Automotive	[38]
Car spoiler	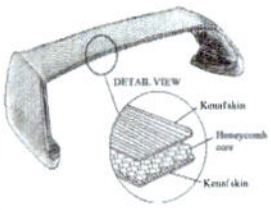	Kenaf fiber polymer composite	Automotive	[37]
Side door impact beam		Natural fiber composite	Auto-motive	[39]
Car front hood	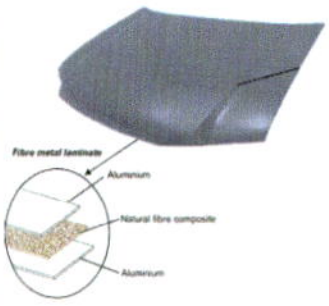	Natural fiber-aluminium laminate	Automotive	[40]
Anti-roll bar		Hybrid natural-carbon fibre reinforced composite	Automotive	[35]
Automotive pedals	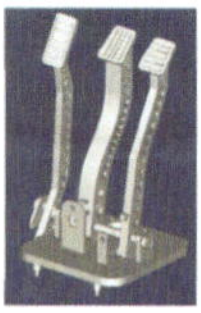	Conventional fibre reinforced with thermoset matrix	Automotive	[41]

Table 10.2 Applications of polymeric composite in automobile [43–46]

Automotive carmakers	Car parts implementing polymer composites as core material
BMW	Door panels, headliner panel, boot lining, seat backs, noise insulation panels molded foot, and well linings
Audi	Seat backs, side and back door panel, boot lining, hat rack, and spare tire lining
Ford	Door panels, B-pillar, and boot liner
Mercedes-Benz	Internal engine cover, engine insulation, sun visor, interior insulation, bumper, wheel box, and roof cover
Toyota	Door panels, seat backs, and spare tire cover
Volkswagen	Door panel, seat back, boot lid finish panel, and boot liner

Globally, all farmers from various countries cultivate and harvest millions of tons of natural fibres from a wide range of plant fibers every years. Somewhere another, various commodity trades between each countries in every continents to fulfill their supply and demand chains. For instance, European car makers mostly exploits hemp and flax while sisal is mainly imported from Brazil, the United States, and South Africa. Meanwhile, bananas are mainly imported from the Philippines; and kenaf and jute are mainly imported from India and Bangladesh. The application of various natural fibres for European carmaker industry can be seen in Fig. 10.1. It can be seen that wood fibre grasps the highest share which 38% of market and followed by cotton fiber, flax, kenaf, hemp, and others [47]. The total usage of natural fibres in automotive composites products in the European automotive industry in 2012 was around 80,000 tons. This can be seen that only 10–15% of the total European composites market which shows a significant moves toward greener technology.

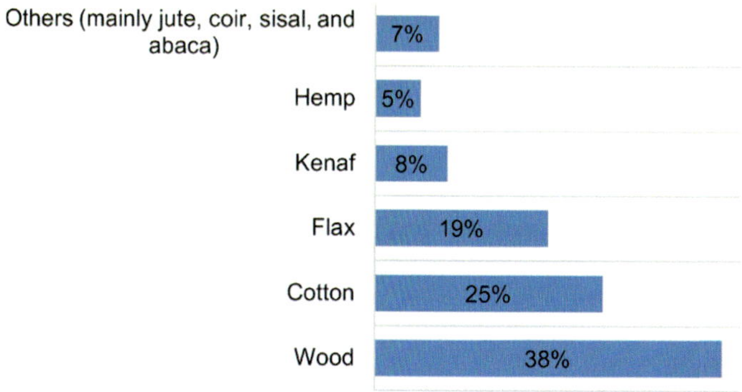

Fig. 10.1 Use of wood and natural fibers for composites in the European automotive industry in 2012. Reproduced from [47]

4 Potential Utilization of Natural/Synthetic Reinforced Polymer Composites as Lightweight Material in Automotive Components

Hybrid composites is referred of two or more different natural and synthetic fibre reinforced together in suitable matrices to form high performance composites which becoming a vital for economic and ecological compatibility view. This idea has stimulated mankind to applied hybrid composite materials in applications for daily life. The beneficial properties of these composites are including recyclability, cost-effectiveness and biodegradability. This could made these hybrid materials as a highly potential to replace the current lightweight synthetic materials which can be used in automotive sectors [50]. Table 10.3 shows the composite preparation of hybrid fibres from previous literatures.

The development of hybrid composites could establish a desirable mechanical, thermal and physical properties, which comparable to that of man-made materials. Moreover, the vital advantage of hybrid composites is that one fibre can complement with other fibre to attain a balance in cost and performance of this composite with appropriate material design considerations [49, 66, 67]. Optimum results can be achieved in hybrid composites by depending on various conditions such as structure of fiber, fiber content and its orientation, fiber arrangement, fibre bonding, individual fiber length, and finally the failure strain of each fiber in the composite [68–71]. Hybrid FRPCs are manufactured by various techniques such as the hand lay-up method, cold press method, hydraulic press, compression molding method, etc. [72–74] to produce various automotive components. Hybrid composites also find their applications in automobile and automotive sectors like in the production of car interior parts, doors, headliners, decking, dash boards, pallets, spare tyre covers, seat backs and spare-wheel panels.

5 Perspectives

The intention of hybridizing natural and synthetic fibres in thermoplastic biocomposites is to enhance the composite's properties. Generally, the failures commonly occurred due to low strength and load bearing properties. In order to improve the properties of composites, more than one fibre have to reinforce in matrix to allow good structural integrity on fibre reinforced polymer composite. The contribution of several fibres in the polymer matrix could eradicate any issues associated by its varying nature via giving a consistent performance. Moreover, it also helps in improving the lifetime of the material.

Table 10.3 Hybrid fiber-reinforced polymer composites preparation methods

Hybrid fibre	Polymer	Curing agent	Accelerator	Manufacturing methods	Refs.
Glass-glass	Epoxy	HY951 Hardener	–	Hand lay-up	[51]
Woven jute-glass	Polyester	–	–	Hand lay-up	[52]
Sisal-silk	Polyester	–	–	Hand lay-up	[53]
Banana-sisal	Polyester	–	–	Hand lay-up, compression moulding	[54]
Jute-glass	Polyester	–	–	Hand lay-up	[55]
Roselle-sisal	Polyester	–	–	Hand lay-up	[56]
Carbon-glass	Epoxy	HY225 Hardener	–	Hand lay-up	[57]
Kenaf-glass	Polyester	–	–	Hand lay-up, cold press	[58]
Pineapple-sisal-glass	Polyester	MEKP methyl ethyl ketone peroxide	Cobalt napthenate	Hydraulic press	[58]
Oil palm-jute	Epoxy	Hardener	–	Compression molding	[59]
Glass-palmyra	Polyester	–	–	Hydraulic compression molding process	[60]
Banana-kenaf	Polyester	–	–	Hydraulic compression molding process	[61]
Banana-sisal	Epoxy	–	–	Hydraulic compression molding process	[62]
Basalt-hemp	Polypropylene	–	–	Hot pressing	[63]
Flax, hemp, and jute	Polypropylene	–	–	Hydraulic press	[64]
Banana-glass	Polypropylene	–	–	Twin screw extrusion	[65]

6 Hybrid Composites in Various Parts and Models of Top Automobiles

NFCs are widely implemented in various models and parts of cars as displayed in Table 10.2. In the near future, hybrid composites will play a significant role for automobile products since it has eye-catching physical and mechanical properties and strength to weight ratio consideration. Vast common automobile parts would be replaced with hybrid composites to increase the structure effectively with better performance. Efficient weight reduction of the existing metal/steel components in an automotive parts without compromising its strength will ensurely decrease the fuel consumption which helps in pollution control. Thus, the production of hybrid composites with specific manufacturing and processing methods will be reaching new levels in the coming years.

7 Natural/Synthetic Hybrid Polymer Composites

Various behaviors obtained based on the significant of constituent material in a polymer hybrid composite is too many for detailing. Most previous literatures manage to the most usual type of hybrid composites such as carbon/Kevlar, carbon/ultrahigh modulus polyethylene (UHMPE), carbon/glass hybrid fibres reinforced polymer composites. Other hybrid fibres composites such as UHMPE/glass/bio fiber, aramid/UHMPE and carbon/nylon shows positive results of a certain property have been reported [75, 76]. The hybrid of fibres would improve the tensile properties as an enhancement of the first failure strain of the low elongation fibre component of hybrid composite.

8 Types of Hybrid Composites

There are several types of hybrid fibre composites as they are differentiate based on manufacturing methods as main criterion. In this case, the way constituent materials are arranged can be classified as follow [77]:

1. Interply hybrid two or more fibre layers by not interfering of hybrid fibres with same reinforcement.
2. Super hybrid composites which contain metal foils or metal composite plies stacked in a specified orientation.
3. Intraply hybrids, in which two (or more) fibers are mixed in the same layers.
4. Selective hybrids where placements in which reinforcements are placed wherever additional strength is required, over the base reinforcing laminate layer.
5. Intermingled hybrids where the constituent fibers are mixed randomly as possible so that no concentrations of either type are present in the material.

9 Properties of Hybrid Polymer Composites

Polymers are considered matrix material which found for preparing composites due to its versatility. The polymers usually divided into thermoset and thermoplastic polymers. In general, Table 10.4 displays the physical and mechanical properties of

Table 10.4 Properties of thermoset polymer matrix material for hybrid composite [77]

Polymers	Tensile strength (MPa)	Tensile modulus (GPa)	Flexural strength (MPa)	Flexural modulus (GPa)	Glass transition temperature (°C)	Specific gravity
Epoxies	55–130	2.7–4.1	110–150	3–4	170–300	1.2–1.3
Phenolic	50–60	4–7	80–135	2–4	175	1.2–1.3
Polyesters	34–105	2.1–3.5	70–110	2–4	130–160	1.1–1.4
Vinyl esters	73–81	3–3.5	130–140	3	–	1.1–1.3

polymer resins used in composites. Many researches established that the mechanical properties of natural/synthetic fibres composites is improved due to the addition of synthetic reinforcement. For instance, the addition of glass fibres in polymer matrix shows significant increase in mechanical strength and stiffness of natural fiber reinforced polymer composites. Moreover, those chemical resistance, flexural and tensile properties of hybrid sisal/carbon fibre reinforced polyester composites is significantly enhanced as compared to both single carbon and sisal composites. Table 10.5 displays the mechanical performance of fibrous material from synthetic and natural sources respectively.

Table 10.5 Mechanical properties of mineral and synthetic fibers [78]

Fibres	Part of plants	Diameter (μm)	Density (g/cm^3)	Tensile strength (MPa)	Tensile modulus (GPa)	Elongation (%)
Abaca	Leaf	10–30	1.5	430–813	31.1–33.6	2.9
Jute	Bast	25–250	1.3–1.49	393–800	13–26.5	1.16–1.5
Sisal	Leaf	50–200	1.34	610–710	9.4–22.2	2–3
Cotton	Seed	–	1.5–1.6	287–597	5.5–12.6	7.0–8.0
Kenaf	Stem	–	1.45	930	53	1.6
Soft Wood	Stem	–	1.5	600–1020	18–40	4.4
Ramie	Stem	20–80	1.5	400–938	61.4–128	3.6–3.8
Coir	Fruit	150–250	1.2	175	4–6	30
Flax	Stem	25	1.5	500–1500	27.6	2.7–3.2
Hemp	Stem	25–600	1.47	690	70	2.0–4.0
E-glass	–	10	2.54	3450	72.4	4.8
S-glass	–	10	2.49	4300	86.9	5.7
Kevlar	–	49	11.9	1.45	3620	131
Carbon (PAN)	–	10	2.00	4900	230	2

10 Hybrid Natural/Synthetic Composites for Automotive Applications

The natural fibres especially cellulosic fibres can be used together with synthetic fibres without disrupting its mechanical behavior. Hybrid fibres composites fabricated from thermoplastic resin illustrates better mechanical properties. This hybridization technique can develop the mechanical properties of single fiber-reinforced polymer composites and increase its versatility due to due to the inclusion of synthetic fiber having comparably high compatibility with the matrix. However, the implementation of natural fibre in polymeric composite due to low cost and density, recyclable, eco-friendlier material, and abundant in resources. Table 10.6 depicts mechanical properties of hybrid natural/synthetic fibres reinforced polymer biocomposites.

Table 10.6 Mechanical properties of hybrid natural/synthetic fibres composites [79]

Hybrid fibre	Polymer	Fibre content (%)	Fibre length (mm)	Treatment	Tensile		Flexural	
					Strength (MPa)	Modulus (GPa)	Strength (MPa)	Modulus (GPa)
Sisal-silk	UPE	–	10	NaOH treatment	20.9	–	50.5	–
Sisal-silk	UPE	–	10	–	16.6	–	33.5	–
PALF-glass	PE	25	–	–	72.0	–	101.3	–
Bamboo-glass	PP	30	–	–	17.5	3	34	3.4
Bamboo-glass	PP	30	–	Maleic anhydride polypropylene	18.5	3.3	42	4.5
Roystonea regia-glass	Epoxy	20	5–8	–	31.9	2.4	40.1	3.9
Palmyra-glass	Rooflite	32	40	–	26.1	1.4	26.6	1.6
Palmyra-glass	Rooflite	41	30	–	26.2	1.4	44.5	1.4
Coir-glass	Phenolic	–	20	–	22.3	3.8	53.4	4.8
Coir-glass	Phenolic	–	20	NaOH treatment	25.6	4.3	68.3	5.6
Sisal-glass	PP	30	–	–	29.6	2.3	66.7	4.0
Glass-sisal	PE	–	30	–	176.2	–	–	–
Sisal-glass	Epoxy	–	35	–	68.6	–	–	–
Sisal-glass	PE	30	–	5% NaOH solution	130.0	–	150.0	–

Notes UPE—Unsaturated polyester, PE—Polyester, PP—Polypropylene

11 Hybridization of Coconut Coir with Glass Fiber Composite

Karikalan et al. [80] studied the hybridizing of coconut coir fibre with glass fibre in polymer matrix composites using hand layup method. They conducted the study to evaluate the dissimilar mechanical properties under various conditions. In the study, they discovered that combination of coconut fibre with glass fibre exhibit compatibility in polymer matrix which bond firmly with one another in macroscopic analyses. The mechanical outcomes shows that tensile strength, hardness and impact strength significantly improved which indicate the hybrid composites form high specific strength. Thus, the hybrid coconut/glass fibre composites found useful for automotive components which results in high strength and low weight value.

12 Hybridization of Kenaf with Glass Fiber Reinforced Polyester Composite

Atiqah et al. [81] conducted a research on characterization of hybrid kenaf/glass fibres reinforced unsaturated polyester composites for automotive component usage. From the study, the flexural strength of kenaf/glass fibres reinforced polymer composites shows that a higher value of tensile strength, Young's modulus, impact and flexural properties at 15% of treated kenaf fibres loading. This could be happened due to 6% of NaOH solution treatment of kenaf fibre at 3 h durations allows enhanced the adhesion between the surface of fibre and the matrix. This established that the hybridization of natural and synthetic fibres in polymer matrix requires chemical treatment to enhance its mechanical performance. The mercerization treated kenaf with 15% volume fraction showed highest tensile modulus which enhanced the stiffness of hybrid composite. The kenaf fibre alone (30% volume fraction) or higher percentage (22.5% volume fraction) cannot withstand higher impact load leading to brittleness and less toughness in hybrid composite. Moreover, the addition of treated kenaf fibres is not only reduced the raw material costs of overall composites but the hybrid composites display they have lightweight properties suitable to replace current conventional materials in automotive components.

After several years past, Nazim et al. [82] have conducted the potential of implementation hybrid kenaf/glass fibres reinforced polyester composites for impact attenuator in automobile application. Figure 10.2 displays the schematic diagram of impact attenuator in front of car. The function of impact attenuator is to provide protection for engine from accident crash as it is fixed right behind the bumper component of the car or vehicle. Currently, the component is made up of steel, aluminum and metal alloys and expected that hybrid kenaf/glass reinforced polyester composites to replace them since these conventional material of steel, aluminum and metal alloys. At the end of

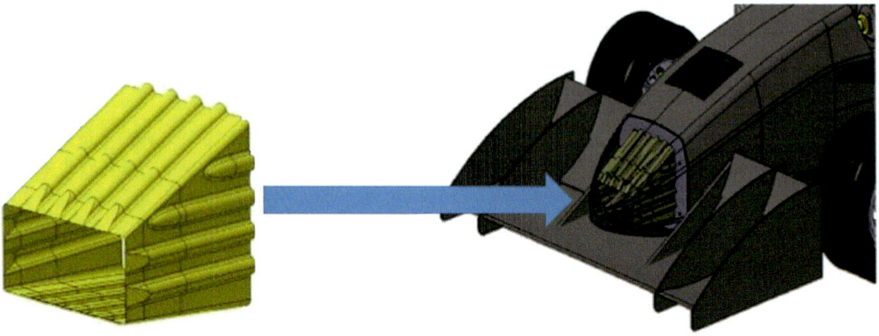

Fig. 10.2 CAD drawing of impact attenuator in front part of racing car. Adapted from Ref. [83]

the study, they explored that impact strength of the hybrid natural/synthetic composites with alkaline treatment was increased by 3.88% than untreated hybrid composites and 9.25% higher than untreated kenaf reinforced polyester composites. Hence, energy absorbing properties of the treated kenaf/glass hybrid composite direct the way to apply this material composite for the attenuator in the frontal assembly of the automotive cars.

13 Hybridization of Sugar Palm with Glass Fiber Reinforced Polyurethane Composite

Afzaluddin et al. [84] had explored the physico-mechanical performance of hybridized sugar palm and glass fibres reinforced polyurethane composites. Based on their results, the incorporation of 30% of treated sugar palm fibre along with 10% of glass fibre reinforced polyurethane exhibited increase in density with reduce water absorption and thickness swelling. On top of that, the tensile and impact performance of this hybrid composites as compared to glass fiber reinforced composites (0/40 SP/G) because of excellent hybrid performance of the two fibres. All in all, the addition glass fiber to sugar palm fiber composites can improve the physical and mechanical properties which suitable for automotive applications.

From this research, a subsequent study has been conducted by Yusof et al. [85] to introduce conceptual design for sugar palm polymer composite Automotive Crash Box (ACB). They expected to replace the current conventional ACB to reduce the weight and effective cost for future production. They implemented integration of several concurrent engineering methods such as theory of inventive problem solving (TRIZ), How-How diagram, Morphological Chart (MC) and Ishikawa methods. At the end of the study, important requirements of the function specification and failure mode analysis as well as geometry specification are identified and new concept design was produced as shown in Fig. 10.3. In this case, they are considered those limitation

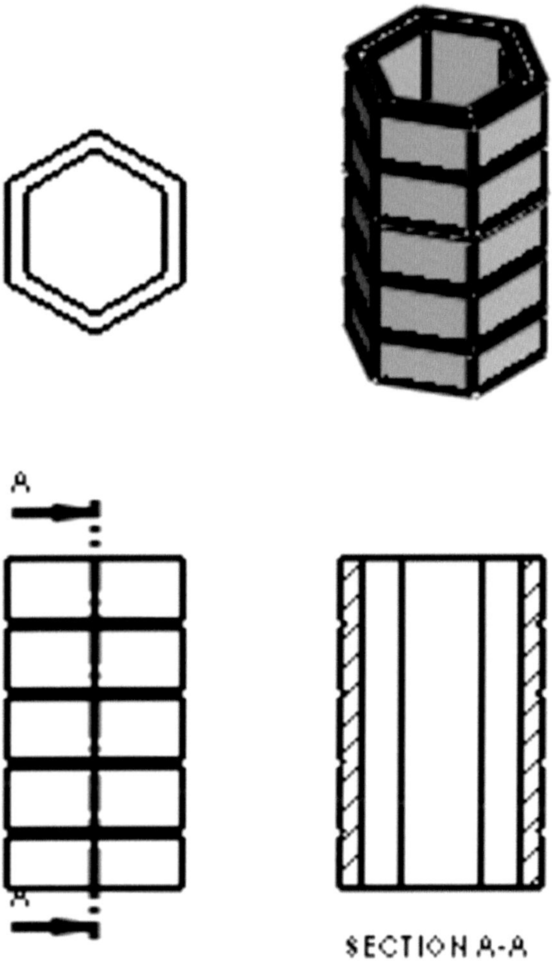

Fig. 10.3 Automotive crush box conceptual design. Reproduced from Ref. [85]

of hybrid composites which at the end differ with the conventional steel ACB. The final conceptual design of sugar palm polymer composite ACB has covered element of low cost, avoid complex manufacturing process to reduce the cost and efficient performance of ACB to absorb energy during collision.

14 Conclusion

This chapters summarize the importance of hybrid synthetic/natural fibres in automobile applications as lightweight materials. In this article, a brief summary of the prominence of various natural and synthetic fibres, their classification, their production, and the use of them for automotive applications. It also recommends future researchers to apply various natural with synthetic fibres to fabricate more green products that can efficiently replace different conventionally-made products. Several profits of hybrid composites such as high fatigue and corrosion properties, high strength-to weight ratio, and most importantly high performance characteristics over metallic materials have made them a suitable material for automotive structures as they are employed widely and play a very significant role in those applications.

Acknowledgements The authors would like to express their gratitude for the financial support received from Universiti Teknologi Malaysia through the project "Characterizations of Hybrid Kenaf Fibre/Fibreglass Meshes Reinforced Thermoplastic ABS Composites for Future Use in Aircraft Radome Applications" under grant number PY/2022/03758—Q.J130000.3824.31J25.

References

1. Azman MA, Asyraf MRM, Khalina A, Petrů M, Ruzaidi CM, Sapuan SM, Wan Nik WB, Ishak MR, Ilyas RA, Suriani MJ (1917) Natural fiber reinforced composite material for product design: a short review. Polym (Basel) 13. https://doi.org/10.3390/polym13121917
2. Asyraf MRM, Ishak MR, Sapuan SM, Yidris N, Rafidah M, Ilyas RA, Razman MR (2020) Potential application of green composites for cross arm component in transmission tower: a brief review. Int J Polym Sci. https://doi.org/10.1155/2020/8878300
3. Alsubari S, Zuhri MYM, Sapuan SM, Ishak MR, Ilyas RA, Asyraf MRM (2021) Potential of natural fiber reinforced polymer composites in sandwich structures: a review on its mechanical properties. Polymers (Basel) 13:423. https://doi.org/10.3390/polym13030423
4. Kundu B, Kurland NE, Bano S, Patra C, Engel FB, Yadavalli VK, Kundu SC (2014) Silk proteins for biomedical applications: bioengineering perspectives. Prog Polym Sci 39:251–267. https://doi.org/10.1016/j.progpolymsci.2013.09.002
5. Ilyas RA, Sapuan MS, Norizan MN, Norrrahim MNF, Ibrahim R, Atikah MSN, Huzaifah MRM, Radzi AM, Izwan S, Azammi AMN, et al (2020) Macro to nanoscale natural fiber composites for automotive components: research, development, and application. In: Sapuan MS, Ilyas RA (eds) Biocomposite and synthetic composites for automotive applications. Woodhead Publishing Series: Amsterdam, Netherland
6. Ranade A, Nayak K, Fairbrother D, D'Souza NA (2005) Maleated and non-maleated polyethylene-montmorillonite layered silicate blown films: creep, dispersion and crystallinity. Polymer (Guildf) 46:7323–7333. https://doi.org/10.1016/j.polymer.2005.04.085
7. Amir AL, Ishak MR, Yidris N, Zuhri MYM, Asyraf MRM (2021) Advances of composite cross arms with incorporation of material core structures: manufacturability, recent progress and views. J Mater Res Technol 13:1115–1131. https://doi.org/10.1016/j.jmrt.2021.05.040
8. Amir AL, Ishak MR, Yidris N, Zuhri MYM, Asyraf MRM (2021) Potential of honeycomb-filled composite structure in composite cross-arm component: a review on recent progress and its mechanical properties. Polymers (Basel) 13:1341. https://doi.org/10.3390/polym13081341
9. Asyraf MRM, Ishak MR, Sapuan SM, Yidris N, Ilyas RA (2020) Woods and composites cantilever beam: a comprehensive review of experimental and numerical creep methodologies. J Mater Res Technol 9:6759–6776. https://doi.org/10.1016/j.jmrt.2020.01.013
10. Ilyas R, Sapuan S, Atikah M, Asyraf M, Rafiqah SA, Aisyah H, Nurazzi NM, Norrrahim M (2021) Effect of hydrolysis time on the morphological, physical, chemical, and thermal behavior of sugar palm nanocrystalline cellulose (*Arenga pinnata* (Wurmb.) Merr). Text Res J 91:152–167. https://doi.org/10.1177/0040517520932393
11. Asyraf MRM, Rafidah M, Azrina A, Razman MR (2021) Dynamic mechanical behaviour of kenaf cellulosic fibre biocomposites: a comprehensive review on chemical treatments. Cellulose 28:2675–2695. https://doi.org/10.1007/s10570-021-03710-3

12. Asyraf MRM, Rafidah M, Ishak MR, Sapuan SM, Yidris N, Ilyas RA, Razman MR (2020) Integration of TRIZ, morphological chart and ANP method for development of FRP composite portable fire extinguisher. Polym Compos 41:2917–2932. https://doi.org/10.1002/pc.25587
13. Feng NL, Malingam SD, Ping CW (2021) Mechanical characterisation of kenaf/PALF reinforced composite-metal laminates: effects of hybridisation and weaving architectures. J Reinf Plast Compos 40:193–205. https://doi.org/10.1177/0731684420956719
14. Asyraf MRM, Syamsir A, Supian ABM, Usman F, Ilyas RA, Nurazzi NM, Norrrahim MNF, Razman MR, Zakaria SZS, Sharma S et al (2022) Sugar palm fibre-reinforced polymer composites: influence of chemical treatments on its mechanical properties. Materials (Basel) 15:3852. https://doi.org/10.3390/ma15113852
15. Omran AAB, Mohammed AABA, Sapuan SM, Ilyas RA, Asyraf MRM, Koloor SSR, Petrů M (2021) Micro- and nanocellulose in polymer composite materials: a review. Polymers (Basel) 13:231. https://doi.org/10.3390/polym13020231
16. Bahrain SHK, Masdek NRN, Mahmud J, Mohammed MN, Sapuan SM, Ilyas RA, Mohamed A, Shamseldin MA, Abdelrahman A, Asyraf MRM (2022) Morphological, physical, and mechanical properties of sugar-palm (*Arenga pinnata* (Wurmb) Merr.)-reinforced silicone rubber biocomposites. Mater (Basel) 15:4062. https://doi.org/10.3390/ma15124062
17. Ilyas RA, Sapuan SM, Harussani MM, Hakimi MYAY, Haziq MZM, Atikah MSN, Asyraf MRM, Ishak MR, Razman MR, Nurazzi NM et al (2021) Polylactic acid (PLA) biocomposite: processing, additive manufacturing and advanced applications. Polymers (Basel) 13:1326. https://doi.org/10.3390/polym13081326
18. Ilyas RA, Sapuan SM, Atiqah A, Ibrahim R, Abral H, Ishak MR, Zainudin ES, Nurazzi NM, Atikah MSN, Ansari MNM, et al. (2020) Sugar palm (*Arenga pinnata* [Wurmb.] Merr) starch films containing sugar palm nanofibrillated cellulose as reinforcement: water barrier properties. Polym Compos 41:459–467. https://doi.org/10.1002/pc.25379
19. Asyraf MZ, Suriani MJ, Ruzaidi CM, Khalina A, Ilyas RA, Asyraf MRM, Syamsir A, Azmi A, Mohamed A (2022) Development of natural fibre-reinforced polymer composites ballistic helmet using concurrent engineering approach: a brief review. Sustainability 14:7092. https://doi.org/10.3390/su14127092
20. Bahrain SHK, Rahim NNCA, Mahmud J, Mohammed MN, Sapuan SM, Ilyas RA, Alkhatib SE, Asyraf MRM (2022) Hyperelastic properties of bamboo cellulosic fibre-reinforced silicone rubber biocomposites via compression test. Int J Mol Sci 23:6338. https://doi.org/10.3390/ijms23116338
21. Asyraf MRM, Ishak MR, Norrrahim MNF, Amir AL, Nurazzi NM, Ilyas RA, Asrofi M, Rafidah M, Razman MR (2022) Potential of flax fiber reinforced biopolymer composites for cross-arm application in transmission tower: a review. Fibers Polym. 23:853–877. https://doi.org/10.1007/s12221-022-4383-x
22. Nurazzi NM, Norrrahim MNF, Sabaruddin FA, Shazleen SS, Ilyas RA, Lee SH, Padzil FNM, Aizat G, Aisyah HA, Mohidem NA et al (2022) Mechanical performance evaluation of bamboo fibre reinforced polymer composites and its applications: a review. Funct Compos Struct 4:015009. https://doi.org/10.1088/2631-6331/ac5b1a
23. Alhayek A, Syamsir A, Supian ABM, Usman F, Asyraf MRM, Atiqah MA (2022) Flexural creep behaviour of pultruded GFRP composites cross-arm: a comparative study on the effects of stacking sequence. Polymers (Basel) 14:1330. https://doi.org/10.3390/polym14071330
24. Asyraf MRM, Ishak MR, Syamsir A, Amir AL, Nurazzi NM, Norrrahim MNF, Asrofi M, Rafidah M, Ilyas RA, Rashid MZA et al (2022) Filament-wound glass-fibre reinforced polymer composites: potential applications for cross arm structure in transmission towers. Polym Bull. https://doi.org/10.1007/s00289-022-04114-4
25. Haris NIN, Hassan MZ, Ilyas RA, Suhot MA, Sapuan SM, Dolah R, Mohammad R, Asyraf MRM (2022) Dynamic mechanical properties of natural fiber reinforced hybrid polymer composites: a review. J Mater Res Technol 19:167–182. https://doi.org/10.1016/j.jmrt.2022.04.155
26. Thiruganasambanthan T, Ilyas RA, Norrrahim MNF, Kumar TSM, Siengchin S, Misenan MSM, Farid MAA, Nurazzi NM, Asyraf MRM, Zakaria SZS et al (2022) Emerging developments

on nanocellulose as liquid crystals: a biomimetic approach. Polymers (Basel) 14:1546. https://doi.org/10.3390/polym14081546
27. Asyraf MRM, Ishak MR, Sapuan SM, Yidris N (2021) Comparison of static and long-term creep behaviors between balau wood and glass fiber reinforced polymer composite for cross-arm application. Fibers Polym 22:793–803. https://doi.org/10.1007/s12221-021-0512-1
28. Johari AN, Ishak MR, Leman Z, Yusoff MZM, Asyraf MRM (2020) Influence of $CaCO_3$ in pultruded glass fibre/unsaturated polyester composite on flexural creep behaviour using conventional and TTSP methods. Polimery 65:46–54. https://doi.org/10.14314/polimery.2020.11.6
29. Ilyas RA, Sapuan SM, Asyraf MRM, Dayana DAZN, Amelia JJN, Rani MSA, Norrrahim MNF, Nurazzi NM, Aisyah HA, Sharma S et al (2021) Polymer composites filled with metal derivatives: a review of flame retardants. Polymers (Basel) 13. https://doi.org/10.3390/polym13111701
30. Mohd Nurazzi NM, Muhammad Asyraf MRM, Khalina A, Abdullah N, Sabaruddin FA, Kamarudin SH, Ahmad S, Mahat AM, Lee CL, Aisyah HA et al (2021) Fabrication, functionalization, and application of carbon nanotube-reinforced polymer composite: an overview. Polymers (Basel) 13:1047. https://doi.org/10.3390/polym13071047
31. Nurazzi NM, Asyraf MRM, Khalina A, Abdullah N, Aisyah HA, Rafiqah SA, Sabaruddin FA, Kamarudin MNF, Ilyas RA, Sapuan SM (2021) A review on natural fiber reinforced polymer composite for bullet proof and ballistic applications. Polym 13:646. https://doi.org/10.3390/polym13040646
32. Shaharuzaman MA, Sapuan SM, Mansor MR (2018) Prioritizing the product design specification of side-door impact beam using analytic hierarchy process method. Mech Eng Res Day 2018:34–35
33. Mansor MR, Sapuan SM, Zainudin ES, Nuraini AA (2014) Conceptual design of kenaf fiber polymer composite automotive parking brake lever using integrated TRIZ-morphological chart-analytic hierarchy process method. Mater Des 54:473–482. https://doi.org/10.1016/j.matdes.2013.08.064
34. Azammi AMN, Sapuan SM, Ishak MR, Sultan MTH (2018) Conceptual design of automobile engine rubber mounting composite using TRIZ-Morphological chart-analytic network process technique. Def Technol 14:268–277. https://doi.org/10.1016/j.dt.2018.05.009
35. Mastura MT, Sapuan SM, Mansor MR, Nuraini AA (2017) Conceptual design of a natural fibre-reinforced composite automotive anti-roll bar using a hybrid approach. Int J Adv Manuf Technol 91:2031–2048. https://doi.org/10.1007/s00170-016-9882-8
36. Okahisa Y, Yoshida A, Miyaguchi S, Yano H (2009) Optically transparent wood-cellulose nanocomposite as a base substrate for flexible organic light-emitting diode displays. Compos Sci Technol 69:1958–1961. https://doi.org/10.1016/j.compscitech.2009.04.017
37. Mansor MR, Sapuan SM, Hambali A (2015) Conceptual design of kenaf polymer composites automotive spoiler using TRIZ and morphology chart methods. Appl Mech Mater 761:63–67. https://doi.org/10.4028/www.scientific.net/AMM.761.63
38. Davoodi MM, Sapuan SM, Yunus R (2008) Conceptual design of a polymer composite automotive bumper energy absorber. Mater Des 29:1447–1452. https://doi.org/10.1016/j.matdes.2007.07.011
39. Shaharuzaman MA, Sapuan SM, Mansor MR, Zuhri MYM (2020) Conceptual design of natural fiber composites as a side-door impact beam using hybrid approach. J Renew Mater 8:549–563. https://doi.org/10.32604/jrm.2020.08769
40. Ishak NM, Sivakumar D, Mansor MR (2018) The application of TRIZ on natural fibre metal laminate to reduce the weight of the car front hood. J Brazilian Soc Mech Sci Eng 40:1–12. https://doi.org/10.1007/s40430-018-1039-2
41. Sapuan SM (2009) A conceptual design of the concurrent engineering design system for polymeric-based composite automotive pedals. Am J Appl Sci 2:514–525. https://doi.org/10.3844/ajassp.2005.514.525
42. Asyraf MRM, Rafidah M (2022) Mechanical and thermal performance of sugar palm fibre thermoset polymer composites: a short review. J Nat Fibre Polym Compos 1:2

43. Hill K, Swiecki B, Cregger J (2012) The bio-based materials automotive value chain. Cent Automot Res 1–92
44. Cristaldi G, Cicala G, Recca G, Latteri A (2010) Composites based on natural fibre fabrics. Woven Fabr Eng 317–342. https://doi.org/10.5772/10465
45. Koronis G, Silva A, Fontul M (2013) Green composites: a review of adequate materials for automotive applications. Compos Part B Eng 44:120–127. https://doi.org/10.1016/j.compositesb.2012.07.004
46. Ahmad F, Choi HS, Park MK (2015) A review: natural fiber composites selection in view of mechanical, light weight, and economic properties. Macromol Mater Eng 300:10–24. https://doi.org/10.1002/mame.201400089
47. Peças P, Carvalho H, Salman H, Leite M (2018) Natural fibre composites and their applications: a review. J Compos Sci 2. https://doi.org/10.3390/jcs2040066
48. Ilyas RA, Sapuan SM, Asyraf MRM, Atikah MSN, Ibrahim R, Norrahim MNF, Yasim-Anuar TAT, Megashah LN (2021) Mechanical and dynamic mechanical analysis of bio-based composites. In: Krishnasamy S, Nagarajan R, Thiagamani SMK (eds) Mechanical and dynamic properties of biocomposites. WILEY-VCH GmbH, Weinheim, Germany
49. Ilyas RA, Asyraf MRM, Sapuan SM, Afiq TMN, Suhrisman A, Atikah MSN, Ibrahim R (2021) Development of roselle fiber reinforced polymer biocomposites mug pad using hybrid design for sustainability and Pugh method. In: Sapuan SM, Razali N, Radzi AM, Ilyas RA (eds) Roselle: production, processing, products and biocomposites. Elsevier Academic Press, Amsterdam, Netherland, pp 1–300. ISBN 978-0323852135
50. Alkbir MFM, Sapuan SM, Nuraini AA, Ishak MR (2016) Fibre properties and crashworthiness parameters of natural fibre-reinforced composite structure: a literature review. Compos Struct
51. Selmy AI, Elsesi AR, Azab NA, Abd El-Baky MA (2012) In-plane shear properties of unidirectional glass fiber (U)/random glass fiber (R)/epoxy hybrid and non-hybrid composites. Compos Part B Eng 43:431–438. https://doi.org/10.1016/j.compositesb.2011.06.001
52. Ahmed KS, Vijayarangan S, Kumar A (2007) Low velocity impact damage characterization of woven jute-glass fabric reinforced isothalic polyester hybrid composites. J Reinf Plast Compos 26:959–976. https://doi.org/10.1177/0731684407079414
53. Noorunnisa Khanam P, Mohan Reddy M, Raghu K, John K, Venkata Naidu S (2007) Tensile, flexural and compressive properties of sisal/silk hybrid composites. J Reinf Plast Compos 26:1065–1070. https://doi.org/10.1177/0731684407079347
54. Idicula M, Malhotra SK, Joseph K, Thomas S (2005) Dynamic mechanical analysis of randomly oriented intimately mixed short banana/sisal hybrid fibre reinforced polyester composites. Compos Sci Technol 65:1077–1087. https://doi.org/10.1016/j.compscitech.2004.10.023
55. Sabeel Ahmed K, Vijayarangan S (2006) Elastic property evaluation of jute-glass fibre hybrid composite using experimental and CLT approach. Indian J Eng Mater Sci 13:435–442
56. Athijayamani A, Thiruchitrambalam M, Natarajan U, Pazhanivel B (2009) Effect of moisture absorption on the mechanical properties of randomly oriented natural fibers/polyester hybrid composite. Mater Sci Eng A 517:344–353. https://doi.org/10.1016/j.msea.2009.04.027
57. Sayer M, Bektaş NB, Demir E, Çallioğlu H (2012) The effect of temperatures on hybrid composite laminates under impact loading. Compos Part B Eng 43:2152–2160. https://doi.org/10.1016/j.compositesb.2012.02.037
58. Mishra S, Mohanty AK, Drzal LT, Misra M, Parija S, Nayak SK, Tripathy SS (2003) Studies on mechanical performance of biofibre/glass reinforced polyester hybrid composites. Compos Sci Technol 63:1377–1385. https://doi.org/10.1016/S0266-3538(03)00084-8
59. Jawaid M, Abdul Khalil HPS (2011) Cellulosic/synthetic fibre reinforced polymer hybrid composites: a review. Carbohydr Polym 86:1–18. https://doi.org/10.1016/j.carbpol.2011.04.043
60. Velmurugan R, Manikandan V (2005) Mechanical properties of glass/palmyra fiber waste sandwich composites. Indian J Eng Mater Sci 12:563–570
61. Thiruchitrambalam M, Alavudeen A, Athijayamani A, Venkateshwaran N, Perumal AE (2009) Improving mechanical properties of banana/kenaf polyester hybrid composites using sodium laulryl sulfate treatment. Mater Phys Mech 8

62. Thwe MM, Liao K (2003) Environmental effects on bamboo-glass/polypropylene hybrid composites. J Mater Sci 38:363–376. https://doi.org/10.1023/A:1021130019435
63. Czigány T (2006) Special manufacturing and characteristics of basalt fiber reinforced hybrid polypropylene composites: mechanical properties and acoustic emission study. Compos Sci Technol 66:3210–3220. https://doi.org/10.1016/j.compscitech.2005.07.007
64. Wambua P, Vangrimde B, Lomov S, Verpoest I (2007) The response of natural fibre composites to ballistic impact by fragment simulating projectiles. Compos Struct 77:232–240. https://doi.org/10.1016/j.compstruct.2005.07.006
65. Nayak SK (2009) Degradation and flammability behavior of PP/banana and glass fiber-based hybrid composites. Int J Plast Technol 13:47–67. https://doi.org/10.1007/s12588-009-0006-2
66. Asyraf MRM, Ishak MR, Sapuan SM, Yidris N (2019) Conceptual design of creep testing rig for full-scale cross arm using TRIZ-morphological chart-analytic network process technique. J Mater Res Technol 8:5647–5658. https://doi.org/10.1016/j.jmrt.2019.09.033
67. Asyraf MRM, Ishak MR, Sapuan SM, Yidris N (2020) Conceptual design of multi-operation outdoor flexural creep test rig using hybrid concurrent engineering approach. J Mater Res Technol 9:2357–2368. https://doi.org/10.1016/j.jmrt.2019.12.067
68. Nurazzi NM, Khalina A, Sapuan SM, Ilyas RA (2019) Mechanical properties of sugar palm yarn/woven glass fiber reinforced unsaturated polyester composites: effect of fiber loadings and alkaline treatment. Polimery 64:12–22. https://doi.org/10.14314/polimery.2019.10.3
69. Nurazzi NM, Khalina A, Chandrasekar M, Aisyah HA, Rafiqah SA, Ilyas RA, Hanafee ZM (2020) Effect of fiber orientation and fiber loading on the mechanical and thermal properties of sugar palm yarn fiber reinforced unsaturated polyester resin composites. Polimery 65:115–124. https://doi.org/10.14314/polimery.2020.2.5
70. Jaafar CNA, Rizal MAM, Zainol I (2018) Effect of kenaf alkalization treatment on morphological and mechanical properties of epoxy/silica/kenaf composite. Int J Eng Technol 7:258–263. https://doi.org/10.14419/ijet.v7i4.35.22743
71. Johari AN, Ishak MR, Leman Z, Yusoff MZM, Asyraf MRM (2020) Creep behaviour monitoring of short-term duration for fiber-glass reinforced composite cross-arms with unsaturated polyester resin samples using conventional analysis. J Mech Eng Sci 14:7361–7368. https://doi.org/10.15282/jmes.14.4.2020.04.0578
72. Asyraf MRM, Syamsir A, Zahari NM, Supian ABM, Ishak MR, Sapuan SM, Sharma S, Rashedi A, Razman MR, Zakaria SZS et al (2022) Product development of natural fibre-composites for various applications: design for sustainability. Polymers (Basel) 14:920. https://doi.org/10.3390/polym14050920
73. Asyraf MRM, Ishak MR, Norrrahim MNF, Nurazzi NM, Shazleen SS, Ilyas RA, Rafidah M, Razman MR (2021) Recent advances of thermal properties of sugar palm lignocellulosic fibre reinforced polymer composites. Int J Biol Macromol 193:1587–1599. https://doi.org/10.1016/j.ijbiomac.2021.10.221
74. Senthilkumar K, Saba N, Rajini N, Chandrasekar M, Jawaid M, Siengchin S, Alotman OY (2018) Mechanical properties evaluation of sisal fibre reinforced polymer composites: a review. Constr Build Mater 174:713–729. https://doi.org/10.1016/j.conbuildmat.2018.04.143
75. Duc F, Bourban PE, Plummer CJG, Månson JAE (2014) Damping of thermoset and thermoplastic flax fibre composites. Compos Part A Appl Sci Manuf 64:115–123. https://doi.org/10.1016/j.compositesa.2014.04.016
76. Banerjee S, Sankar BV (2012) Mechanical properties of hybrid composites using finite element method based micromechanics. In: Proceedings of the 27th annual technical conference of the American society for composites 2012, held jointly with 15th joint US-Japan conference on composite materials and ASTM-D30 meeting, pp 1258–1276
77. Hwang J-H, Seo D-W, Park K-T, You Y-J (2014) Experimental study on the mechanical properties of FRP bars by hybridizing with steel wires. Engineering 06:365–373. https://doi.org/10.4236/eng.2014.67039
78. Ravishankar B, Nayak SK, Kader MA (2019) Hybrid composites for automotive applications—a review. J Reinf Plast Compos 38:835–845. https://doi.org/10.1177/0731684419849708

79. Puttegowda M, Rangappa SM, Jawaid M, Shivanna P, Basavegowda Y, Saba N (2018) Potential of natural/synthetic hybrid composites for aerospace applications. In: Jawaid M, Sultan MTH (eds) Sustainable composites for aerospace applications. Elsevier, Amsterdam, Netherland, pp 315–351
80. Karikalan L, Chandrasekran M, Ramasubramanian S, Baskar S (2019) Hybridization of composites using natural and synthetic fibers for automotive application. IJSRST 3:916–920
81. Atiqah A, Maleque MA, Jawaid M, Iqbal M (2014) Development of kenaf-glass reinforced unsaturated polyester hybrid composite for structural applications. Compos Part B Eng 56. https://doi.org/10.1016/j.compositesb.2013.08.019
82. Nazim ARM, Ansari MNM, Atiqah A (2019) Impact strength and morphological properties of Kenaf/glass fibre/polyester hybrid composite for attenuator application. Mater Today Proc 29:119–122. https://doi.org/10.1016/j.matpr.2020.05.683
83. Wang J, Yang N, Zhao J, Wang D, Wang Y, Li K, He Z, Wang B (2016) Design and experimental verification of composite impact attenuator for racing vehicles. Compos Struct 141:39–49. https://doi.org/10.1016/j.compstruct.2016.01.013
84. Afzaluddin A, Jawaid M, Salit MS, Ishak MR (2019) Physical and mechanical properties of sugar palm/glass fiber reinforced thermoplastic polyurethane hybrid composites. J Mater Res Technol 8:950–959. https://doi.org/10.1016/j.jmrt.2018.04.024
85. Yusof NSB, Sapuan SM, Sultan MTH, Jawaid M (2018) Concept generation of sugar palm/glass fiber reinforced thermoplastic polyurethane hybrid composite automotive crash box. J Adv Res Mater Sci 49:10–17

Potential of Natural/Synthetic Hybrid Composites for Automotive Applications

Jeyaguru Sangilimuthukumar, Thiagamani Senthil Muthu Kumar, Krishnasamy Senthilkumar, Muthukumar Chandrasekar, and Suchart Siengchin

Abstract Natural/synthetic fiber reinforced hybrid composite materials are attractive because of their unique characteristics like, high mechanical and thermal properties, low cost and low specific weight. When compared to single fiber-reinforced materials, hybrid composites possessed better characteristics (i.e., mechanical, thermal). Many engineering fields use hybrid composites and various interior and exterior parts in the automobile industry. This chapter discusses the mechanical and thermal properties of the hybrid natural/synthetic fiber reinforced thermoset and thermoplastic polymer-based composites and their automotive application.

Keywords Hybrid composites · Natural/synthetic fiber · Thermoset polymer · Thermoplastic polymer · Automotive applications

---

J. Sangilimuthukumar · T. Senthil Muthu Kumar (✉)
Department of Automobile Engineering, Kalasalingam Academy of Research and Education, Krishnankoil 626126, Tamil Nadu, India
e-mail: tsmkumar@klu.ac.in

K. Senthilkumar
Departmet of Mechanical Engineering, PSG Institute of Technology and Applied Research, Coimbatore 641062, Tamil Nadu, India

M. Chandrasekar
School of Aeronautical Sciences, Hindustan Institute of Technology and Science, Padur, Kelambakkam, Chennai 603103, India

S. Siengchin
Department of Materials and Production Engineering, The Sirindhorn International Thai-German Graduate School of Engineering (TGGS), King Mongkut's University of Technology North Bangkok (KMTNB), Bangkok, Thailand

Institute of Plant and Wood Chemistry, Technische Universität Dresden, Pienner Str. 19, 01737 Tharandt, Germany

T. Khan and M. Jawaid (eds.), *Green Hybrid Composite in Engineering and Non-Engineering Applications*, Composites Science and Technology,
https://doi.org/10.1007/978-981-99-1583-5_11

1 Introduction

Nowadays, Polymer matrix composites are used in many applications such as construction, marine, aerospace, and automotive industries [1–4]. They possess high specific strength, abundant availability, eco-friendly nature, renewability, and high stiffness [5–9]. The polymer matrix composites are developed using natural fibers and/or synthetic fibers. The mechanical strengths are lesser when comparing the natural fiber reinforced composites with synthetic fiber composites. Though they have lesser strengths, the natural fiber composites have many advantages such as biodegradability nature, low density, low cost, environmentally friendly, etc. Some disadvantages of natural fibers are poor stability, variations in properties, high susceptibility to moisture absorption and incompatibility, and low thermal resistance [10, 11].

Glass fibers are used widely, while carbon and aramid fibers also increased [10]. The requirements of Glass fiber reinforced composites in structural applications are higher due to their excellent adhesion property, well-developed manufacturing methods, and good strength [1]. Carbon fibers are used when the demand for strength and durability is higher [12, 13]. Sports cars, engine covers, and electric cars are used more often. Some major drawbacks of synthetic fibers are non-biodegradability, the need for fossil fuels, environmental accumulation, and higher costs [14, 15].

A practical method to overcome these disadvantages is to develop hybrid composites. Hybrid composites can be developed by combining natural fiber/natural fiber, natural fiber/synthetic fiber, and synthetic fiber/synthetic fiber in the matrix [10, 16]. Over a decade, hybridization has increased because of its unique properties. It is used in various applications, especially in the automotive industry, because hybrid composites help reduce the vehicle's overall weight. They are resulting in a reduction of CO_2 emissions to the environment.

This chapter summarizes recent works carried out on the mechanical and thermal behavior of natural-synthetic fiber reinforced hybrid polymer composites. Furthermore, the potential use of natural/synthetic hybrid composites in automotive industries are discussed.

2 Mechanical Properties of Natural/Synthetic Fiber Reinforced Polymer Composites

2.1 *Thermoset Polymer Composites*

The mechanical behavior of jute-glass-epoxy matrix composites were analyzed by varying the fiber layering sequences. When the glass fibers were positioned at the skin layer, the respective fiber layered composites exhibited improved properties than the other combinations [17]. In another study, the mechanical behavior of glass/flax/vinyl ester matrix hybrid composites was studied. Results reported that of hybrid composites showed the highest tensile properties compared to pure glass composites. Furthermore, glass fiber as skin layer possess highest flexural and impact strength [18]. The effect of mechanical characteristics of jute and carbon fiber reinforced epoxy composites was investigated. The tensile strength, tensile modulus, flexural strength, and flexural modulus of jute and carbon fiber with epoxy hybrid composite is nearly identical to that of a carbon-epoxy composite and much better than of pure jute fiber reinforced epoxy composites [19].

Ramesh et al. [20] studied the mechanical properties of hemp-carbon fiber based polyester composites. The results observed that the alkali treated hemp-carbon hybrid composites was better mechanical strength compared to untreated one [20]. Azrin et al. [21] evaluated the flexural and impact properties of kevlar-coir fiber-reinforced epoxy-based hybrid composites. Results revealed that kevlar-coir fiber reinforced hybrid composites showed the highest flexural and impact properties compared to pure coir composites. However, coir-kevlar hybrid combinations performed lesser than the pure kevlar composites in terms of flexural and impact properties [21].

In another work, Yahaya et al. [22] explored the mechanical behavior of kenaf-kevlar fiber reinforced with epoxy composites. It indicated that the hybrid composites containing kevlar as skin layers have greater mechanical properties. Moreover, treated hybrid composites have better tensile and flexural characteristics than untreated hybrid composites [22]. Amuthakkannan et al. [23] have reported the hybrid basalt (B)-jute (J) fiber reinforced polyester composites, and the results showed that BBJ layering sequences have superior tensile strength. Then, the improved flexural properties were observed when the fibers were stacked alternatively [23].

Petrucci et al. [24] investigated the mechanical performance of basalt fibers with flax-hemp-glass fiber-reinforced epoxy-based hybrid composites. They reported that Glass-flax-basalt type hybrid composites outperformed in mechanical properties than Glass-hemp-basalt and flax-hemp-basalt hybrid composites. The researchers reported that the composites made using the combination of hemp fibers exhibited a lesser quality interface [24].

Investigation on mechanical properties of jute abaca glass fiber reinforced with epoxy matrix hybrid composites. The findings revealed that the hybrid composites made with two layers of abaca fiber, one layer of jute fiber and two layers of glass fiber reinforced hybrid composites shows better mechanical properties [25]. In another study, mechanical properties of glass/flax/vinyl ester matrix composites. The pure and hybrid composites were fabricated using a vacuum-assisted resin transfer molding process (VARTM). Results evident that the tensile properties of pure woven glass fiber composites were lower than the hybrid composites. The composites with a glass ply on the bottom surface had higher flexural strength, and composites with glass plies on both ends had higher impact strength than the other composites [18].

Pandey et al. [26] studied the physical and mechanical properties of hemp, flax, and glass fibers reinforced polyurethane (PU) hybrid composites. They reported that the performance of the hybrid composites was significantly higher than the pure fiber-reinforced composites. In another work, Yahaya et al. [27] analyzed the mechanical properties of kenaf/aramid/epoxy matrix composites. The results showed that the woven kenaf/aramid composite has greater tensile and impact strength than other hybrid composites. Ramesh et al. [28] evaluated mechanical and interfacial properties of sisal/jute/glass hybrid fibers reinforced epoxy matrix composites. The findings showed that both sisal and jute fibers with glass fiber hybrid composites outperformed in mechanical properties of their single fiber-reinforced composites.

The effect of various fiber loadings on flexural properties of banana/pineapple leaf (PALF)-glass fiber reinforced with epoxy hybrid composite was investigated [29]. Flexural strength increased with an increase in fiber volume. For instance, at 40 wt% of fiber loading, the PALF-glass hybrid composites given higher flexural strength [29]. Another work studied the mechanical performance of hybrid woven jute, vetiver, and glass fibers reinforced vinyl ester composites. Furthermore, the untreated fiber composites were compared with alkali-treated fiber composites. Results reported that the alkali-treated vetiver fiber-reinforced composites showed improved properties [30].

Amico et al. [31] prepared hybrid composites with sisal-glass fiber reinforced polyester composites by a compression molding technique. Results revealed that the stacking of sisal-glass hybrid composites were improved in mechanical performance. Furthermore, the top and bottom surfaces of glass fiber in the hybrid composites showed enhanced flexural behavior [31]. In another study, the mechanical behavior of sisal-glass-polyester and pineapple leaf fiber-glass-polyester hybrid composites was investigated. Both the polyester-based hybrid composites obtained enhanced tensile, impact and flexural properties. Further, better results were observed by incorporating glass fibers for 8.6 and 5.7 wt% with their hybrid combinations, such as pineapple leaf fiber-glass and sisal-glass fiber composites, respectively [32].

Abdullah et al. [33] fabricated jute–glass–polyester hybrid composites by hand layup method. The result indicated that positive hybridization was found. Further, hybrid jute/glass fiber (1:3) ratio had highest mechanical properties compared to pure composites [33]. The mechanical performance of sisal-glass fiber reinforced polyester composites were investigated by Heitor et al. [34]. The addition of glass fibres improved the flexural and impact properties of the hybrid composites. The slight variations in fibre length have no discernible influence on flexural properties.

The effect of mechanical behaviors of bamboo and glass fiber-reinforced polyester composites by using the compression molding technique was investigated by Tran et al. [35]. Results revealed that the mechanical behaviors of bamboo-glass fibers (25:75) volume ratio of hybrid composites showed improved properties than the other composites. In another work, Sabeel et al. [36] examined the effects of tensile, flexural, and interlaminar shear strength (ILLS) of jute and glass fibers reinforced polyester composites. The results revealed that the two glass fiber was positioned as the skin layer and the jute fibers in the middle produced better mechanical properties. Researchers examined the mechanical characteristics of kapok-glass fiber-reinforced polyester hybrid composites. From the results, kapok-glass hybrid composite exhibited enhanced flexural strength, compressive strength, and ILLS strength. The improved results were owing to the stronger bond between the fibers and the matrix. In flexural and compressive strength, the kapok-glass hybrid composites gave an optimum result compared to kapok-sisal and sisal-glass fiber-reinforced composites [37].

Audibert et al. [38] reported the tensile properties of kevlar fiber and flax fiber-reinforced epoxy hybrid composites. Results revealed better tensile properties for Kevlar/flax hybrid composites [38]. The effect of mechanical properties of sisal-aramid-phenolic resin hybrid composites was studied by Zhong et al. [39]. The results showed that the tensile and compression properties of hybrid composites were significantly improved. Bakar et al. [40] explored the effect of different stacking sequences of kenaf-kevlar fiber-reinforced epoxy hybrid composites. The maximum impact energy was observed for Kevlar-kenaf-kevlar hybrid composites. Moreover, higher impact strength and hardness were found with the addition of the weight percentage of Kevlar fiber in the kenaf-kevlar hybrid composites [40].

In another work, the effect of mechanical properties of coir fiber and Kevlar fiber reinforced with epoxy composites. The results showed that the pure Kevlar composites gave the highest impact strength. Further, kevlar-coir-epoxy hybrid composites exhibited the highest flexural and impact properties [21]. Hasim et al. examined the effect of ply orientation of pineapple leaf fiber-carbon-epoxy composites [41]. The maximum tensile strength and tensile modulus values were obtained for (0°, 90°) ply orientation of carbon-pineapple leaf fiber hybrid composites. Moreover, the outer layer of carbon fiber showed the highest flexural strength and flexural modulus of the composites [41].

The influence of the stacking sequence of the jute-carbon fiber reinforced with epoxy composites was studied. Results reported that the jute-carbon-carbon-jute hybrid composites showed the highest tensile strength while the carbon-jute-jute-carbon hybrid composites possessed the highest impact properties [42]. Venkatasudhakar et al. [43] examined the effect of different layering sequences of jute/banana/carbon fiber reinforced with epoxy matrix hybrid composites. Results showed that the carbon/jute/jute/jute/carbon fiber reinforced epoxy composites exhibited the highest mechanical properties compared to pure fiber reinforced composites [43]. El-baky et al. [44] explored the flax-basalt-glass-epoxy matrix hybrid composites. The pure glass or pure basalt fiber-reinforced composites showed the highest mechanical properties, while the pure flax fiber reinforced composites exhibited the lowest mechanical properties. However, the hybrid composites such as flax-basalt-glass hybrid combinations exhibited intermediate performance in mechanical performance.

2.2 *Thermoplastic Polymer Composites*

The mechanical characteristics of date palm, wood flour, and glass fiber reinforced polypropylene (PP) hybrid composites were reported. Results reported that mechanical characteristics were observed to be higher with the addition of glass fiber in the wood flour-polypropylene matrix composites. It could be due to the good bonding between the fiber and matrix. Likewise, incorporating the glass fiber reinforced composites outperformed in hardness properties [45].

In another work, the mechanical properties of wood flour-glass fiber reinforced with polyvinyl chloride hybrid composites were studied. The L glass-wood flour fibers were attributed to the significant increase in impact strength of hybrid composites. Furthermore, the incorporation of 5% of L glass fiber in the hybrid composites exhibited higher flexural and impact properties [46]. Massoud et al. [47] studied the flax-glass-PP-based hybrid composites. It was revealed that the addition of glass fiber in the flax-PP composites showed higher tensile, impact, and hardness properties [47].

The mechanical performance of wood/carbon/PP composites was studied [48]. Results reported that the wood fiber reinforced composites exhibited improved performance with the addition of carbon fiber in PP matrix. However, the elongation at break was found to reduce by incorporating the carbon fibers with wood.

The effect of mechanical characteristics of kenaf-carbon fiber reinforced thermoplastic rubber hybrid composites was investigated experimentally. Results observed that the addition of fiber loading possess increase in flexural properties. However, when compared to the hybrid composite, the composite with a single type of reinforcement had better flexural properties [49]. In another work, the mechanical properties of kenaf-glass fiber reinforced thermoplastic rubber hybrid composites were examined by Wan et al. [50]. Results observed that 30 wt% of kenaf fiber and 70 wt% of glass fiber reinforced with thermoplastic rubber hybrid composites possessed improved mechanical properties than the other combinations.

Sivakumar et al. [51] examined the tensile strength of kenaf-kevlar-PP hybrid composites with various fiber configurations. The hybrid combinations exhibited the positive hybridization effect, whereby the Kevlar fiber positioned as a skin layer exhibited better tensile properties. Another study conducted by Lin et al. [52] the effect of PALF-kevlar reinforced polypropylene hybrid composites. The tensile properties of hybrid composites were characterized and it was found that the higher tensile properties were achieved for kevlar-PALF-kevlar stacking of hybrid composites [52].

The mechanical behavior of kenaf-basalt fiber reinforced with UHMWPE (ultra-high molecular weight polyethylene)—HDPE (high density polyethylene) hybrid composites were investigated and they results had improved the mechanical properties. When comparing all type of composites, the pure UHMWPE/HDPE composites possess better results with flexural and impact strength [53]. The effect of fiber length and chemical treatment of sisal-glass fiber reinforced low density polyethylene composites were examined by kalaprasad et al. [54]. Results are evident that the enhanced tensile properties of all chemical treated sisal-glass hybrid composites. However, the benzoyl peroxide treated hybrid composites exhibited better tensile properties [54].

In the study of hybrid composites with bast-basalt reinforced PP matrix were analyzed. The bast/basalt hybrid composites have a greater mechanical property than the bast/GF hybrid composites, according to a comparison of their mechanical properties [55]. Anshida et al. [56] investigated the mechanical performance of banana-glass fiber reinforced polystyrene hybrid composites and they have reported an increase in the weight percentage of glass fiber based on total fiber content enhances all mechanical properties [56].

The mechanical properties of hemp-glass-PP hybrid composites were produced by injection molding method and their mechanical properties were analyzed. The addition of glass fibers to short hemp fiber composites improves mechanical properties. Furthermore, the hybrid composite with 15% short glass fiber content had greater flexural strength and modulus [57].

The effect of mechanical properties on the flax/glass PP-based hybrid composites was studied. The 30 wt% of glass fiber in the flax/glass hybrid composites demonstrated higher mechanical strength. On the other hand, the addition of flax fiber in the hybrid composites showed reduce the mechanical properties [58]. Anuar et al. [59] explored the mechanical behavior of glass/oil palm empty fruit bunch hybrid fiber reinforced with thermoplastic rubber composites. It was found that the maximum tensile and impact strengths were observed at 10 wt% glass fiber and 10 wt% of oil palm empty fruit bunch fiber content in both treated and untreated hybrid composites [59].

The effect of different stacking sequence on mechanical properties of the glass/flax/carbon/kenaf fiber reinforced epoxy hybrid composites. The results showed that the carbon/flax/carbon stacking type hybrid composite has the highest flexural strength, while the glass/flax/glass stacking type hybrid composite has the highest impact strength. As a result, natural/synthetic fiber reinforced polymer hybrid composites could have uses in automobile structural parts [60]. Mohanavel et al. [61] investigated the influence on different stacking sequence of madar-glass-jute fiber reinforced epoxy composites. According to the studies, the glass fiber mats as the core and skin layers had a greater impact on the mechanical properties of the hybrid composites than other composites [61]. The effect of mechanical properties of carbon-flax fiber reinforced with poly butyl succinate matrix hybrid composites. It was found that the carbon-flax hybrid composites exhibited better tensile, flexural and impact properties compared to other composites [62].

3 Thermal Properties of Natural/Synthetic Fiber Reinforced Polymer Composites

The thermal behavior of (sisal/glass) fiber-reinforced with PP hybrid composites were reported. According to the findings, the addition of glass fiber to the PP composites increased their thermal stability. The temperature of thermal disintegration of the composites increased as the glass fiber concentration increased [63]. Sushanta et al. [64] explored the thermal properties of bamboo-glass fiber reinforced PP hybrid composites. The results indicated that the incorporation of bamboo-glass hybrid composites exhibited better thermal stability. Further the dynamic mechanical analysis of bamboo-glass hybrid composites has improved the storage modulus owing to the higher stiffness of composites [64].

Suhara et al. [57] reported the thermal properties of hemp-glass-PP hybrid composites were investigated. Results found that the addition of glass fibers produced enhanced thermal stability of hemp-glass-PP hybrid composites, according to thermogravimetric analysis. All of these findings suggested that injection-molded hemp-glass-PP hybrid composites had improved performance features and might be used in automobile applications like interior parts of the car [57]. Researchers examined the dynamic mechanical characteristics of oil palm-glass fiber-reinforced phenol formaldehyde hybrid composites. From the results, hybrid composites have a better damping factor and a lower storage modulus compared to pure oil palm fiber-phenol formaldehyde composite [65]. The investigation of the dynamic mechanical analysis (DMA) of curaua and glass hybrid fiber reinforced with polylactic acid found that the maximum storage modulus and loss modulus were obtained with the increase in glass fiber content. All-glass composite has the highest activation energy and glass transition temperature [66].

In another study, the influence of different volume fractions on thermal properties of curaua and glass fibers reinforced with polyester hybrid composites was examined by Jose et al. [67]. The hybridization with curaua-glass fiber reinforcement had better thermal stability. However, 30 wt% curaua fiber replacing the glass fiber has similar properties to the mono glass composite [67]. Sanjay et al. [68] selected bamboo-glass fiber reinforced hybrid composites using PP matrix. Furthermore, the researchers explored the thermal and DMA properties of hybrid composites. According to the findings, the incorporation of bamboo-glass fibers demonstrated good thermal stability due to improving fiber and matrix reinforcement. The DMA results revealed that storage modulus increases in the addition of fibers and PP while hybridization with glass fibers [68].

Glass and sugar palm fibers reinforced with polyurethane (PU) composites were introduced and analyzed [69]. From the DMA results, the maximum storage and loss modulus was obtained by increasing glass fiber content, whereas the lowest damping factor was observed for maximum sugar palm fiber content. According to thermogravimetric analysis (TGA), the amount of residue reduced as the glass fiber content lowered. Moreover, combining glass-sugar palm-PU hybrid composites improve thermal performance for automotive applications [69]. The effect of dynamic mechanical properties of various stacking sequences of pineapple leaf fiber (PALF) and glass fiber in the polyester matrix was investigated experimentally. Dynamic mechanical study revealed that intimately mixed and glass/PALF/glass hybrid composites had the highest storage modulus values compared to PALF/glass/PALF hybrid composites. Although, the 0.26 volume fraction of glass/PALF/glass hybrid composites demonstrated more rigidity at higher temperature [70].

Heitor et al. [34] examined the DMA analysis of glass fiber and sisal fiber reinforced polyester composites with various volume fraction. The results showed that, by increasing the glass fiber percentage, the storage modulus and loss modulus of hybrid composites were increased. This could be due to the higher stiffness of the fiber and the matrix. Further the glass transition temperature was shifted to higher as the overall fiber loading and glass fiber percentage increased [34]. Another study conducted by Anuar et al. [49] sulfuric acid treated kenaf fiber and carbon fiber incorporated thermoplastic rubber hybrid composites. The DMA properties of hybrid composites were characterized, and it was found that the DMA properties of untreated hybrid composites were better than the treated hybrid composites. The treated hybrid composite's glass transition temperature was lower than the untreated hybrid composite's, indicating poor damping capabilities [49].

The thermal properties of hemp and glass fibers reinforced with polypropylene hybrid composites were studied. Results are evident that the hybridization with glass fibers increased the thermal characteristics of hemp fiber composites. These findings suggested that hemp fiber and glass fiber reinforced with polypropylene hybrid composites had improved thermal performance characteristics and might be used in automotive structural applications [57]. In the study of polypropylene composites fabricated using twin screw extruder, the banana and glass fibers were categorized based on the ratio of 15:15. Further, the treated banana fiber and glass fiber hybrid composites thermal characteristics were analyzed. In DMA results, the highest storage modulus value was obtained for treated banana fiber with glass fiber hybrid composites. Moreover, the thermal studies using differential scanning calorimetry (DSC) and TGA analysis revealed decreasing thermal stability and crystallization temperature with the addition of maleic anhydride grafted polypropylene treated banana and glass fiber hybrid composites [71].

Daiane et al. [72] investigated the dynamic mechanical analysis of glass-ramie-polyester hybrid composites. Composites containing 75 wt% glass fiber have a higher activation energy. They also reported that the 75:25 (glass-ramie) had better storage modulus, loss modulus and tan delta of the hybrid composites [72]. Aisyah et al. [73] examined the thermal characteristics of kenaf-carbon fibers reinforced with epoxy composites. Hybrid composite with more woven kenaf fiber content was found to have improved thermal stability, whereas a mono carbon fiber composite had the best thermal stability. According to DSC measurements, the addition of woven kenaf increased the degradation temperature [73]. Table 1 illustrate the mechanical properties of natural/synthetic fiber reinforced thermoset and thermoplastic polymer composites.

4 Automotive Applications of Natural/Synthetic Fiber Reinforced Hybrid Composites

Several studies have investigated the use of natural-synthetic based hybrid composites in automobile parts such as instrument panels, door panels, vehicle spall liner, interior and exterior parts, seat back, brake pads, engine cover, battery tray, motor bike silencer applications [74–79]. Glass fiber reinforced composites are the most extensively used composites in the automotive sector. In hybridization, we could provide an opportunity to infuse a glass fiber in natural fiber partially. In automotive brake lever applications, hybrid kenaf/glass fiber composites have showed better characteristics in comparable to glass fiber composites [80]. Furthermore, when compared to natural fiber composites, natural-glass fiber hybrid composites have shown lower moisture absorption, enhancing their potential in structural applications [5]. Also, a

Table 1 Mechanical properties of natural/synthetic fiber reinforced thermoset and thermoplastic polymer composites

Hybrid composites	Mechanical properties				
Thermoset polymer	Tensile strength (MPa)	Tensile modulus (GPa)	Flexural strength (MPa)	Flexural modulus (GPa)	References
Jute/glass/epoxy	42	2.95	241	13	[17]
Flax/glass/vinyl ester	140	11	240	9	[18]
Kevlar/kenaf/epoxy	140	3	80	7	[22]
Basalt/jute/polyester	130	–	410	4	[23]
Sisal/jute/glass/epoxy	81	–	347	–	[28]
Thermoplastic polymer					
Wood flour/glass/polyvinyl chloride	–	–	45	5	[46]
Carbon/wood/polypropylene	55	2.5	70	4.3	[48]
Carbon/kenaf/thermoplastic rubber	–	–	38	4.5	[49]
Kenaf/Kevlar/polypropylene	85	3	–	–	[51]
Sisal/glass/polyethylene	35	1.4	–	–	[54]

hybrid composite of glass fiber and coconut coir have proven structures in automobile industries [81].

For car bumper beam applications, mechanical characteristics of jute and glass fiber composites were investigated. The hybrid composites demonstrated higher hardness and impact strength, making it a viable replacement for commercial bumper materials [74]. Glass/chicken feather fiber (CFF) based on the reinforced epoxy composite were studied for the development of a prototype bike silencers [75].

Hybrid composites with kenaf and glass fiber reinforced epoxy matrix used the bumper beam material of passenger cars was investigated (Fig. 1). In comparison with GMT bumper and glass mat thermoplastic, glass thermoplastic possesses higher mechanical properties. Further, the effect of impact property is still below the desired level. However, this suggests the kenaf-glass hybrid composites could be used in vehicle bumper beams [77]. Another application is the use of sisal-glass fiber reinforced polypropylene hybrid composites in the fabrication of car interior and exterior parts [78].

In coming years, hybrid composites will play a significant part in the automotive sector owing to their physical and mechanical characteristics, as well as their strength-to-weight ratio. Most of the automotive components were changed into hybrid composite materials to improve the automotive structure and performance. In an automobile industry, the effective reduction of weight in conventional steel components will reduce fuel consumption, hence assisting in pollution control. As a result, process of hybrid materials along with various process will reach more heights [81]. Table 2 gives the various natural/synthetic hybrid composites in automobile industry.

Fig. 1 Hybrid composite components for automobiles [76] (open access)

5 Conclusion

This chapter discusses on mechanical behavior of different natural fiber and synthetic fiber reinforced hybrid composites have been seen. And it attains a better property both the natural and synthetic should be combined. The different structures of automotive applications were highlighted. The general observations were as follows:

- In hybrid composites with two different fibers, load bearing capability, stiffness and flexural rigidity were dependent on the fiber weight% and volume fraction of fiber. The increasing proportion of high strength fiber in fiber weight was found to be beneficial. However, the optimum strength under various mechanical loads, stiffness and rigidity can occur at a particular fiber weight proportion or at different proportions for each type of load.
- Stacking sequence of the fiber layers in the composite laminate can significantly influence the mechanical properties. Composites with the high strength fiber in the outer layer was found to have better tensile properties as it can withstand more tensile load before failure. The tensile failure initially starts with matrix crack and further propagate into the outer fabric layer to the core fabric layers. Thus, the composite with high strength fiber on the outer layer possess superior tensile properties. In case of the impact load, failure occurs by fiber pull-out, fiber-matrix de-bonding and delamination, thus, the impact resistance depends on the fiber stacking sequence as well as the fiber pre-treatment due to their influence

Table 2 Hybrid composites used in automobile application

Hybrid composites	Automotive applications	References
Kenaf/glass fibers reinforced polypropylene (PP) composites	Parking brake lever component	[80]
Hybridization of cellulose fiber with several inorganic fibers to reinforce a PP matrix	Automotive "under-the-hood" and body interior components	[79]
Banana/glass fiber reinforced with polypropylene (PP) hybrid composites	Body panels such as seat cushions, cabin linings, etc.	[82]
Hybrid composite with coir and carbon fiber reinforced epoxy	Helmet shells in automobile application and sports applications	[83]
Glass/chicken feather fibers/hydroxyapatite reinforced epoxy hybrid composites	Motorbike silencer	[75]
Hybrid composite with sisal and glass fiber reinforced PP	Car interior and exterior parts	[78]
Jute/glass fiber reinforced with epoxy composites	Car bumper beam	[74]
Coir and glass fiber reinforced with epoxy composites	Outdoor structure in automobiles	[81]
Hemp and glass fiber reinforced with PP composites	Automotive interior parts	[57]
Carbon and flax fiber reinforced with epoxy composites	Automotive structural parts	[84]

on the interfacial bonding characteristics between the natural fibers and polymer matrix.

- From the research, it has been evident that natural and synthetic fiber-based hybrid composites possessed better results from the literature. Nowadays the mix of natural fiber along with glass fiber has been increased in automobile applications.

References

1. Holbery J, Houston D (2006) Natural-fiber-reinforced polymer composites in automotive applications. Jom 58(11):80–86. https://doi.org/10.1007/s11837-006-0234-2
2. Sangilimuthukumar J, Kumar TSM, Santulli C, Chandrasekar M, Senthilkumar K, Siengchin S (2020) The use of pineapple fiber composites for automotive applications: a short review. J Mater Sci Res Rev 39–45
3. Chandrasekar M et al (2019) Flax and sugar palm reinforced epoxy composites: effect of hybridization on physical, mechanical, morphological and dynamic mechanical properties. Mater Res Express
4. Thiagamani SMK et al (2019) Investigation into mechanical, absorption and swelling behaviour of hemp/sisal fibre reinforced bioepoxy hybrid composites: effects of stacking sequences. Int J Biol Macromol 140:637–646

5. Saleem A, Medina L, Skrifvars M, Berglin L (2020) Hybrid polymer composites of bio-based bast fibers with glass, carbon and basalt fibers for automotive applications—a review. Molecules 25(21):4933
6. Gupta MK, Srivastava RK (2016) Mechanical properties of hybrid fibers-reinforced polymer composite: a review. Polym Plast Technol Eng 55(6):626–642
7. Chandrasekar M, Siva I, Kumar TSM, Senthilkumar K, Siengchin S, Rajini N (2020) Influence of fibre inter-ply orientation on the mechanical and free vibration properties of banana fibre reinforced polyester composite laminates. J Polym Environ 28(11):2789–2800
8. Senthilkumar K et al (2020) Dual cantilever creep and recovery behavior of sisal/hemp fibre reinforced hybrid biocomposites: effects of layering sequence, accelerated weathering and temperature. J Ind Text 1528083720961416
9. Krishnasamy S et al (2019) Effects of stacking sequences on static, dynamic mechanical and thermal properties of completely biodegradable green epoxy hybrid composites. Mater Res Express 6(10):105351
10. Shahzad A, Nasir SU (2017) Mechanical properties of natural fiber/synthetic fiber reinforced polymer hybrid composites. In: Green biocomposites. Springer, pp 355–396
11. Atiqah A, Chandrasekar M, Kumar TSM, Senthilkumar K, Ansari MNM (2019) Characterization and interface of natural and synthetic hybrid composites
12. Yao S-S, Jin F-L, Rhee KY, Hui D, Park S-J (2018) Recent advances in carbon-fiber-reinforced thermoplastic composites: a review. Compos Part B Eng 142:241–250
13. Alam P, Mamalis D, Robert C, Floreani C, Brádaigh CMÓ (2019) The fatigue of carbon fibre reinforced plastics—a review. Compos Part B Eng 166:555–579
14. Dér A, Kaluza A, Kurle D, Herrmann C, Kara S, Varley R (2018) Life cycle engineering of carbon fibres for lightweight structures. Procedia CIRP 69:43–48
15. Zhang J, Chaisombat K, He S, Wang CH (2012) Hybrid composite laminates reinforced with glass/carbon woven fabrics for lightweight load bearing structures. Mater Des 36:75–80
16. Nasimudeen NA et al (2021) Mechanical, absorption and swelling properties of vinyl ester based natural fibre hybrid composites. Appl Sci Eng Prog 14(4):680–688
17. Acharya SDS, Dalbehera S (2014) Study on mechanical properties of natural fiber reinforced woven jute-glass hybrid epoxy composites. Adv Polym Sci Technol 4(1):1–6
18. Kumar CN, Prabhakar MN, Song J (2019) Effect of interface in hybrid reinforcement of flax/glass on mechanical properties of vinyl ester composites. Polym Test 73:404–411
19. Ramana MV, Ramprasad S (2017) Experimental investigation on jute/carbon fibre reinforced epoxy based hybrid composites. Mater Today Proc 4(8):8654–8664
20. Ramesh M, Deepa C, Arpitha GR, Gopinath V (2019) Effect of hybridization on properties of hemp-carbon fibre-reinforced hybrid polymer composites using experimental and finite element analysis. World J Eng
21. Azrin Hani Abdul R, Roslan A, Jaafar M, Roslan MN, Ariffin S (2011) Mechanical properties evaluation of woven coir and kevlar reinforced epoxy composites. In: Advanced materials research, vol 277, pp 36–42
22. Yahaya R, Sapuan SM, Jawaid M, Leman Z, Zainudin ES (2015) Effect of layering sequence and chemical treatment on the mechanical properties of woven kenaf–aramid hybrid laminated composites. Mater Des 67:173–179
23. Amuthakkannan P, Manikandan V, Jappes JTW, Uthayakumar M (2012) Influence of stacking sequence on mechanical properties of basalt-jute fiber-reinforced polymer hybrid composites. J Polym Eng 32(8–9):547–554
24. Petrucci R, Santulli C, Puglia D, Sarasini F, Torre L, Kenny JM (2013) Mechanical characterisation of hybrid composite laminates based on basalt fibres in combination with flax, hemp and glass fibres manufactured by vacuum infusion. Mater Des 49:728–735
25. Ramnath BV, Manickavasagam VM, Elanchezhian C, Krishna CV, Karthik S, Saravanan K (2014) Determination of mechanical properties of intra-layer abaca–jute–glass fiber reinforced composite. Mater Des 60:643–652
26. Pandey P, Bajwa D, Ulven C, Bajwa S (2016) Influence of hybridizing flax and hemp-agave fibers with glass fiber as reinforcement in a polyurethane composite. Materials (Basel) 9(5):390

27. Yahaya R, Sapuan SM, Jawaid M, Leman Z, Zainudin ES (2016) Effect of fibre orientations on the mechanical properties of kenaf–aramid hybrid composites for spall-liner application. Def Technol 12(1):52–58
28. Ramesh M, Palanikumar K, Reddy KH (2016) Evaluation of mechanical and interfacial properties of sisal/jute/glass hybrid fiber reinforced polymer composites. Trans Indian Inst Met 69(10):1851–1859
29. Zin MH, Abdan K, Norizan MN (2019) The effect of different fiber loading on flexural and thermal properties of banana/pineapple leaf (PALF)/glass hybrid composite. In: Structural health monitoring of biocomposites, fibre-reinforced composites and hybrid composites. Elsevier, pp 1–17
30. Vinayagamoorthy R, Rajeswari N (2014) Mechanical performance studies on Vetiveria zizanioides/Jute/glass fiber-reinforced hybrid polymeric composites. J Reinf Plast Compos 33(1):81–92
31. Amico SC, Angrizani CC, Drummond ML (2010) Influence of the stacking sequence on the mechanical properties of glass/sisal hybrid composites. J Reinf Plast Compos 29(2):179–189
32. Mishra S et al (2003) Studies on mechanical performance of biofibre/glass reinforced polyester hybrid composites. Compos Sci Technol 63(10):1377–1385
33. Abdullah-Al-Kafi, Abedin MZ, Beg MDH, Pickering KL, Khan MA (2006) Study on the mechanical properties of jute/glass fiber-reinforced unsaturated polyester hybrid composites: effect of surface modification by ultraviolet radiation. J Reinf Plast Compos 25(6):575–588
34. Ornaghi HL Jr, Bolner AS, Fiorio R, Zattera AJ, Amico SC (2010) Mechanical and dynamic mechanical analysis of hybrid composites molded by resin transfer molding. J Appl Polym Sci 118(2):887–896
35. Dieu TV, Liem NT, Mai TT, Tung NH (2004) Study on fabrication of BMC laminates based on unsaturated polyester resin reinforced by hybrid bamboo/glass fibers. JSME Int J Ser A Solid Mech Mater Eng 47(4):570–573
36. Ahmed KS, Vijayarangan S (2008) Tensile, flexural and interlaminar shear properties of woven jute and jute-glass fabric reinforced polyester composites. J Mater Process Technol 207(1–3):330–335
37. Venkata Reddy G, Shobha Rani T, Chowdoji Rao K, Venkata Naidu S (2009) Flexural, compressive, and interlaminar shear strength properties of kapok/glass composites. J Reinf Plast Compos 28(14):1665–1677
38. Audibert C, Andreani A-S, Lainé É, Grandidier J-C (2018) Mechanical characterization and damage mechanism of a new flax-Kevlar hybrid/epoxy composite. Compos Struct 195:126–135
39. Zhong LX, Fu SY, Zhou XS, Zhan HY (2011) Effect of surface microfibrillation of sisal fibre on the mechanical properties of sisal/aramid fibre hybrid composites. Compos Part A Appl Sci Manuf 42(3):244–252
40. Bakar NH, Hyie KM, Ramlan AS, Hassan MK, Jumahat A (2014) Mechanical properties of kevlar reinforcement in kenaf composites. Appl Mech Mater 465:847–851
41. Hashim MKR, Abdul Majid MS, Jamir MRM, Kasim FH, Sultan MTH (2021) The effect of stacking sequence and ply orientation on the mechanical properties of pineapple leaf fibre (Palf)/carbon hybrid laminate composites. Polymers (Basel) 13(3):455
42. Sathiyamoorthy M, Senthilkumar S (2020) Mechanical, thermal, and water absorption behaviour of jute/carbon reinforced hybrid composites. Sādhanā 45(1):1–12
43. Venkatasudhahar M, Kishorekumar P, Dilip Raja N (2020) Influence of stacking sequence and fiber treatment on mechanical properties of carbon-jute-banana reinforced epoxy hybrid composites. Int J Polym Anal Charact 25(4):238–251
44. Abd El-baky MA, Attia MA, Abdelhaleem MA, Hassan MA (2020) Mechanical characterization of hybrid composites based on flax, basalt and glass fibers. J Compos Mater 54(27):4185–4205
45. AlMaadeed MA, Kahraman R, Khanam PN, Madi N (2012) Date palm wood flour/glass fibre reinforced hybrid composites of recycled polypropylene: mechanical and thermal properties. Mater Des 42:289–294

46. Jiang H, Pascal Kamdem D, Bezubic B, Ruede P (2003) Mechanical properties of poly (vinyl chloride)/wood flour/glass fiber hybrid composites. J Vinyl Addit Technol 9(3):138–145
47. Ghasemzadeh-Barvarz M, Duchesne C, Rodrigue D (2015) Mechanical, water absorption, and aging properties of polypropylene/flax/glass fiber hybrid composites. J Compos Mater 49(30):3781–3798
48. Kada D, Migneault S, Tabak G, Koubaa A (2016) Physical and mechanical properties of polypropylene-wood-carbon fiber hybrid composites. BioResources 11(1):1393–1406
49. Anuar H, Ahmad SH, Rasid R, Ahmad A, Busu WNW (2008) Mechanical properties and dynamic mechanical analysis of thermoplastic-natural-rubber-reinforced short carbon fiber and kenaf fiber hybrid composites. J Appl Polym Sci 107(6):4043–4052
50. Wan Busu WN, Anuar H, Ahmad SH, Rasid R, Jamal NA (2010) The mechanical and physical properties of thermoplastic natural rubber hybrid composites reinforced with Hibiscus cannabinus, L and short glass fiber. Polym Plast Technol Eng 49(13):1315–1322
51. Malingam SD, Feng NL, Sean NC, Subramaniam K, Razali N, Mustafa Z (2018) Mechanical properties of hybrid kenaf/kevlar fibre reinforced thermoplastic composites. Sci Technol Res Inst Def 11(2):181–209
52. Ng LF, Malingam SD, Subramaniam K, Selamat M, Juan W (2020) The investigation of the tensile and quasi-static indentation properties of pineapple leaf/kevlar fibre reinforced hybrid composites. Def ST Tech Bull 13:117–129
53. Mokhtar I, Yahya MY, Kadir MRA, Kambali MF (2013) Effect on mechanical performance of UHMWPE/HDPE-blend reinforced with kenaf, basalt and hybrid kenaf/basalt fiber. Polym Plast Technol Eng 52(11):1140–1146
54. Kalaprasad G et al (2004) Effect of fibre length and chemical modifications on the tensile properties of intimately mixed short sisal/glass hybrid fibre reinforced low density polyethylene composites. Polym Int 53(11):1624–1638
55. Saleem A, Medina L, Skrifvars M (2020) Influence of fiber coating and polymer modification on mechanical and thermal properties of bast/basalt reinforced polypropylene hybrid composites. J Compos Sci 4(3):119
56. Haneefa A, Bindu P, Aravind I, Thomas S (2008) Studies on tensile and flexural properties of short banana/glass hybrid fiber reinforced polystyrene composites. J Compos Mater 42(15):1471–1489
57. Panthapulakkal S, Sain M (2007) Injection-molded short hemp fiber/glass fiber-reinforced polypropylene hybrid composites—mechanical, water absorption and thermal properties. J Appl Polym Sci 103(4):2432–2441
58. Arbelaiz A, Fernández B, Cantero G, Llano-Ponte R, Valea A, Mondragon I (2005) Mechanical properties of flax fibre/polypropylene composites. Influence of fibre/matrix modification and glass fibre hybridization. Compos Part A Appl Sci Manuf 36(12):1637–1644
59. Anuar H, Ahmad SH, Rasid R, Daud NSN (2006) Tensile and impact properties of thermoplastic natural rubber reinforced short glass fiber and empty fruit bunch hybrid composites. Polym Plast Technol Eng 45(9):1059–1063
60. Saroj S, Nayak S, Jesthi DK (2022) Effect of hybridization of carbon/glass/flax/kenaf fibre composite on flexural and impact properties. Mater Today Proc 49:502–506
61. Mohanavel V, Suresh Kumar S, Vairamuthu J, Ganeshan P, NagarajaGanesh B (2021) Influence of stacking sequence and fiber content on the mechanical properties of natural and synthetic fibers reinforced penta-layered hybrid composites. J Nat Fibers 1–13
62. Bahrami M, Enciso B, Gaifami CM, Abenojar J, Martinez MA (2021) Characterization of hybrid biocomposite poly-butyl-succinate/carbon fibers/flax fibers. Compos Part B Eng 221:109033
63. Jarukumjorn K, Suppakarn N (2009) Effect of glass fiber hybridization on properties of sisal fiber–polypropylene composites. Compos Part B Eng 40(7):623–627
64. Samal SK, Mohanty S, Nayak SK (2009) Polypropylene—bamboo/glass fiber hybrid composites: fabrication and analysis of mechanical, morphological, thermal, and dynamic mechanical behavior. J Reinf Plast Compos 28(22):2729–2747

65. Sreekala MS, Thomas S, Groeninckx G (2005) Dynamic mechanical properties of oil palm fiber/phenol formaldehyde and oil palm fiber/glass hybrid phenol formaldehyde composites. Polym Compos 26(3):388–400
66. Ornaghi HL Jr, da Silva HSP, Zattera AJ, Amico SC (2011) Hybridization effect on the mechanical and dynamic mechanical properties of curaua composites. Mater Sci Eng A 528(24):7285–7289
67. Júnior JHSA, Júnior HLO, Amico SC, Amado FDR (2012) Study of hybrid intralaminate curaua/glass composites. Mater Des 42:111–117
68. Nayak SK, Mohanty S, Samal SK (2009) Influence of short bamboo/glass fiber on the thermal, dynamic mechanical and rheological properties of polypropylene hybrid composites. Mater Sci Eng A 523(1–2):32–38
69. Atiqah A, Jawaid M, Sapuan SM, Ishak MR, Alothman OY (2018) Thermal properties of sugar palm/glass fiber reinforced thermoplastic polyurethane hybrid composites. Compos Struct 202:954–958
70. Devi LU, Bhagawan SS, Thomas S (2010) Dynamic mechanical analysis of pineapple leaf/glass hybrid fiber reinforced polyester composites. Polym Compos 31(6):956–965
71. Samal SK, Mohanty S, Nayak SK (2009) Banana/glass fiber-reinforced polypropylene hybrid composites: fabrication and performance evaluation. Polym Plast Technol Eng 48(4):397–414
72. Romanzini D, Ornaghi HL Jr, Amico SC, Zattera AJ (2012) Influence of fiber hybridization on the dynamic mechanical properties of glass/ramie fiber-reinforced polyester composites. J Reinf Plast Compos 31(23):1652–1661
73. Aisyah HA et al (2019) Thermal properties of woven kenaf/carbon fibre-reinforced epoxy hybrid composite panels. Int J Polym Sci 2019
74. Olorunnishola AAG, Adubi EG (2018) A comparative analysis of a blend of natural jute and glass fibers with synthetic glass fibers composites as car bumper materials. IOSR J Mech Civ Eng 15(3):67–71
75. Nguyen DD et al (2020) Fabrication and modeling of prototype bike silencer using hybrid glass and chicken feather fiber/hydroxyapatite reinforced epoxy composites. Prog Org Coat 148:105871
76. Mohammed L, Ansari MNM, Pua G, Jawaid M, Islam MS (2015) A review on natural fiber reinforced polymer composite and its applications. Int J Polym Sci 2015. https://doi.org/10.1155/2015/243947
77. Davoodi MM, Sapuan SM, Ahmad D, Ali A, Khalina A, Jonoobi M (2010) Mechanical properties of hybrid kenaf/glass reinforced epoxy composite for passenger car bumper beam. Mater Des 31(10):4927–4932. https://doi.org/10.1016/j.matdes.2010.05.021
78. Birat KC, Panthapulakkal S, Kronka A, Agnelli JAM, Tjong J, Sain M (2015) Hybrid biocomposites with enhanced thermal and mechanical properties for structural applications. J Appl Polym Sci 132(34)
79. Reale Batista MD, Drzal LT, Kiziltas A, Mielewski D (2020) Hybrid cellulose-inorganic reinforcement polypropylene composites: lightweight materials for automotive applications. Polym Compos 41(3):1074–1089
80. Mansor MR, Sapuan SM, Zainudin ES, Nuraini AA, Hambali A (2013) Hybrid natural and glass fibers reinforced polymer composites material selection using analytical hierarchy process for automotive brake lever design. Mater Des 51:484–492
81. Ravishankar B, Nayak SK, Kader MA (2019) Hybrid composites for automotive applications—a review. J Reinf Plast Compos 38(18):835–845
82. Zin MH, Razzi MF, Othman S, Liew K, Abdan K, Mazlan N (2016) A review on the fabrication method of bio-sourced hybrid composites for aerospace and automotive applications. In: IOP conference series: materials science and engineering, vol 152, no 1, p 12041

83. Singh Y, Singh J, Sharma S, Lam T-D, Nguyen D-N (2020) Fabrication and characterization of coir/carbon-fiber reinforced epoxy based hybrid composite for helmet shells and sports-good applications: influence of fiber surface modifications on the mechanical, thermal and morphological properties. J Mater Res Technol 9(6):15593–15603
84. Fiore V, Valenza A, Di Bella G (2012) Mechanical behavior of carbon/flax hybrid composites for structural applications. J Compos Mater 46(17):2089–2096

Investigation of Natural/Synthetic Hybrid Composite for Marine Application

Mohammad Azad Alam, H. H. Ya, Mohammad Azeem, Faisal Masood, Tauseef Ahmad, S. M. Sapuan, Rehan Khan, and Mohammad Yusuf

Abstract Watercraft, submersibles, offshore structures, and other marine structural components are exposed to relevant environmental challenges. Therefore, materials exhibiting elevated resistance and requiring few or no maintenance for extended periods of time are generally under consideration. To date, composites are employed in all areas of the marine sector and for a variety of components and structures, namely hulls, bearings, propellers, hatch covers, exhausts, topside structures, radomes, sonar domes, railings, vessels of all types, valves and other subsea structures. The advantages of composite materials for marine structures, notably their high specific properties, resistance to degradation in water, and flexible fabrication to produce special shapes, become even more valuable when underwater applications are considered. This chapter presents the various natural and synthetic reinforcements utilized to produce composite materials, the mechanical and thermal properties of synthetic hybrid composites and their different processing techniques to produce the composite materials are discussed in detail.

M. A. Alam (✉) · H. H. Ya · M. Azeem · T. Ahmad
Department of Mechanical Engineering, Universiti Teknologi PETRONAS, Seri Iskandar, Perak, Malaysia
e-mail: azadalam@zhcet.ac.in

F. Masood
Department of Electrical Engineering, University of Engineering and Technology Taxila, Rawalpindi, Pakistan

S. M. Sapuan
Laboratory of Biocomposite Technology, Institute of Tropical Forestry and Forest Products, Universiti Putra Malaysia, UPM, 43400 Serdang, Selangor, Malaysia

Department of Mechanical and Manufacturing Engineering, Universiti Putra Malaysia, UPM, 43400 Serdang, Selangor, Malaysia

R. Khan
College of Electrical and Mechanical Engineering, National University of Sciences and Technology, Rawalpindi, Pakistan

M. Yusuf
Department of Petroleum Engineering, Universiti Teknologi PETRONAS, Seri Iskandar, Malaysia

T. Khan and M. Jawaid (eds.), *Green Hybrid Composite in Engineering and Non-Engineering Applications*, Composites Science and Technology,
https://doi.org/10.1007/978-981-99-1583-5_12

Keywords Natural fiber · Synthetic fiber · Composites · Marine applications

1 Introduction

The traditional material for the construction of ship and submarine hulls is steel, with aluminum having been used for particularly weight sensitive applications. The technology of composite materials in the marine industry, in the form of glass-reinforced plastic, is now mature, and the time is ripe to develop the use of the next generation of composite materials to improve the cost-effectiveness of marine structures. Structural problems in surface ships, submarines and some other marine structures are reviewed in this light and new applications are proposed. Polymeric composite materials have been used in boats, ships, submersibles, offshore structures and other marine structural applications for about 50 years. Considerable progress has been made in this period on understanding the behavior of these materials and the tailored structures under mechanical, thermal and fire-induced load scenarios. Processing and production considerations too have received much attention leading to a capability of constructing quite complex, multi-material, large, three-dimensional assemblies capable of sustaining extreme loads. Nevertheless, there is still an air of conservatism and even hesitation in specifying polymer composite-based solutions for several applications. This is owed to doubts about new ways to use existing materials or to use new and existing materials in new applications. This implies the need for further work for enhanced application and use of composite materials in marine applications.

2 Natural Fibers/Synthetic Fibres

The use of natural fibers is growing due to their ecological and economic benefits, however there are some limitations regarding their properties. On the other hand, synthetic fibers possess remarkable properties but are not environmentally friendly. Therefore, combining the two can have advantageous effect on the properties of the resultant composites. The properties of natural and synthetic fibers that are mostly utilized as reinforcing materials in composites are mentioned in Table 1.

Table 1 Properties of natural and synthetic fibres

Material	Density (g/cm^3)	Fiber diameter (μm)	Youngs modulus (GPa)	Tensile strength (GPa)	References
Sisal	1.46	50–300	9–20	2.27–4.00	[4]
Kenaf	1.4	81	4.30	2.50	
Flax	1.54	–	0.3–20	28.85	[5]
Bamboo	1.32	–	11–17	140–230	
Coir	1.15	100–460	4.00–6.00	1.08–2.52	[6]
Banana	1.35	80–250	5.29–7.59	8.2	
S-Glass	2.6	10	90	4.60	[7]
E-Glass	2.5	10	70	1.5–2.0	
Carbon A	1.9	7–10	220	3.20	
Carbon HM	1.8	7–10	400	2.0–2.8	
Kevlar 49	1.45	12	130	3.60	[7]

3 Types of Natural Fibers

Sisal fibres

Sisal is a type of agave grown primarily in East Africa and Brazil. Synthetic substitutes and harvesting systems using no or less twine have proceeded to undercut the traditional fibre industry. Using various polymer matrices, sisal fibre has been employed to manufacture variety of structural and non-structural industrial goods. The researchers also experimented with sisal fibres. It has been discovered that altering the orientation of sisal fibres within the specimen has an effect on the nature of the composite [1].

Coir fiber

Coir is a lignocellulosic natural fibre derived from the coconut's outer shell or husk. Coir has the potential to be used as a reinforcing material in the manufacture of low load-bearing thermoplastic composites [2]. Coconut cultivation is concentrated in Asia's and East Africa's tropical belts. Coconut fibres are classified into two types: brown fibre and white fibres obtained from mature and immature coconuts respectively. Coconut fibres are tough and stiff, with a low thermal conductivity. These are commercially available in three varieties: long fibres (bristle), relatively short fibres (mattress), and mixed fibers (decorticated). Depending on the application, these various fibres can be used in a variety of ways. Brown fibres are commonly used in engineering.

Kenaf fibre

Kenaf is one of the most commonly used natural fibres for reinforcement. It is a herbaceous plant that can be produced in a wide range of climatic conditions and environments reaching heights of more than 3 m in 3 months, even in temperate climates [3].

Flax fibre

Flax fibre is a type of bast fibre that is extracted through scutching and retting operations. These cellulose-based fibres have high aspect ratio, low density, good tensile, and stiffness [8]. The mechanical property of flax fibre varies based on its located in the stem.

Banana fibre

Banana fibre is a bast fibre that is a byproduct of banana cultivation. It is a lingocellulosic fibre obtained from the pseudo-stem of the banana plant with improved mechanical properties [9]. The volume fraction of short banana fibre has proved to have significant impact on the dynamic mechanical properties of polymer composites [10].

4 Types of Synthetic Fibers

Kevlar 29 fiber

Kevlar is a man-made fibre that is classified as an organic fibre in the aromatic polyamide family that was developed for high-strength industrial and innovative-technology applications due to their high toughness, modulus, strength and thermal stability. Presently, different types of Kevlar are manufactured to meet a wide demand for high strength fibre in several applications [11].

Glass fibers

Glass fibres are the most commonly used reinforcements in large composite structures. The performances and properties achieved depends on their size, filament diameter, and chemistry. The physical and mechanical properties of different types of glass fibres are mentioned in Table. Among these E-glass fibre is mostly preferred for marine applications due to their higher tensile strength (2200 MPa) and ultimate tensile strain (2.5%), as well as excellent electrical insulation properties and resistance to chemicals and moisture [11].

Other types that are referred as high-strength glass fibres are R-glass in Europe, T-glass in Japan and S-glass in the United States. S glass contains more Al_2O_3, SiO_2, and MgO than E glass and is usually 40–70% stronger. When the temperature rises from 70 °F (room temperature) to 1000 °F (approx. 540 °C), both E-glass and S-glass lose up to half of their tensile strength, even though both fibre types retain good resistance [12].

Carbon fibers

Carbon fibres are stronger and stiffer than glass. Depending upon the type their properties vary, as mentioned in the Table 1. Due to their higher cost as compared to glass fibres, structures made of entirely carbon fibre are unaffordable, thus and ship designers prefer to use hybrid laminates of carbon and glass fibres [13, 14].

5 Properties of Various Natural Fibers/Synthetic Fibres Hybrid Composites

5.1 Mechanical Properties

Tensile Properties

The tensile properties of epoxy hybrid composites made with three different combinations of fibres including jute, sisal and glass with a fiber-to-matrix weight ratio of 30:70 was evaluated. In comparison to other combinations, sisal/glass/epoxy hybrid composite had the highest tensile and impact strength [15]. By using hand lay-up technique, the jute/glass fibre reinforced polyester composites with varying fibre lengths were fabricated. The fibre lengths greatly influenced the tensile strength of composites [16].

The kenaf/glass fibre reinforced unsaturated polyester hybrid composites were developed by Sheet moulding compound process. The tensile strength and modulus of the treated kenaf-hybrid composite was higher than untreated hybrid composites as shown in Fig. 1. This can be attributed to the better bonding between fibre and matrix in case of treated kenaf fibers [17]. The jute fibre/Kevlar reinforced epoxy composites were fabricated, and their properties were evaluated. The results demonstrated that the properties of hybrid composites were significantly enhanced by combining kevlar in jute fibre [18]. The hybridization of flax fiber/glass fibres reinforced composites were developed using the compression moulding technique concluded that the addition of of glass fibre increased the young's modulus and tensile strength of laminates of flax fibre due to glass fibre being stiffer than flax fibre [19]. Furthermore, the banana/flax and glass fibre reinforced polymer (GFRP) composites fabricated by hand layup method, exhibited higher ultimate tensile strength (39 N/mm^2) as compared to banana-GFRP (30 N/mm^2) and flax-GFRP(32 N/mm^2) composites [20].

Flexural Properties

The effect of glass fabric loading on the flexural strength of silk/glass hybrid epoxy composites revealed 35.52%. increase in the flexural strength of the composite at

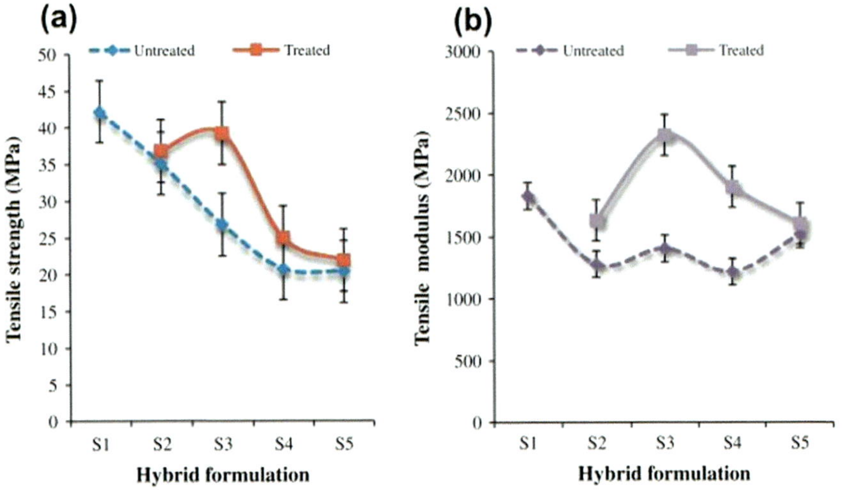

Fig. 1 Tensile properties of Kenaf/glass fibre epoxy hybrid composite: **a** tensile strength and **b** tensile modulus [17]

10 wt% glass fiber. In addition, 50 wt% of the fiber content in the composites increased the strength about 46.89%. The flexural modulus of hybrid composites also increased as the percentage of glass fabric in the composites increases thereby increasing the load bearing efficiency of the composites [21]. The flexural strength of unsaturated polyester hybrid composites reinforced with sisal and glass fibres were investigated with varying fibre content. The hybrid composite exhibited greater flexural strength than that of the sisal fibre reinforced composite although less than that of the glass fibre reinforced composite [22]. The studies on flexural properties of oil palm empty fruit bunch/glass fiber reinforced polypropylene composites revealed the decrease in flexural strength with the addition of both fibers. In contrast, flexural modulus increased with increasing fiber loading. Additionally, maleated polypropylene and 3-(trimethoxysilyl)-propymethacrylate as coupling agents improved flexural properties of the hybrid composites [23].

The reinforcement of the polypropylene composites with sisal/glass fibres improved the flexural strength of the composites due to stiffer and stronger characteristics of the glass fibers, while having no effect on the flexural module. [11]. The effect of carbon fabric layer on the mechanical properties of flax fabric/epoxy composites developed using a vacuum bagging process was investigated. There was an improvement of 221.7 and 110% in flexural modulus and strength of hybrid structures, respectively as compared to flax fiber reinforced composites. Furthermore, the hybrid structure exhibited better flexural (average values of young's modulus and failure stress approximately 23.8 GPa, and 160.4 MPa respectively) (Table 2) [24].

Table 2 Natural fiber/synthetic fiber reinforced composites

Matrix	Reinforcement (natural fiber/synthetic fiber)	Technique	Study	References
Epoxy resin	Flax fiber/basalt fiber	Vacuum bagging	– Water absorption – Dynamic mechanical properties – Impact strength	[25]
Polyester resin	Oil palm EFB/glass fiber	Hand layup	– Water absorption – Flexural properties	[26]
Epoxy resin	Flax fiber/carbon fiber	Vacuum assisted resin transfer molding	– Flexural properties – Tensile properties – Impact properties	[28]
Epoxy resin	Kenaf fiber/Kevlar fiber	Hand layup and hydraulic cold press	– Morphology – Tensile properties – Compressive strength	[29]
Natural rubber	Kenaf fiber/E-glass	Hot pressing	– Impact strength – Flexural properties	[30]
Epoxy resin	Jute fiber/glass fiber	Hand layup	– Tribological properties	[31]
Epoxy resin	Sisal fiber/glass fiber	Hot pressing	– Tensile strength – Impact strength	[32]
Polypropylene	Bamboo fiber/glass fiber	Compression moulding	– Thermal properties – Impact energy – Dynamic impact properties	[33]
Unsaturated polyester	Kenaf-glass	Sheet moulding compound process	– Flexural properties – Impact strength – Tensile properties	[17]
Epoxy	Jute/E-glass fibre	Pultrusion technique	– Flexural properties – Tensile properties – Water absorption	[34]
Unsaturated polyester	Banana/glass fibre	Hand layup and compression moulding	– Flexural properties – Tensile properties	[35]

Impact Strength

The impact strength of ramie/glass fibre reinforced polyester hybrid composites fabricated by using resin transfer moulding was evaluated as a function of fibre content. Higher fibre content improved the impact strength of the composites as a result of high interfacial strength between the fibre and the matrix [36]. Panthapulakkal and Sain [37] observed that the addition of hemp fiber (40 wt%) increased the impact strength by 86%, further the impact strength increased by 35% by the addition of glass fibers. This can be attributed to the better resistance offered by the glass fibers in the composites.

The evaluation of impact behaviour of the banana/E-glass fabrics reinforced polyester hybrid composites revealed that their impact strength was 5 times than that of banana reinforced polyester composites [38]. Živković [39] investigated the effect of moisture absorption on the impact properties of basalt, flax, and flax/basalt hybrids. The moisture absorption affects the impact behaviour of flax and basalt fibres; however, there was no significant change in the impact performance exhibited by flax-basalt hybrid composite in dry or wet conditions.

Water absorption

Natural fiber-synthetic composite of kenaf-Kevlar and jute-Kevlar laminated with epoxy resin were immersed in water to investigate the effect of water absorption on their mechanical properties containing approximately 30 wt% of fibre content. The tensile properties of the humid composite samples reduced as compared to the dry sample and with increase in fibre content As the weight the percentage of moisture uptake increased as a result of their higher high cellulose content [40].

Also, the influence of moisture absorption on the mechanical properties of kenaf/Kevlar hybrid composites concluded that the water ageing had a significant effect on their tensile strength but negligible effect on their impact strength [41].

The moisture absorption capacity of mono-hemp based propylene composites was found to be higher and was reduced significantly by 40% on addition of the glass fiber in hemp/glass fiber based fiber composites [42]. The flax/glass fiber reinforced epoxy hybrid composite revealed that the hydrophilic/hydrophobic behavior of the composites influenced their water uptake and wettability indicating that the glass fibers play a crucial role in promoting water absorption stability of the composites [13].

5.2 *Thermal Properties*

The thermal properties of short hemp-glass fibre reinforced propylene composites developed by injection moulding. The study revealed that the hybrid composites with 25% hemp and 15% glass-fibre have better thermal properties than mono hemp propylene composites [43]. The thermal properties of palmyra fiber (raw and alkali treated) and glass fibers reinforced epoxy composites prepared using hand layup technique were evaluated. The thermal stability and the glass transition temperature of the composite specimens were better than the untreated ones, though the addition of natural fibers reduced the thermal conductivity [44].

6 Different Processing Techniques to Fabricate Composites

6.1 Hand Layup

The matrix material reinforced with layers of aligned fibres makes up high-performance composites. These fibres provide excellent structural properties to composites but are complex to manufacture because they are built up layer by layer. Hand layup is a manufacturing process that involves pouring resin on reinforcement material. The resin is forced into the reinforcement by using roller as shown in Fig. 2. By applying roller over the resin ensures enhancement of interaction between the successive layers of the reinforcement and resin material [45].

This technique was employed to fabricate hybrid composites of Glass Fiber Reinforced Plastic (GFRP) with sisal fibre. The top and bottom layers of the composite were made of GFRP, while the intermediate layers were made of glass and sisal fibres. Three types of composites were developed in this study. The fibres in the second layer in the first category were orthogonal to the top and bottom layers, the second category fibers were placed parallel to each other, while the fibres in the third category were placed at a 45-degree angle to the end layers [47]. This technique was also used to evaluate the penetration energy and energy absorption capacities of kenaf/Kevlar fibre reinforced epoxy composites [48]. The randomly oriented unsaturated polyester-based sisal/carbon fibres with were also developed by this technique and the fluctuation in the mechanical properties due to treated fibers were studied. Alkali treatment improved the tensile and flexural properties of the hybrid composites significantly [49].

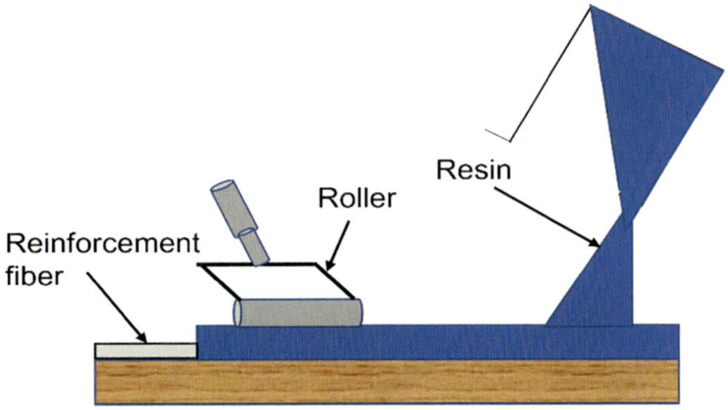

Fig. 2 Schematic diagram of hand layup technique (adapted and modified from [46])

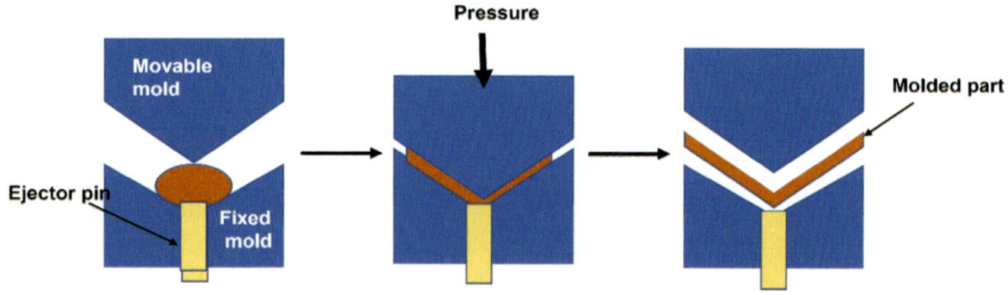

Fig. 3 Schematic diagram of compression moulding (adapted and modified from [46])

6.2 Compression Molding

Compression moulding can be used to create hybrid composites in two ways: hot press or autoclave. In this technique the compounded materials are placed in vacuum atmosphere created inside the pressure chamber at a temperature above the softening point of the polymer matrix in order to ensure adequate flow of the fibres and matrix. A composite is formed after several heat–pressure cycles. In case of autoclave, mould should be closed on the other hand in hot pressing technique a mould may or may not be closed. It is critical to ensure that the fibres do not break under a specific load in the hot press. Bulk moulding compounds (BMCs) and Sheet moulding compounds (SMCs) are the two most common initial charges in the compression moulding process. In comparison to compression moulding, the injection molding moves the material through a screw and a hopper (Fig. 3).

Injection-molded hemp/glass fibre hybrid polypropylene composite performance exhibited improved performance properties. A hybrid composite with flexural strength of 101 MPa and a flexural modulus of 5.5 GPa were acquired. The impact strength of hybrid composites improved significantly (34%). In addition, the glass fibres enhanced the thermal stability of hemp fibre PP composites [37].

6.3 Vacuum Bagging

Vacuum bagging or Vacuum bag molding is a very flexible process for combining laminates mostly made of fibre-reinforced polymers in wider ranges of sizes and shapes. Laminates are first made by means of hand lay-up technique followed by placing them between mold and vacuum bag. The air preset between vacuum bag and mold is evacuated by the vacuum pump. The consolidation process takes place under atmospheric pressure leading to compressing the part. This technique was employed to develop flax fiber/carbon fiber reinforced epoxy resin composites. The evaluation of mechanical properties of the two hybrid configurations (flax fiber-carbon fiber-flax fiber (FCF) and carbon fiber-flax fiber- carbon fiber (CFC)) concluded that better mechanical and impact absorption performances were exhibited by FCF specimens as compared to CFC ones. Post-impact flexural tests indicated that hybridization can improve the damage tolerance as compared to carbon fibres reinforced configurations while retaining acceptable stiffness and residual flexural strength [27].

6.4 Resin Transfer Molding

Resin Transfer Molding (RTM) process is one of the popular manufacturing techniques due to their cost-effectiveness and high-volume production capability. In the RTM process, either a dry fibre preforms (impregnating) or a porous fibrous preform is positioned in the mould cavity. Two identical mould halves are tightly held together to prevent resin leakage during the process. Then, using allocating equipment, a pressurised molten plastic is inserted into the heated mould through single or multiple inlet ports until the complete mould is filled. After cooling the part is removed from the mould. Post-curing is usually required to ensure that the resin is completely cured.

Vacuum infusion, also known as vacuum assisted resin transfer moulding (VARTM), is a relatively new technology in which preform fibres are placed on a mold and a perforated tube is placed between the vacuum bag and the resin container. As shown in Fig. 4, vacuum force causes the resin to be sucked through the perforated tubes over the fibres to consolidate the laminate structure. This process eliminates the need for excess air in the composite structure, making it popular for the production of large objects such as boat hulls and wind turbine blades [48]. The flax/glass fibre reinforced vinyl ester laminates were fabricated using this technique. The low-velocity impact behaviour and the moisture absorption influence on the hybrid laminates were investigated. The hybrid composite exhibited higher load and energy and lower moisture uptake as compared to flax/vinyl ester composite [50].

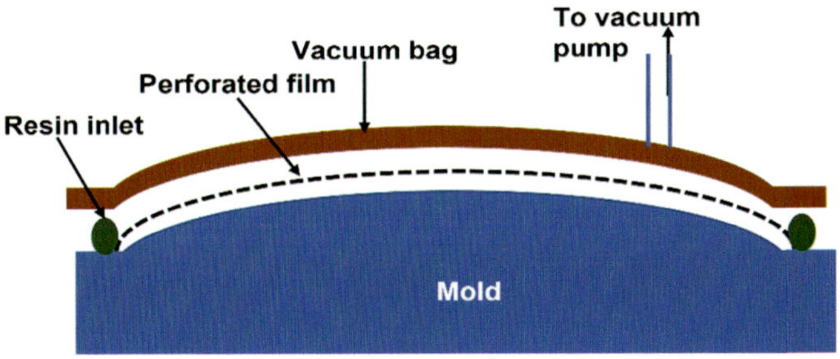

Fig. 4 Schematic diagram of vacuum infusion process (adapted and modified from [46])

7 Summary

Natural fibres are sources of renewable natural materials. These fibres can be used to make lightweight and strong composites. Composites reinforced with natural fibres allows to produce sustainable, economic and environmentally friendly technology. There are some drawbacks associated with natural fibres reinforced composites such as low mechanical properties. To address this, the addition of synthetic fibres to natural fibre reinforced composites has proven to be a viable strategy for improving structural performance of such composites, making them suitable for the development of structures in marine applications. To improve their performance further, the processing of these hybrid materials requires better understanding particularly factors affecting their microstructure, mechanical, thermal and water absorption capacities. Further by developing advanced manufacturing techniques and replacing the traditional ones. There is a critical need for the development of a flexible and sustainable manufacturing technology.

References

1. Senthilkumar K, Saba N, Rajini N et al (2018) Mechanical properties evaluation of sisal fibre reinforced polymer composites: a review. Constr Build Mater 174:713–729. https://doi.org/10.1016/j.conbuildmat.2018.04.143
2. Harish S, Michael DP, Bensely A et al (2009) Mechanical property evaluation of natural fiber coir composite. Mater Char 60:44–49. https://doi.org/10.1016/j.matchar.2008.07.001
3. Nishino T, Hirao K, Kotera M et al (2003) Kenaf reinforced biodegradable composite. Compos Sci Technol 63:1281–1286. https://doi.org/10.1016/S0266-3538(03)00099-X
4. Malec K, Gasiński M, Kuchta K, Kozok A (2013) Unsuccessful suicidal attempt with use of self-prepared bullet—case report. Przegląd Lek 70:754–756
5. Shahinur S, Hasan M (2020) Natural fiber and synthetic fiber composites: comparison of properties, performance, cost and environmental benefits. Elsevier Ltd.
6. Ho MP, Wang H, Lee JH et al (2012) Critical factors on manufacturing processes of natural fibre composites. Compos Part B Eng 43:3549–3562. https://doi.org/10.1016/j.compositesb.2011.10.001
7. Chalmers DW (1991) Experience in design and production of FRP marine structures. Mar Struct 4:93–115. https://doi.org/10.1016/0951-8339(91)90015-4
8. Charlet K, Eve S, Jernot JP et al (2009) Tensile deformation of a flax fiber. Procedia Eng 1:233–236. https://doi.org/10.1016/j.proeng.2009.06.055
9. Mukhopadhyay S, Fangueiro R, Arpaç Y, Şentürk Ü (2008) Banana fibers—variability and fracture behaviour. J Eng Fiber Fabr 3:155892500800300. https://doi.org/10.1177/155892500800300207
10. Reed KE (1980) Dynamic mechanical analysis of fiber reinforced composites. Polym Compos 1:44–49. https://doi.org/10.1002/pc.750010109
11. Jarukumjorn K, Suppakarn N (2009) Effect of glass fiber hybridization on properties of sisal fiber-polypropylene composites. Compos Part B Eng 40:623–627. https://doi.org/10.1016/j.compositesb.2009.04.007
12. Stickel JM, Nagarajan M (2012) Glass fiber-reinforced composites: from formulation to application. Int J Appl Glas Sci 3:122–136. https://doi.org/10.1111/j.2041-1294.2012.00090.x

13. Calabrese L, Fiore V, Scalici T, Valenza A (2019) Experimental assessment of the improved properties during aging of flax/glass hybrid composite laminates for marine applications. J Appl Polym Sci 136:1–12. https://doi.org/10.1002/app.47203
14. Mouritz AP, Gellert E, Burchill P, Challis K (2001) Review of advanced composite structures for naval ships and submarines. Compos Struct 53:21–42. https://doi.org/10.1016/S0263-8223(00)00175-6
15. Sathish S, Ganapathy T, Bhoopathy T (2014) Experimental testing on hybrid composite materials. Appl Mech Mater 592–594:339–343. https://doi.org/10.4028/www.scientific.net/AMM.592-594.339
16. Mandal A, Nigam MK (2018) E-R E-R E-R E-R 1:46–54
17. Atiqah A, Maleque MA, Jawaid M, Iqbal M (2014) Development of kenaf-glass reinforced unsaturated polyester hybrid composite for structural applications. Compos Part B Eng 56:68–73. https://doi.org/10.1016/j.compositesb.2013.08.019
18. Patil SS, Praveen BA, Kempaiah UN, Adarsha H (2017) Fabrication and characterization of Kevlar/jute reinforced epoxy. Int Res J Eng Technol 4:1441–1444
19. Saidane EH, Scida D, Assarar M et al (2016) Hybridisation effect on diffusion kinetic and tensile mechanical behaviour of epoxy based flax-glass composites. Compos Part A Appl Sci Manuf 87:153–160. https://doi.org/10.1016/j.compositesa.2016.04.023
20. Srinivasan VS, Rajendra Boopathy S, Sangeetha D, Vijaya Ramnath B (2014) Evaluation of mechanical and thermal properties of banana-flax based natural fibre composite. Mater Des 60:620–627. https://doi.org/10.1016/j.matdes.2014.03.014
21. Priya SP, Rai SK (2006) Mechanical performance of biofiber/glass-reinforced epoxy hybrid composites. J Ind Text 35:217–226. https://doi.org/10.1177/1528083706055754
22. Road B, Pradesh A (2011) Tensile and flexural properties of sisal/glass fibre reinforced hybrid composites. Int J Macromol Sci Univers Res Publ 1:19–22
23. Rozman HD, Tay GS, Kumar RN et al (2001) Polypropylene-oil palm empty fruit bunch-glass fibre hybrid composites: a preliminary study on the flexural and tensile properties. Eur Polym J 37:1283–1291. https://doi.org/10.1016/S0014-3057(00)00243-3
24. Fiore V, Valenza A, Di Bella G (2012) Mechanical behavior of carbon/flax hybrid composites for structural applications. J Compos Mater 46:2089–2096. https://doi.org/10.1177/0021998311429884
25. Fiore V, Scalici T, Calabrese L et al (2015) Durability of flax-basalt hybrid composites for marine applications. In: ICCM international conference on composite materials 2015, pp 19–24
26. Karina M, Onggo H, Dawam Abdullah AH, Syampurwadi A (2008) Effect of oil palm empty fruit bunch fiber on the physical and mechanical properties of fiber glass reinforced polyester resin. J Biol Sci 8:101–106. https://doi.org/10.3923/jbs.2008.101.106
27. Sarasini F, Tirillò J, D'Altilia S et al (2016) Damage tolerance of carbon/flax hybrid composites subjected to low velocity impact. Compos Part B Eng 91:144–153. https://doi.org/10.1016/j.compositesb.2016.01.050
28. Flynn J, Amiri A, Ulven C (2016) Hybridized carbon and flax fiber composites for tailored performance. Mater Des 102:21–29. https://doi.org/10.1016/j.matdes.2016.03.164
29. Salman SD, Sharba MJ, Leman Z et al (2016) Hybrid composites failure. BioResources 11:3575–3586
30. Muhammad YH, Ahmad S, Abu Bakar MA et al (2015) Mechanical properties of hybrid glass/kenaf fibre-reinforced epoxy composite with matrix modification using liquid epoxidised natural rubber. J Reinf Plast Compos 34:896–906. https://doi.org/10.1177/0731684415584431
31. Dalbehera S, Acharya SK (2015) Impact of stacking sequence on tribological wear performance of woven Jute-glass hybrid epoxy composites. Tribol Mater Surf Interfaces 9:196–201. https://doi.org/10.1080/17515831.2015.1121343
32. Hashmi SAR, Naik A, Chand N et al (2011) Development of environment friendly hybrid layered sisal-glass-epoxy composites. Compos Interfaces 18:671–683. https://doi.org/10.1163/156855412X626252
33. Zuhudi NZM, Lin RJT, Jayaraman K (2016) Flammability, thermal and dynamic mechanical properties of bamboo-glass hybrid composites. J Thermoplast Compos Mater 29:1210–1228. https://doi.org/10.1177/0892705714563118

34. Akil HM, Santulli C, Sarasini F et al (2014) Environmental effects on the mechanical behaviour of pultruded jute/glass fibre-reinforced polyester hybrid composites. Compos Sci Technol 94:62–70. https://doi.org/10.1016/j.compscitech.2014.01.017
35. Hanifawati IN, Azmah Hanim MA, Sapuan SM, Zainudin ES (2011) Tensile and flexural behavior of hybrid banana pseudostem/glass fibre reinforced polyester composites. Key Eng Mater 471–472:686–691. https://doi.org/10.4028/www.scientific.net/KEM.471-472.686
36. Romanzini D, Lavoratti A, Ornaghi HL et al (2013) Influence of fiber content on the mechanical and dynamic mechanical properties of glass/ramie polymer composites. Mater Des 47:9–15. https://doi.org/10.1016/j.matdes.2012.12.029
37. Panthapulakkal S, Sain M (2007) Injection-molded short hemp fiber/glass fiber-reinforced polypropylene hybrid composites-mechanical, water absorption and thermal properties. J Appl Polym Sci 103:2432–2441. https://doi.org/10.1002/app.25486
38. Sanjay MR, Arpitha GR, Laxmana Naik L et al (2016) Studies on mechanical properties of Banana/E-Glass fabrics reinforced polyester hybrid composites. J Mater Environ Sci 7:3179–3192
39. Živković I, Fragassa C, Pavlović A, Brugo T (2017) Influence of moisture absorption on the impact properties of flax, basalt and hybrid flax/basalt fiber reinforced green composites. Compos Part B Eng 111:148–164. https://doi.org/10.1016/j.compositesb.2016.12.018
40. Aktas L, Hamidi Y, Altan MC (2002) Effect of moisture absorption on mechanical properties of resin. 2–7
41. Zhang Y, Li Y, Ma H, Yu T (2013) Tensile and interfacial properties of unidirectional flax/glass fiber reinforced hybrid composites. Compos Sci Technol 88:172–177. https://doi.org/10.1016/j.compscitech.2013.08.037
42. Panthapulakkal S, Sain M (2007) Studies on the water absorption properties of short hemp-glass fiber hybrid polypropylene composites. J Compos Mater 41:1871–1883. https://doi.org/10.1177/0021998307069900
43. Yang Z, Peng H, Wang W, Liu T (2010) Crystallization behavior of poly(ε-caprolactone)/layered double hydroxide nanocomposites. J Appl Polym Sci 116:2658–2667. https://doi.org/10.1002/app
44. Mohanasundaram S, Sivakumar M (2021) Thermal properties of natural fiber reinforced hybrid composites. 9:125–128
45. Elkington M, Bloom D, Ward C et al (2015) Hand layup: understanding the manual process. Adv Manuf Polym Compos Sci 1:138–151. https://doi.org/10.1080/20550340.2015.1114801
46. Rajak DK, Pagar DD, Menezes PL, Linul E (2019) Fiber-reinforced polymer composites: manufacturing, properties, and applications. Polymers (Basel) 11. https://doi.org/10.3390/polym11101667
47. Yuvaraj G, Vijaya Ramnath B, Abinash A et al (2016) Investigation of mechanical behaviour of sisal epoxy hybrid composites. Indian J Sci Technol 9. https://doi.org/10.17485/ijst/2016/v9i34/101004
48. Yahaya R, Sapuan SM, Jawaid M et al (2014) Quasi-static penetration and ballistic properties of kenaf-aramid hybrid composites. Mater Des 63:775–782. https://doi.org/10.1016/j.matdes.2014.07.010
49. Aslan M, Tufan M, Küçükömeroğlu T (2018) Tribological and mechanical performance of sisal-filled waste carbon and glass fibre hybrid composites. Compos Part B Eng 140:241–249. https://doi.org/10.1016/j.compositesb.2017.12.039
50. Barouni AK, Dhakal HN (2019) Damage investigation and assessment due to low-velocity impact on flax/glass hybrid composite plates. Compos Struct 226:111224. https://doi.org/10.1016/j.compstruct.2019.111224

Advanced Natural/Synthetic Composite Materials for Marine Applications

Ashwini Karrupaswamy, Jayavel Sridhar, D. Aravind, K. Senthilkumar, T. Senthil Muthu Kumar, M. Chandrasekar, and N. Rajini

Abstract Composite material research has witnessed ubiquitous contributions to sustainable development. Environmental impact assessment of composite materials has revealed the non-biodegradability and environmental menace causing health hazards. Green hybrid composite materials pose as an effective alternative for alleviating the utility of synthetic composites. The compilation corresponds to green hybrid composite materials and significant parametric characters for increased marine applications in engineering and Non-Engineering arena.

Keywords Green hybrid composites · Natural and synthetic fiber-reinforced composites · Biodegradability · Environmental-friendly · Marine applications · Macroalgal derived biopolymers

A. Karrupaswamy · J. Sridhar
Department of Biotechnology (DDE), Madurai Kamaraj University, Palkalai Nagar, Madurai 625021, Tamil Nadu, India

D. Aravind
University Science Instrumentation Centre, Madurai Kamaraj University, Palkalai Nagar, Madurai 625021, Tamil Nadu, India

D. Aravind · T. Senthil Muthu Kumar · N. Rajini
Department of Mechanical Engineering, Kalasalingam Academy of Research and Education, Krishnankoil 626126, Tamil Nadu, India

K. Senthilkumar (✉)
Department of Mechanical Engineering, PSG Institute of Technology and Applied Research, Coimbatore 641062, Tamil Nadu, India
e-mail: kmsenthilkumar@gmail.com

M. Chandrasekar
Department of Aeronautical Engineering, Hindustan Institute of Technology & Science, Padur, Kelambakkam, Chennai 603103, Tamil Nadu, India

T. Khan and M. Jawaid (eds.), *Green Hybrid Composite in Engineering and Non-Engineering Applications*, Composites Science and Technology,
https://doi.org/10.1007/978-981-99-1583-5_13

1 Introduction

Composite materials correspond to more than two materialistic combinations of fibers that possess uniqueness through neither blending nor dissolving, resulting in large-scale applications [27]. Composite materials in the ancient era were primarily confined to warfare productions, fiber-reinforced polymers, radars, electric appliances, sports materials, automotive and aerospace industries [59]. The incumbent composition of composite materials can be corroborated by the continuous matrix phase, discontinuous reinforcements, and an interface region [15, 109]. Green composite materials are widely utilized in several industrial applications owing to their Eco-friendly nature and large-scale benefits [10]. Comprehensive reviews revealing the significance and prominent applications of natural composite materials are summarized for the inherent characteristics, preparation methodology, and robust engineering and non-engineering realms [110, 121]. Green composite materials are defined as renewable composites prepared by blending natural and synthetic composites. A meager addition of the renewable resource in the composite preparation renders the green composite nature of any composite [40]. The field of composites has foreseen explicit applications in day-to-day life usage for augmented utilization. The world witnesses a plethora of deleterious and devastating natural calamities that are stressed by Scientists for dire attention ubiquitously. Further, the illustrations of 17 Sustainable Development Goals (SDG's) by the United Nations necessitate adequate infrastructure in attaining the pinnacle of global welfare [29]. The present compilation encompasses the usage of advanced green hybrid composite materials incorporating both engineering and non-engineering applications for an upheaval of marine environmental health. Thus, the present review summarizes the plant, animal fibers reinforced to the formation of green hybrid composite materials and their significant applications that determine the copious marine applications. The marine applications are alone concentrated based on the Engineering and Non-Engineering counterparts.

2 Green Composite Materials and Their Applications

Polymers and their associated composite materials provide magnanimous utilities for industrial applications and domestic purposes in regular life. The reinforced material used in preparation renders the constituent phase/matrix together with the matrix interface linkage contributing to escalated tenacity [111]. Synthetic composites with polymeric ingredients are advantageous to use except that they are non-biodegradable and not Eco-friendly, posing lesser effective use in non-renewable resources management [81]. Green composites are preferred over the commonly available composite materials due to their economical feasibility, technological availability, reduced cost, easy recyclable properties, and ease of transport [21]. Thus the realm of green composite materials has drawn the immense attention of the technocrats globally for effective fiber reinforcement protocols. Natural fibers encompassing ordinary plants like jute, wheat, sugarcane, oil palm, silk, cotton, coconut, flax, and sisal have been included in Engineering applications to refined manufacturing products [6, 107]. The natural fiber reinforcement strategy has gained momentum due to its environmental friendliness, cost-effectiveness, and increased usage. A wide arena for natural fiber reinforced matrices in manufacturing applications has witnessed the utility of polylactic acid (PLA) in thermoplastics and thermoset industries, biopolymers for packaging industries, and automobile industries [123]. The methodologies adopted for biocomposites incorporation comprise the utilization of apt technology like injection molding, compression molding, resin transfer molding, pultrusion, thermoforming, etc. [34]. Hence, the use of green composite materials in Engineering applications and other fields has envisaged adequate fabrication methodologies and appropriate intrinsic fiber reinforcements. Thus the interdisciplinary applications of the innovative green composites with Eco-friendly prospects has foreseen copious applications in packaging industries, biomedical industry, automobiles, leisure, sports, musical instruments manufacture and Energy industries [70]. However, there has been a dire need to address several technological innovations, fabrication strategies and complexity of agglomerating natural and synthetic composite materials for enhanced properties along with the emphasis for biodegradability, compatibility, cost-effectiveness and environment friendly approaches. The present assessment aims to abridge the optimal significance of blending the natural composite fibers along with synthetic composites for enhanced green hybrid composite materials that play pivotal roles in marine applications involving both Engineering and Non-Engineering applications.

3 Natural Green Composite Material

Biofibers correspond to natural fiber reinforcements that are agglomerated over a polymer matrix phase of the primary composite that exists in a dispersed layer. This congregation provides tenacity, robustness, and escalated challenging characteristics of the composite preparations [51, 121]. Biopolymers that are easily degradable, like bamboo, jute, and hemp, are increasingly preferred over synthetic fibers involving plastics made of polyvinyl chloride and polystyrene due to renewable nature, lesser emission of Green House Gases (GHG's), and biodegradability [47, 94]. The increased pollution hazards rendered by synthetic composite materials necessitated the stress for alternate natural fiber reinforcements, which pose lesser environmental damage. Environmental friendly resins replace a variety of conventional polymers that are petroleum-based for a plethora of applications in Science and Technology [99]. Biopolymers resemble synthetic polymers as the matrix withstands physical harm and resists the appropriate burden implied from the surroundings. Biodegradable biopolymers are usually fractionated from the natural biomass of polysaccharides and proteins. Whereas polyesters like polyhydroxyalkanoates (PHAs), polyester amide (PEA), and polycaprolactone (PCL) represent the commercially available polyesters derived from chemical processes [11]. Nevertheless, polylactic acid (PLA) biopolymers corroborating to the optimal bio-compatible and biodegradable (microbial bio degradation) with high tensile strength are usually derived from renewable energy resources shows versatile applications in thermoset and thermoplastics industries [40, 78, 103]. Microbial Biodegradation of naturally reinforced fiber composites have been evident in several sewage treatment plants efficiently [74]. The above research reports depict the basis for PLA biopolymers and their versatile uses in regular Engineering and Non-Engineering applications. However, the mode of action onto how the PLA exerts the eco-friendly nature of biopolymer characteristics dates back to 1993. PLA-based biopolymers pose as a better alternative to petroleum based polymers due to enzymatic ring-opening polymerization (eROP) that involves esterases like lipases for hydrolytic reactions catalyzing esterification reactions for Green composites manufacturing [13, 65, 137]. The chemical entities envisage the wide applications in tissue engineering and regeneration potentials in Biomedical applications. Natural nanocomposites based on PLA correspond to escalated drug delivery, anti-tumor effects, construction industry, automobiles, packaging industries, textile technology, agriculture and electronics manufacture with enhanced biodegradability. Nevertheless the monomer and polymer synthesis emancipates further adequate research for increased applications surpassing the intricate limitations [13]. Energy industries comprise the usage of biotex flax for wind blades manufacture in wind energy industries, and bamboo has been recognized as a potential component in natural bio-composites [40]. Other significant applications of plant fiber-based biocomposites include musical instruments due to lightweight properties, toys, sports industry materials, skateboards, and marine applications like boat hulls and canoes for easing transportation [70]. Thus versatile applications of green composite bio-materials are revealed for their optimistic benefits in many applications.

4 Synthetic Composite Materials

Synthetic fibers are widely used based on their physical and resistance properties and are utilized for reinforcement materials comprising glass, kevlar, and carbon components. Unique characteristics of glass fibers (E-glass, AR-glass, R-glass, and S-glass) comprehending the tenacity, stiffness, high strength, enhanced flexible nature, resistance to chemical damages, and high insulation properties render their composite applications [22]. Glass fiber-reinforced polymer composites (GFRPCs) have been increasingly utilized for construction, transportation, electronics, sports and leisure applications [22]. Similarly, carbon fibers reinforcements in synthetic composites are used in plastics, carbon–carbon composites, and carbon-based cement [85, 97]. The precursors for carbon fiber reinforcements belong to polyacrylonitrile (PAN), rayon, and petroleum pitch-based carbon fibers, resulting in higher brittle characteristics than the glass and aramid fibers [18]. Aramid fibers named Kevlar for commercial usage as Kevlar 29, Kevlar 49, and Kevlar 149 possess high tenacity, low density, high resistance to physical impact, and enhanced tensile properties like that of glass fibers [88]. However, limitations are attributed to low compression, sensitivity to heat and moisture [88]. Comprehensive reports regarding the summarization of synthetic composites and potential future directions for alternative bio-composites like green hybrid composites have been put forth in Engineering and Non-Engineering realms [51, 101, 126, 73, 125]. Hence, our focus was to establish the positive attributes of green hybrid composites composing both natural and synthetic composites. Further, the compilation describes the appropriate marine applications and possible future directions to enhance the biomass augmentation and biodegradability of green hybrid composite materials.

5 Significance of Green Hybrid Composite Materials and Prominent Applications

Polymer composite matrix composites (PMC) are the revolutionizing composites overpowering the conventional composite materials [60, 122]. Fiber reinforcement falls primarily into either glass fiber or carbon fiber plastics, along with natural fiber incorporation, resulting in resin formation through hybridization [2, 61, 68]. However, carbon-based nanocomposites or nanofillers upon fabrication strategies resulted in enhanced mechanical properties of hybrid composites [14]. The arena of hybridization is evolving consistently for developing augmented biodegradability and mechanical compatibility for a wide range of applications proving the escalated versatility [63].

Hybrid composites materials are manufactured using the combinations of either natural/synthetic, natural/natural, or synthetic/synthetic fiber incorporation [100]. Moreover, the biodegradability of polymer matrices involved in manufacturing green hybrid composite materials rely on the basic principle of Reduce, Reuse, and Recycle for human and environmental health perspectives [124, 19]. Natural fibers are basically obtained from either plants, animals or minerals. Plant fibers are rich in cellulose components. However, synthetic composites accounted for lesser biodegradability have been harnessed for hindrances in bio-degradation. Thus, nanocomposites of biological origin are focused by marine researchers in marine applications management for sustainable development. On the contrary, animal fibers are incumbent components of protein-rich parts such as silk, wool, and hair. Plant fibers are classified into bast or stem comprising Flax, Hemp, Jute, Kenaf, Roselle, and Ramie. Abaca, Banana, Cantala, Caroa, Curaua, Date palm, Henequen, Pineapple, and Sisal corroborates to leaf or hard fiber components. Cotton seeds, grass fibers like Bagasse, Bamboo, cereal, wood, and fruit fibers comprising Coir, Kapok, Oil palm, Sponge gourd account for the comprehensive categorization for varied mechanical strength [108, 112, 116]. GFRP—Glass-fibre-reinforced polymer consists of finer glass fibers used for thermoset and thermoplastics manufacturing with high strength, robustness, and lightweight. The majority of applications ranging from boats to roofing materials are attributed with non-recyclable properties as a potential limitation [41, 43, 95]. Carbon fiber reinforcement based on nanofillers, carbon nanotubes, graphene particles provide enhanced electrical and thermal conductivity for use as nanocomposites [44, 77]. Of the above, Green hybrid composites encompass natural/synthetic, natural/natural hybridization along with nanocomposites fiber incorporation that has versatile applications. Natural and synthetic fiber hybridization has been proved for escalated mechanical characteristics that render high structural utilization with chemical modifications and treatment strategies [57]. An extensive review revealing the green hybrid composites for wide applications corroborates resistance to moisture and fire, aging of natural fibers, and interface research on hybridization has been focused on increased green hybrid composites application [20, 58]. Thus, the applications of green hybrid composites and their mechanical, structural attributes are discussed. However, marine applications in a holistic way necessitate the appropriate marine applications of green hybrid composites.

6 Marine Applications of Green Hybrid Composites (Engineering and Non-engineering)

6.1 Green Hybrid Composites and Synthetic Fiber-Reinforced Polymers

Green hybrid composites are representative composites in attaining Environmental sustenance composing polymers that are biodegradable for sailing applications using the mineral, basalt and plant, flax composites [28]. Animal fiber Reinforced hybrid fiber composite for marine applications rely on high tensile strength and malleability. Epoxy resins were used for maximal synthetic fiber reinforcements rendering animal fiber utility in marine applications [52]. Carbon hybrid composite materials have been established for effective and high tensile marine propellers than glass fiber reinforcements [89]. CFRP—Carbon fiber-reinforced polymer composites pose cost-effective, high mechanical properties for marine propeller applications [62]. Further applications of CFRP include the development of hulls and marine crafts [128]. Hence, alternatives to reinforcing composites to prevent the aging factor due to sea water were recommended for improvising marine applications [91]. Figure 1 depicts the varied marine applications of green hybrid composite materials as a holistic view. The summary of futuristic values of natural fiber reinforcements is self-explanatory in Fig. 1. However, further authentic research with reproducible results is required for escalated applications in this regard. Thus, future research in compatible fiber reinforcements that can combat aging in seawater environments can have potential implications. Marine coral reefs, microalgae, and various macroalgae substituting plant fibers can have substantiating benefits. Lichens that represent a phycobiont and a mycobiont can yield high tensile strength and mechanical properties for marine

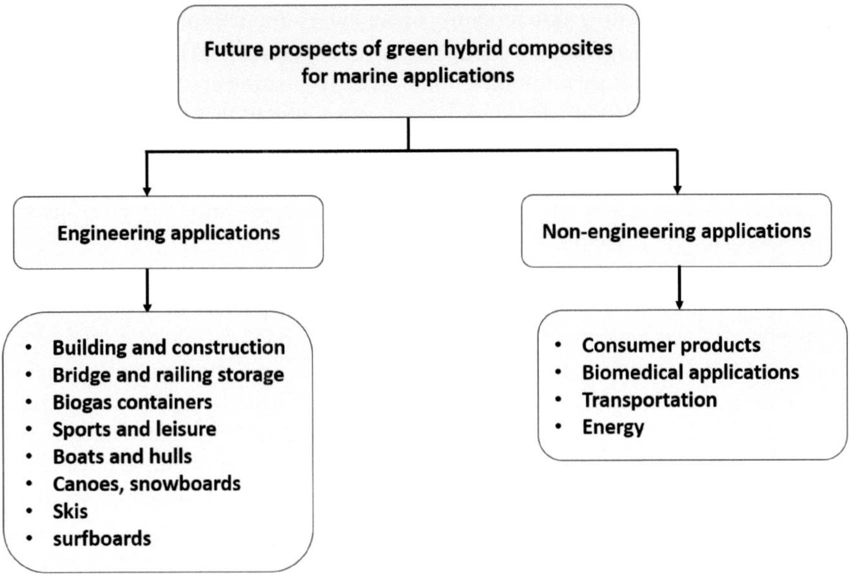

Fig. 1 Future prospects of marine applications of green hybrid composites

applications. Nevertheless, incorporating synthetic composites in agglomerating the natural fiber composites and embedding experiments will provide provocative benefits. Along with structural contouring based on bonding patterns, eco-friendly nature, biodegradability, compatibility can have a holistic phenomenon in developing green hybrid composites for various marine applications.

Fiber-reinforced polymers have been utilized for marine applications over several decades. The principal applications of composite materials in the marine environment encompass hulls, propellers, radomes, railings, valves, exhausts, hatch covers, and bearings for structural integrity. The composites integration for marine utilities impart longevity and safer construction of powerboats used for racing at sea [98]. Glass fiber reinforced plastics (GFRP) have been reported for shipbuilding for naval applications and are highly resistant to salt, corrosion, fire resistance and physical impacts [71]. Ample reports have documented the use of fiber-reinforced composite materials for military shipbuilding and submarine construction [26, 75]. Among civil applications in the marine environment, lifeboats, hovercrafts, yachts, and catamarans have been utilized for GRP based composites incorporation [55, 75, 119, 120] Fiber-reinforced composites made of glass or carbon fiber reinforced plastics (CFRP) have been assessed for replacing the nickel-aluminum-bronze (NAB) components in marine propellers, bulkheads, heat exchangers, discharge funnels, rudders, decks, hatches, engine components and protection systems [98]. The offshore sector marine applications are attributed to Oil and gas industries comprising fiberglass tubes, fiber spar line pipe, composite reinforced linear tube, risers, and caissons [60]. GFRP and CFRP are used in marine energy production industries [46, 80]. Fiber-reinforced composites have been utilized in repairing underwater structures and wooden pillars, revealing corrosion resistance and strengthening of concretes [105]. Thus a bird's eye view of approaches in finer reinforced composites for marine application shows the involvement of natural/synthetic hybrid composites largely than natural fiber reinforced hybrid composites due to sea water aging and corrosion resistance. Nonetheless research on nanocomposites and blending along to green hybrid composites will foresee numerous and enormous intricate marine applications of green hybrid composites in the near future.

GFRP composites, upon blending with glass or basalt, have been proved effective in marine naval applications [36]. GFRP and CFRP composites used for manufacturing fabrics in marine applications showed higher mechanical properties, improved tenacity, and high resistance against seawater aging [48]. In a similar study, the laminates made of flax and glass hybrid composites proved structurally intact laminates with mechanical properties, resistance to seawater aging, cost-effectiveness, and environmental friendly [16]. Metal hybrid composites involving Grade 5 Titanium alloy (Ti-6Al-4 V) blended with Kevlar®, and carbon fiber reinforcements show predominant marine applications owing to cost-effectiveness, high tensile strength, resistance to corrosion, lightweight for marine applications in hulls and structures in the marine environment [31]. A similar study for assessing natural/synthetic hybrid composites preparation employing kevlar (K29), Eggshell, and coir incorporation proved effective due to lightweight, high mechanical strength for enhanced marine applications [30]. Lightweight aluminum alloys usually used in marine applications can have high mechanical strength, corrosion resistance, reduced water absorption, and increased structural properties due to fiber reinforcement protocols. However, more synthetic composites impregnation and blending experiments are present questioning the environmental friendliness of the fiber reinforced composites applications in the marine environment. Nanocomposites embedding with appropriate nanoparticles, epoxy composites and circular bionomy approaches for reducing, reusing and recycling can have potential research and development of eco-friendly composites for marine apllications.

A circular bionomy approach was made use to develop a novel hybrid composite system based on discarded Nylon fishnet, and glass fiber reinforcements revealed the novel methodologies adopted for eco-friendly hybrid composite development [115]. Hybrid composite materials applications in marine turbines under tidal wave currents, when assessed for nozzle improvements, revealed the utility of carbon/glass/glass—CGG stack show high resistance to impact than laminates in combating damage and absorption of energy [64]. Metal matrix composites have proved significant over other composites due to their lightweight, mechanical and structural properties. Aluminum hybrid composite materials impregnated with silicon carbide nanoparticles and fly ash proved improved structural properties for enhanced marine applications in the future [130]. Comparative assessment of corrosion resistance between CF/E composite—carbon fiber/epoxy composite, pure titanium TA2, and Ti FML—titanium-carbon fiber/epoxy fiber-metal laminate for corrosion resistance and mechanical stability was analyzed for marine applications. The results depict that under conditioning under hygrothermal conditions, Ti FML showed less water absorption and corrosion resistance due to the stable film formation of Titanium oxides [4]. An extensive review of hybrid composites in propellers design in marine applications has affirmed the significant positive outcomes of reinforcement strategies. GFRP has been corroborated to better performance and cost-effectiveness, and CFRP has an equally prominent characteristic in marine propellers design. Thus, hybrid composite materials have been attributed to versatile marine applications [127]. Epoxy materials for coating and surface modification properties have shown the utility of nanoparticles for filler applications in marine environments. Epoxy composites composed of silica nanoparticles and PES—Polyethersulfone showed that silica–PES core–shell NPs (PES@silica) composite possess high tenacity, resistance to seawater aging, and high flexure strength for enhanced marine filler applications [23]. When incorporated into the preparation of fiber reinforced composite materials, recycling strategies will have vast applications for using non-biodegradable resins, complex materials like fishnets, and some polymeric plastics of interest. Hence circular bionomy approaches will pave the way for the above phenomenon and apt usage of composite embedding and incorporation in effectuating marine applications of green hybrid composite materials.

7 Epoxy Resin Composites in Marine Applications

Polyethylene epoxy composites have been used for high resistance to seawater aging and pose as an alternative strategy in utilizing plastics that are non-biodegradable and hazardous to the environment [69]. Graphene nanoplatelets upon reinforcement into epoxy composites with basalt have proved the reduced absorption of water, thereby resistant to corrosion and wear/tear [118]. Lightweight sandwich structural composites, including metamaterials [102], and lattices [131], will have profound implications in the construction industries in marine applications [82]. Glass fibre/epoxy (GE) composites with graphene oxide embedding shows high structural complexity owing to mechanical strength, a barrier against seawater aging and escalated viscoelasticity proving high marine applications. The reduced water absorption linking to resistance to corrosion was analyzed by kinetics and was correlated to cationic cross-linking for the positive attributes [7]. Epoxy composites and their respective oxides formation along with necessary nanoparticles and synthetic fibres for improved structural and mechanical properties can have a wide arena of research in marine applications in the years to come.

Polymeric substances derived from marine environments have been categorized for blue biotechnology approaches involving polysaccharides (hyaluronic acid, chitin, chitosan) and proteins (collagen, proteins and polypetides, Enzymes) [25, 76]. Macroalgal biopolymers comprise cell wall and storage polysaccharides belonging to the macroalgal species Chlorophyceae, Rhodophyta, and Phaeophyceae for effective biopolymer biomass compositions [90]. Mycosporine-Like Amino Acids represent zwitterionic components in macroalgae that possess biomedical applications [96].

Copious marine-derived polymeric substances and respective applications for polymeric biomass constituents and composite synthesis with enhanced applications from the marine environment have been deciphered [25]. Figure 2 depicts the blue biotechnology innovative macroalgal constituents for effective biopolymer composite applications. Thus the advent of blue biotechnology can have massive inclusions of rigidity parameters of marine macroalgae for effectuating the versatility of biocomposites preparation from macroalgae in marine applications. Hence future research upon macroalgal-derived composites will provide biocomponent implications in green hybrid composite preparation and potential applications in the future.

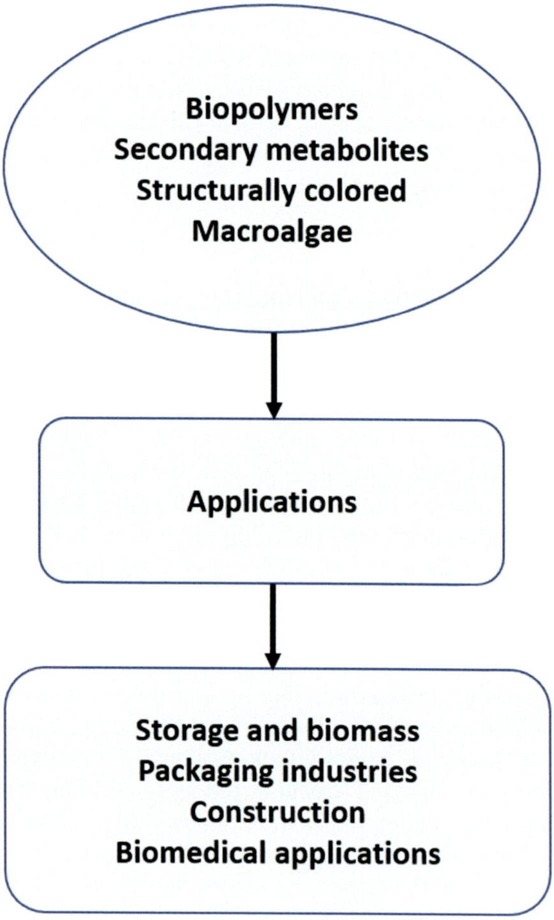

Fig. 2 Blue biotechnology derived macroalgal biopolymeric substances and applications

8 Nanocomposites in Green Hybrid Composites for Marine Applications

Green hybrid composites in developing sustainable sources for marine utilities pose the large-scale usage of synthetic fiber reinforcements. Hence the Eco-friendly nature and biodegradable compatible and cost-effective research witnessed the nanocomposite fiber reinforcements that can have a wider outreach for marine applications. Green nanocomposites have been made use for energy storage applications in the marine environment [53]; Green nanofillers comprising the nano clay that are easily degradable, carbon nanofiller, metal nanoparticles, cellulosic nanofibers, and nanotubes will have potential marine applications perspective [53]. Polyethylene glycol polymer matrices have long been addressed as a prominent alternative owing to physical, structural texture with improved mechanical parameters and antibacterial properties [3, 42, 67]. Earlier dielectric capacitors for energy storage have been adequately reported [134].

Moreover, natural composites composed of hybrid nanocomposites have been documented for sustainable applications for thermosets and thermoplastics [86]. Polylactic acid nanocomposites play a pivotal role in automotive applications [8], that can be harnessed in marine transports, including marine propellers. Nanocomposites of natural origin harboring natural fiber reinforcements have been reported for versatile applications, including marine engineering prospects [45]. Nevertheless, nanocomposites have also been proved worthy due to their antifouling properties in marine environments. Biomimmetics and nanocomposites incorporation have several positive attributes encompassing minimal free energy, antibacterial activities, and superior mechanical properties [104]. Figure 3 summarizes the disadvantages of the green hybrid composite materials and necessary reinforcements. Hence additional research in this arena will aid in advancing the green hybrid nanocomposites-based reinforcement strategies for enhanced marine applications in Engineering and Non-Engineering sectors. Thus future research in augmenting the nanocomposites incorporation with fiber reinforcement surpassing synthetic fiber composites will escalate marine applications of green hybrid nanocomposites.

9 Comprehensive Assessment of Nanocomposites in Marine Applications: Engineering/Non-engineering

Nanocomposites are accounted for the majority of marine applications and have been regarded as a potential revolution in the Environment friendly and cost-effective fiber reinforcements. The alleviation of marine applications using non-biodegradable synthetic fiber reinforcements has foreseen tremendous utility in the marine environment. Since the nanofillers possess good mechanical properties and structural integrity, they are primarily employed in the marine construction industry [106]. Aluminum alloys impregnated with nanocomposites based on

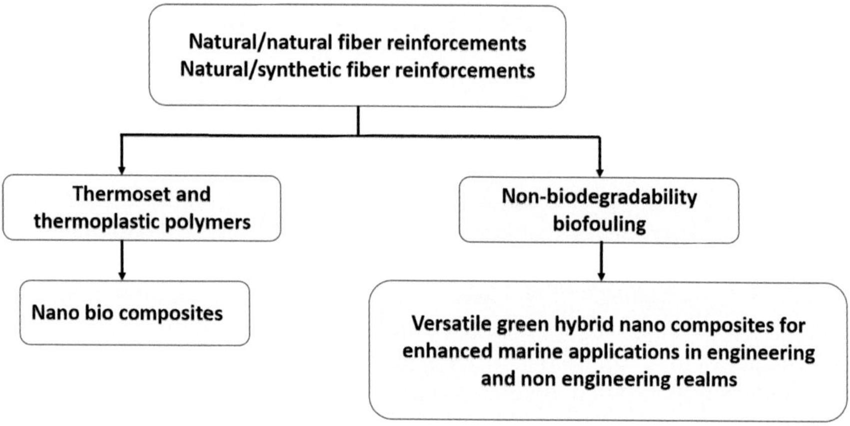

Fig. 3 Illustration of Green hybrid nanobiocomposites and potentials

cerium oxides are being chosen for electrochemical anodes in marine applications enhancing the galvanic cells in the marine surroundings [113]. Nanocomposites and necessary incorporation's have been advanced using appropriate computational strategies in minimizing the cost-effectiveness and time consumption in choosing the apt nanocomposite fiber reinforcements. An algorithm based on TOPSIS-PSI methodology was adopted to affirm nanobiocomposites made of aluminium alloys for hybrid marine applications with material selections in marine engineering applications [72]. The above strategy resulted in both quantitative and qualitative selection of nanocomposite hybrid materials for effective construction of fishing boats, offshore platforms, bridging rails, cylinder blocks, etc. [72]. Biofouling accounts for corrosion in the marine environments leading to a loss in ship construction industries [5, 24]. TBT—tributyltin was used as an effective additive in painting industries for minimizing economic loss due to corrosion but presently neglected for biotoxicity [32, 56]. Thus novel incorporation employing epoxy resins agglomerated with clay nanotubes pose as an effectuating natural nanocomposite material with silver or biocide particles acting as anti-biofouling in marine environments [38]. DCOIT—4,5-dichloro-2-octyl-isothiazolinone has been proved as a potent epoxy resin composite for paint applications and minimizing bacterial colonization combating antibacterial and thus antifouling properties [38]. Hence, using nano clay addition, the silver nanoparticles additives with the painting industries can enhance the bioprospecting of innovative petroleum reserves in the arctic areas [38]. Moreover, the biosynthesis of nanoparticles using marine algae and seagrasses provides effective biogenic nanoparticles in augmenting marine engineering applications in an environment-friendly and with minimal side effects like corrosion, fouling, and economic loss in the marine constructions [35]. Metal nanoparticles synthesized from marine algae and plants in the marine environment. An extensive review has cataloged the efficient marine algal counterparts in metals and metal oxide nanoparticles biogenic synthesis [9, 84].

Metal nanoparticles based on Ag (*Codium capitatum, Spirogyra insignis, Padina tetrastromatica*), Au (*Sargassum wightii Greville, Turbinaria conoides, Stoechospermum marginatum*) and Pd (*Chlorella vulgaris, Laminaria digitata*) have been synthesized in biogenic mode for structural applications [17, 33, 50, 83, 87, 92, 93, 114]. Similarly metal oxide nanoparticles have been utilized for Cu2O, CuO nanoparticles (*Bifurcaria bifurcata*), Cu/Cu_2O nanopartiles (*Kappaphycus alvarezii*), Fe_3O_4 nanoparticles (*Sargassum muticum*) and ZnO metaloxide nanoparticles (*Sargassum muticum, Gracilaria gracilis*) [1, 12, 37, 54]. Thus macroalgal species in the marine environment can be harnessed for biogenic synthesis of metal and metal oxide nanoparticles that can be used for fiber fillers or nanofillers in reinforcement strategies in designing green hybrid nanobiocomposites for engineering and non-engineering applications in the marine environment. Future research targetting the biomass and bioenergy usage from marine macroalgae apart from basic biomedical applications research will revolutionize the field of hybrid composites in marine applications. Several research on nanocomposites has been preceded by copious researchers globally. Nevertheless, the lab-to-field transition and affirmation of the appropriate mechanical properties, structural integrity, cost-effectiveness, and environment-friendly biodegradable green hybrid composite materials remains unraveled.

Comprehensive reviews pertaining to the potential marine applications of nanocomposites have been reported. Nonetheless, the effective applications in the context of practical utility with improved structural characteristics are still under rigorous research. Marine turbines employing carbon nanotubes are depicted as a key strategy in addressing environmental-friendly and clean renewable resources of energy in the marine environment accounting for sustainable development. The marine applications of carbon nanotubes-based marine turbines are attributed to fiber reinforcements in fillers and marine structures, anti-biofouling enhancing the structural integrity of marine engineering context, compatible and highly efficient wiring, and lubrication prospects [79]. Similar research concerned with microbial fuel cells in marine benthic environments has witnessed manganese oxide coatings with graphite or multi-walled carbon nanotubes have confirmed the utility in manufacturing oceanography devices. The effective mode of action has been deciphered as the electron shuttle amongst the interface for enhanced electrochemical properties with abatement of biofilms resulting in anti-biofouling properties [39]. On the contrary, nano cellular incorporation of silicone rubber foams from supercritical CO_2 revealed high mechanical properties with enhanced tensile strength than microcellular silicone rubber foams [132]. Thus high order of intrinsic cellular modification proves that nanocomposites of biological origin or biogenic nanoparticles from marine origin can provide explicit advancements in green hybrid natural nanocomposites replacing synthetic fiber reinforcements for environmental sustainability. Hence biopolymers agglomeration with nanocomposites or incorporation with biogenic nanoparticles will have the versatility of marine applications with augmented biodegradability, environmental-friendliness, improved mechanical properties, and structural texture. However, detailed research of intrinsic chemical bonding properties and necessary redox reactions will address the potential gap of using green hybrid natural composite fibers for marine engineering and non-engineering perspectives.

Advancing strategies in bio-polymer technologies will provide enormous and ubiquitous benefits in designing and developing green hybrid natural fiber reinforcements for marine applications in the Engineering sector than the non-engineering sector due to biodegradability principles. PDMS—polydimethylsiloxane with epoxy resin development for multi-functional nanocomposites-based coatings for marine applications comprising corrosion resistance and anti-biofouling properties. Zinc oxide reinforcements revealed the high quality of performance ability for Eco-friendly marine utilities [129]. Thus a clear and precise picture of avoiding synthetic fiber reinforcement protocols necessitates laborious research on nanocomposites and epoxy resin blending for developing superior green hybrid composite materials for marine applications. Facile blending mechanism using zinc oxide nanoparticles enhancing hydrophobicity for anti-bacterial bioactivity and anti-biofouling characteristics with acrylic polyurethane depicted escalated structural stability, suppression of adhesion properties of marine algae and biofilm disruption [133]. Hence multiple modalities and mechanistic action emphasizing versatile biopolymers for sustainable marine environment applications are the pivotal need of the scenario research. Flame retardation has been stressed for enhanced qualities and superior performance of fiber reinforcement nanocomposites for marine engineering applications. Rigid nature and flammable properties render incorporating nanocomposites with epoxy resins provide the thermoset nano biocomposites as an effective alternative for marine applications [49]. Moreover, marine environments are prone to flammable nature as petrochemical resources and fossil-based polymers are utilized in marine engineering and non-engineering applications. Hence to preserve the renewable resources of energy and environmental sustainability, the dire need for nanocomposites is focused on by researchers globally.

Resistance to corrosion and the maintenance of infrared stealth are contradictory phenomena in the marine environment. Graphene incorporated composite materials were proved for high-performance stealth using epoxy resin coating. Thus the graphene basis showed effective resistance to corrosion and improved emission profiles of stealth properties [136]. In a similar study, hydrothermal biosynthesis of graphene flakes using Ag nanoparticles proved effectuating biofilm disruption and potent anti-biofouling characteristics. Thus the green hybrid nanobiocomposites based on graphene silver conjugates showed inhibition properties against the marine bacteria *H. pacifica* and the marine algae, *D. tertiolecta* and Isochrysis sp. Hence synergistic conjugation of graphene with biogenic nanoparticles proved potential green composite reinforcement nature with biofilm abatement properties [135]. Green hybrid nanocomposites have been increasingly used for marine naval ships and warfare applications. Resistance to seawater stress shocks and atmospheric pressure management has been assessed for mathematical modeling and assessment of theoretical extrapolation in determining the structural texture and mechanical integrity of ship hulls in naval warfare marine applications. Clay Silica hybrid nanocomposites have been shown for shock resistance in sea waters by naval warship hulls [66]. Thus nanobiocomposites have revolutionized the green hybrid natural and synthetic composite reinforcement materials for potential marine applications in engineering and non-engineering applications. The comprehensive compilation addresses the green hybrid composite materials that comprise plant, animal, and mineral fibers that possess biodegradable polymer matrices and environmental friendliness. The green hybrid composite materials are given prime importance for biodegradable potentials that impact the environment and ecosystem and minimize the full hazardous release in the natural systems. The present compendium of research on green hybrid natural and synthetic composite materials effectively indicates the nanobiocomposites and biogenic nanoparticles synthesized from marine algae and marine plants are corroborated as the launchpad research arena. Further, the nanobiocomposites blending to epoxy resins with accessory composite ingredients will have massive research outcomes in enhancing the engineering and non-engineering marine applications for eco-friendly and sustainable ecosystems. Nevertheless, the cost-effectiveness and high performance marine and maritime applications will necessitate the dire need of interdisciplinary and multidisciplinary research areas including mathematical modeling, computational inputs and statistical assessment.

10 Conclusion

The present holistic compilation addresses the marine applications of green hybrid composite materials about Engineering and Non-Engineering applications. Several research reports have redressed the impact of natural/natural, natural/synthetic fiber reinforcement strategies. However, limitations like moisture absorption, aging of natural fiber composites, interface research in deciphering intricate structural conformations are necessitated. Further, the utility of synthetic fibers in augmenting the longevity and quality of the green hybrid composites requires explicit research and validation. Marine biomass like microalgae, macroalgae, corals, and other lichens could provide altered characteristics of green hybrid composites with escalated mechanical properties. Bonding patterns to reveal the esterification and similar hydrophilic reactions will provide a more intrinsic phenomenon for enhanced green hybrid patterns with increased marine applications.

References

1. Abboud Y, Saffaj T, Chagraoui A et al (2014) Biosynthesis, characterization and antimicrobial activity of copper oxide nanoparticles (CONPs) produced using brown alga extract (Bifurcaria bifurcata). Appl Nanosci 4:571–576
2. Agarwal R, Saxena NS, Sharma KB et al (2003) Thermal conduction and diffusion through glass-banana fiber polyester composites
3. Alcantar NA, Aydil ES, Israelachvili JN (2000) Polyethylene glycol–coated biocompatible surfaces. J Biomed Mater Res An Off J Soc Biomater Jpn Soc Biomater Aust Soc Biomater Korean Soc Biomater 51:343–351

4. Ali A, Pan L, Duan L et al (2016) Characterization of seawater hygrothermal conditioning effects on the properties of titanium-based fiber-metal laminates for marine applications. Compos Struct 158:199–207. https://doi.org/10.1016/j.compstruct.2016.09.037
5. Almeida E, Diamantino TC, de Sousa O (2007) Marine paints: the particular case of antifouling paints. Prog Org Coatings 59:2–20
6. Alothman OY, Jawaid M, Senthilkumar K et al (2020) Thermal characterization of date palm/epoxy composites with fillers from different parts of the tree. J Mater Res Technol 9:15537–15546
7. Anand A, Ghosh SK, Fulmali AO, Prusty RK (2021) Enhanced barrier, mechanical and viscoelastic properties of graphene oxide embedded glass fibre/epoxy composite for marine applications. Constr Build Mater 268:121784
8. Arjmandi R, Hassan A, Zakaria Z (2017) Polylactic acid green nanocomposites for automotive applications. In: Green biocomposites. Springer, pp 193–208
9. Asmathunisha N, Kathiresan K (2013) A review on biosynthesis of nanoparticles by marine organisms. Colloids Surf B Biointerfaces 103:283–287
10. Atiqah A (2020) Characterization and interface of natural and synthetic hybrid composites. Elsevier Ltd
11. Avérous L, Halley PJ (2009) Biocomposites based on plasticized starch. Biofuels, Bioprod Biorefin 3:329–343
12. Azizi S, Ahmad MB, Namvar F, Mohamad R (2014) Green biosynthesis and characterization of zinc oxide nanoparticles using brown marine macroalga Sargassum muticum aqueous extract. Mater Lett 116:275–277
13. Balla E, Daniilidis V, Karlioti G et al (2021) Poly (lactic Acid): a versatile biobased polymer for the future with multifunctional properties—from monomer synthesis, polymerization techniques and molecular weight increase to PLA applications. Polymers (Basel) 13:1822
14. Begum S, Fawzia S, Hashmi MSJ (2020) Polymer matrix composite with natural and synthetic fibres. Adv Mater Process Technol 6:547–564
15. Bledzki AK, Jaszkiewicz A, Scherzer D (2009) Mechanical properties of PLA composites with man-made cellulose and abaca fibres. Compos Part A Appl Sci Manuf 40:404–412
16. Calabrese L, Fiore V, Scalici T, Valenza A (2019) Experimental assessment of the improved properties during aging of flax/glass hybrid composite laminates for marine applications. J Appl Polym Sci 136:47203
17. Castro L, Blázquez ML, Muñoz JA et al (2013) Biological synthesis of metallic nanoparticles using algae. IET Nanobiotechnol 7:109–116
18. Chand S (2000) Review carbon fibers for composites. J Mater Sci 35:1303–1313
19. Chandrasekar M, Kumar TSM, Senthilkumar K et al (2021a) Socioeconomic impact of bio-based packaging bags. Bio-based Packag Mater Environ Econ Asp 427–435
20. Chandrasekar M, Kumar TSM, Senthilkumar K et al (2021b) Performance of natural fiber reinforced recycled thermoplastic polymer composites under aging conditions. In: Recycling of plastics, metals, and their composites. CRC Press, pp 127–139
21. Chandrasekar M, Siva I, Kumar TSM et al (2020) Influence of fibre inter-ply orientation on the mechanical and free vibration properties of banana fibre reinforced polyester composite laminates. J Polym Environ 1–12
22. Chawla K, Tekwani B (2013) Studies of glass fiber reinforced concrete composites. Int J Struct Civ Eng Res 2:176–182
23. Cheon J-W, Kim I-J, Kim J-H et al (2021) Silica-polyethersulfone core-shell nanoparticles as multifunctional filler for marine applications. Compos Part A Appl Sci Manuf 106721
24. Ciriminna R, Bright F V, Pagliaro M (2015) Ecofriendly antifouling marine coatings
25. Claverie M, McReynolds C, Petitpas A et al (2020) Marine-derived polymeric materials and biomimetics: an overview. Polymers (Basel) 12:1002
26. Critchfield MO, Judy TD, Kurzweil AD (1994) Low-cost design and fabrication of composite ship structures. Mar Struct 7:475–494
27. Datta LP, Manchineella S, Govindaraju T (2020) Biomolecules-derived biomaterials. Biomaterials 230:119633

28. Dhandapani A, Krishnasamy S, Muthukumar C et al (2020) Plastics in Marine Engineering. Ref Modul Mater Sci Mater Eng 1–12
29. Di Marco M, Baker ML, Daszak P et al (2020) Opinion: Sustainable development must account for pandemic risk. Proc Natl Acad Sci 117:3888–3892
30. Dinesh S, Elanchezhian C, Vijayaramnath B et al (2020) Experimental investigation of natural and synthetic hybrid composite for marine application. Mater Today Proc 22:322–329
31. Elanchezhian C, Ramnath BV, Ramakrishnan G et al (2018) Review on metal matrix composites for marine applications. Mater Today Proc 5:1211–1218
32. Elshaarawy RFM, Mustafa FHA, Sofy AR et al (2019) A new synthetic antifouling coatings integrated novel aminothiazole-functionalized ionic liquids motifs with enhanced antibacterial performance. J Environ Chem Eng 7:102800
33. Eroglu E, Chen X, Bradshaw M et al (2013) Biogenic production of palladium nanocrystals using microalgae and their immobilization on chitosan nanofibers for catalytic applications. RSC Adv 3:1009–1012
34. Faruk O, Bledzki AK, Fink H-P, Sain M (2012) Biocomposites reinforced with natural fibers: 2000–2010. Prog Polym Sci 37:1552–1596
35. Fawcett D, Verduin JJ, Shah M et al (2017) A review of current research into the biogenic synthesis of metal and metal oxide nanoparticles via marine algae and seagrasses. J Nanosci
36. Fiore V, Di Bella G, Valenza A (2011) Glass–basalt/epoxy hybrid composites for marine applications. Mater Des 32:2091–2099
37. Francavilla M, Pineda A, Romero AA et al (2014) Efficient and simple reactive milling preparation of photocatalytically active porous ZnO nanostructures using biomass derived polysaccharides. Green Chem 16:2876–2885
38. Fu Y, Wang W, Zhang L et al (2019) Development of marine antifouling epoxy coating enhanced with clay nanotubes. Materials (Basel) 12:4195
39. Fu Y, Yu J, Zhang Y, Meng Y (2014) Graphite coated with manganese oxide/multiwall carbon nanotubes composites as anodes in marine benthic microbial fuel cells. Appl Surf Sci 317:84–89
40. Georgios K, Silva A, Furtado S (2016) Applications of green composite materials. Biodegrad Green Compos 16:312
41. Gomes A, Matsuo T, Goda K, Ohgi J (2007) Development and effect of alkali treatment on tensile properties of curaua fiber green composites. Compos Part A Appl Sci Manuf 38:1811–1820
42. Gopi S, Amalraj A, Kalarikkal N et al (2019) Preparation and characterization of nanocomposite films based on gum arabic, maltodextrin and polyethylene glycol reinforced with turmeric nanofiber isolated from turmeric spent. Mater Sci Eng C 97:723–729
43. Harish S, Michael DP, Bensely a, et al (2009) Mechanical property evaluation of natural fiber coir composite. Mater Charact 60:44–49. https://doi.org/10.1016/j.matchar.2008.07.001
44. Harris PJF, Harris PJF (2009) Carbon nanotube science: synthesis, properties and applications. Cambridge University Press
45. Hasan KM, Horváth PG, Alpár T (2020) Potential natural fiber polymeric nanobiocomposites: A review. Polymers (Basel) 12:1072
46. Hayman B, Wedel-Heinen J, Brøndsted P (2008) Materials challenges in present and future wind energy. MRS Bull 33:343–353
47. Hu G, Cai S, Zhou Y et al (2018) Enhanced mechanical and thermal properties of poly (lactic acid)/bamboo fiber composites via surface modification. J Reinf Plast Compos 37:841–852. https://doi.org/10.1177/0731684418765085
48. Jesthi DK, Nayak RK (2019) Improvement of mechanical properties of hybrid composites through interply rearrangement of glass and carbon woven fabrics for marine application. Compos Part B Eng 168:467–475
49. Kandola BK, Deli D (2014) Flame-retardant thermoset nanocomposites for engineering applications. In: Polymer green flame retardants. Elsevier, pp 503–549
50. Kannan RRR, Stirk WA, Van Staden J (2013) Synthesis of silver nanoparticles using the seaweed Codium capitatum PC Silva (Chlorophyceae). South Afr J Bot 86:1–4

51. Karim Z, Afrin S, Husain Q, Danish R (2017) Necessity of enzymatic hydrolysis for production and functionalization of nanocelluloses. Crit Rev Biotechnol 37:355–370
52. Karthi N, Kumaresan K, Sathish S et al (2020) An overview: Natural fiber reinforced hybrid composites, chemical treatments and application areas. Mater today Proc 27:2828–2834
53. Kausar A (2021) Green nanocomposites for energy storage. J Compos Sci 5:202
54. Khanehzaei H, Ahmad MB, Shameli K, Ajdari Z (2014) Synthesis and characterization of Cu@ Cu_2O core shell nanoparticles prepared in seaweed Kappaphycus alvarezii Media. Int J Electrochem Sci 9:8189–8198
55. Galanis K (2002) Hull construction with composite materials for ships over 100 m in length by. Ocean Eng 1–119
56. Kotrikla A (2009) Environmental management aspects for TBT antifouling wastes from the shipyards. J Environ Manage 90:S77–S85
57. Krishnasamy S, Muthukumar C, Nagarajan R et al (2019) Effect of fibre loading and $Ca(OH)_2$ treatment on thermal, mechanical, and physical properties of pineapple leaf fibre/polyester reinforced composites. Mater Res Express 6:085545. https://doi.org/10.1088/2053-1591/ab2702
58. Krishnasamy S, Thiagamani SMK, Muthu Kumar C et al (2019) Recent advances in thermal properties of hybrid cellulosic fiber reinforced polymer composites. Int J Biol Macromol 141:1–13. https://doi.org/10.1016/j.ijbiomac.2019.08.231
59. Krishnasamy S, Thiagamani SMK, Muthukumar C et al (2019) Effects of stacking sequences on static, dynamic mechanical and thermal properties of completely biodegradable green epoxy hybrid composites. Mater Res Express 6:105351. https://doi.org/10.1088/2053-1591/ab3ec7
60. Kumar A, Krishna GL, Subramanian VA (2019a) Design and analysis of a carbon composite propeller for podded propulsion. In: proceedings of the fourth international conference in ocean engineering (ICOE2018). Springer, pp 203–215
61. Kumar D (2014) Mechanical characterization of treated bamboo natural fiber composite. Int J Adv Mech Eng 4:551–556
62. Kumar NV, Krishna BS, Chandrika NS (2019) Evaluation of properties of glass-banana-fiber reinforced hybrid fiber polymer composite. Mater today Proc 18:2137–2141
63. Kumar TSM, Senthilkumar K, Chandrasekar M et al (2020) Influence of fillers on the thermal and mechanical properties of biocomposites: an overview. In: Biofibers and biopolymers for biocomposites. Springer, pp 111–133
64. Laaouidi H, Tarfaoui M, Nachtane M, Trihi M (2021) Energy absorption characteristics in hybrid composite materials for marine applications under impact loading: Case of tidal current turbine. Int J Energy Res 45:5894–5911
65. Lee C, Hong S (2014) An overview of the synthesis and synthetic mechanism of poly (lactic acid). Mod Chem Appl 2:1–5
66. Lingaraju D, Ramji K, Pramiladevi M, Rajyalakshmi U (2012) Prediction of post underwater shock properties of polymer-clay/silica hybrid nanocomposites through regression models. Int J Mater Metall Eng 6:1081–1085
67. Ljungberg N, Wesslen B (2003) Tributyl citrate oligomers as plasticizers for poly (lactic acid): thermo-mechanical film properties and aging. Polymer (Guildf) 44:7679–7688
68. Madhusudhana HK, Desai B, Venkatesha CS (2018) Experimental investigation on parameter effects on fracture toughness of hemp fiber reinforced polymer composites. Mater Today Proc 5:20002–20012
69. Manjunath RN, Khatkar V, Behera BK (2020) Investigation on seawater ageing of PET-epoxy composites: an ecological and sustainable approach for marine applications. J Polym Environ 28:2289–2300
70. Mann GS, Singh LP, Kumar P, Singh S (2020) Green composites: a review of processing technologies and recent applications. J Thermoplast Compos Mater 33:1145–1171
71. Mathijsen D (2016) Leading the way in thermoplastic composites. Reinf Plast 60:405–407
72. Mieszkin S, Callow ME, Callow JA (2013) Interactions between microbial biofilms and marine fouling algae: a mini review. Biofouling 29:1097–1113

73. Mochane MJ, Mokhena TC, Mokhothu TH et al (2019) Recent progress on natural fiber hybrid composites for advanced applications: a review
74. Mooney BP (2009) The second green revolution? Production of plant-based biodegradable plastics. Biochem J 418:219–232
75. Mouritz AP, Gellert E, Burchill P, Challis K (2001) Review of advanced composite structures for naval ships and submarines. Compos Struct 53:21–42
76. Muffler K, Sana B, Mukherjee J, Ulber R (2015) Marine enzymes–Production & applications. In: Springer handbook of marine biotechnology. Springer, pp 413–429
77. Muralidharan M, Sathishkumar TP, Rajini N et al (2021) Evaluation of tensile strength retention and service life prediction of hydrothermal aged balanced orthotropic carbon/glass and Kevlar/glass fabric reinforced polymer hybrid composites. J Appl Polym Sci 139:51602
78. Netravali AN, Chabba S (2003) Composites get greener mater today 6: 22–29. http://dx. doi. org/10.1016. S1369-7021 (03) 427–429
79. Ng K-W, Lam W-H, Pichiah S (2013) A review on potential applications of carbon nanotubes in marine current turbines. Renew Sustain Energy Rev 28:331–339
80. Nicholls-Lee RF, Turnock SR (2007) Enhancing performance of a horizontal axis tidal turbine using adaptive blades. In: OCEANS 2007-Europe. IEEE, pp 1–6
81. Nirmal U, Hashim J, Megat Ahmad MMH (2015) A review on tribological performance of natural fibre polymeric composites. Tribol Int 83:77–104. https://doi.org/10.1016/j.triboint.2014.11.003
82. Palomba G, Epasto G, Crupi V (2021) Lightweight sandwich structures for marine applications: a review. Mech Adv Mater Struct 1–26
83. Parker HL, Dodson JR, Budarin VL et al (2015) Direct synthesis of Pd nanoparticles on alginic acid and seaweed supports. Green Chem 17:2200–2207
84. Patel V, Berthold D, Puranik P, Gantar M (2015) Screening of cyanobacteria and microalgae for their ability to synthesize silver nanoparticles with antibacterial activity. Biotechnol Rep 5:112–119
85. Pico D, Steinmann W (2016) Synthetic fibres for composite applications. In: Fibrous and textile materials for composite applications. Springer, pp 135–170
86. Praharaj S, Rout D, Nayak RK (2021) Synergistic effect of natural fiber-reinforced polymer composite. In: Hybrid natural fiber composites. Elsevier, pp 145–165
87. Princy KF, Gopinath A (2015) Eco-friendly synthesis and characterization of silver nanoparticles using marine macroalga Padina tetrastromatica. Int J Sci Res 4:1050–1054
88. Puttegowda M, Rangappa SM, Jawaid M et al (2018) Potential of natural/synthetic hybrid composites for aerospace applications. In: Sustainable composites for aerospace applications. Elsevier, pp 315–351
89. Raheem A, Subbaya D (2021) Material science research India
90. Rahman MA, Halfar J (2014) First evidence of chitin in calcified coralline algae: new insights into the calcification process of Clathromorphum compactum. Sci Rep 4:1–11
91. Rajak DK, Pagar DD, Menezes PL, Linul E (2019) Fiber-reinforced polymer composites: manufacturing, properties, and applications. Polymers (Basel) 11:1667
92. Rajathi FAA, Parthiban C, Kumar VG, Anantharaman P (2012) Biosynthesis of antibacterial gold nanoparticles using brown alga, Stoechospermum marginatum (kützing). Spectrochim Acta Part A Mol Biomol Spectrosc 99:166–173
93. Rajeshkumar S, Malarkodi C, Gnanajobitha G et al (2013) Seaweed-mediated synthesis of gold nanoparticles using turbinaria conoides and its characterization. J Nanostructure Chem 3:1–7
94. Ramamoorthy SK, Skrifvars M, Persson A (2015) A review of natural fibers used in biocomposites: plant, animal and regenerated cellulose fibers. Polym Rev 55:107–162
95. Ramesh M, Palanikumar K, Reddy KH (2013) Mechanical property evaluation of sisal–jute–glass fiber reinforced polyester composites. Compos Part B Eng 48:1–9
96. Rastogi RP, Incharoensakdi A (2014) Characterization of UV-screening compounds, mycosporine-like amino acids, and scytonemin in the cyanobacterium Lyngbya sp. CU2555. FEMS Microbiol Ecol 87:244–256

97. Reinhart TJ, Schneider SJ (1987) Engineered materials handbook. American Society for Metals
98. Rubino C, Bonet Aracil M, Gisbert-Payá J et al (2019) Composite eco-friendly sound absorbing materials made of recycled textile waste and biopolymers. Materials (Basel) 12:4020
99. Rwawiire S, Tomkova B, Militky J et al (2015) Development of a biocomposite based on green epoxy polymer and natural cellulose fabric (bark cloth) for automotive instrument panel applications. Compos Part B Eng 81:149–157
100. Safri A, Thariq M, Sultan H, Jawaid M (2018) Impact behaviour of hybrid composites for structural applications: a review. 133:112–121. https://doi.org/10.1016/j.compositesb.2017.09.008
101. Sanjay MR, Arpitha GR, Naik LL et al (2016) Applications of natural fibers and its composites: an overview. Nat Resour 7:108–114
102. Santos FA, Rebelo H, Coutinho M et al (2021) Low velocity impact response of 3D printed structures formed by cellular metamaterials and stiffening plates: PLA vs. PETg. Compos Struct 256:113128
103. Satyanarayana KG (2015) Recent developments in'green'composites based on plant fibers-preparation: Structure property studies. J Bioprocess Biotech 5:1
104. Selim MS, El-Safty SA, Shenashen MA et al (2020) Progress in biomimetic leverages for marine antifouling using nanocomposite coatings. J Mater Chem B 8:3701–3732
105. Sen R, Mullins G (2007) Application of FRP composites for underwater piles repair. Compos Part B Eng 38:751–758
106. Senthil Kumar P, Gunasundari E (2018) Nanocomposites: recent trends and engineering applications. Nano Hybrids Compos 20:65–80
107. Senthilkumar K, Saba N, Chandrasekar M et al (2021) Compressive, dynamic and thermo-mechanical properties of cellulosic pineapple leaf fibre/polyester composites: influence of alkali treatment on adhesion. Int J Adhes Adhes 106:102823
108. Senthilkumar K, Saba N, Rajini N et al (2018) Mechanical properties evaluation of sisal fibre reinforced polymer composites: a review. Constr Build Mater 174:713–729
109. Senthilkumar K, Siva I, Rajini N et al (2018b) Mechanical characteristics of tri-layer eco-friendly polymer composites for interior parts of aerospace application. In: Sustainable composites for aerospace applications. Elsevier, pp 35–53
110. Senthilkumar K, Siva I, Rajini N, Jeyaraj P (2015) Effect of fibre length and weight percentage on mechanical properties of short sisal/polyester composite. Int J Comput Aided Eng Technol 7:60–71
111. Senthilkumar K, Ungtrakul T, Chandrasekar M et al (2020) Performance of sisal/hemp bio-based epoxy composites under accelerated weathering. J Polym Environ 1–13
112. Shahroze RM, Chandrasekar M, Senthilkumar K et al (2021) Mechanical, interfacial and thermal properties of silica aerogel-infused flax/epoxy composites. Int Polym Process 36:53–59
113. Shibli SMA, Archana SR, Ashraf PM (2008) Development of nano cerium oxide incorporated aluminium alloy sacrificial anode for marine applications. Corros Sci 50:2232–2238
114. Singaravelu G, Arockiamary JS, Kumar VG, Govindaraju K (2007) A novel extracellular synthesis of monodisperse gold nanoparticles using marine alga, Sargassum wightii Greville. Colloids Surf B Biointerfaces 57:97–101
115. Singh AAMM, Raj FM, Franco PA, Binoj JS (2018) Evaluation of mechanical behavior of multifilament discarded fishnet/glass fiber and polyester composites for marine applications. Mar Struct 58:361–366
116. Singh SP, Jawaid M, Chandrasekar M et al (2021) Sugarcane wastes into commercial products: Processing methods, production optimization and challenges. J Clean Prod 328:129453
117. Subramanian K, Krishnasamy S, Thiagamani SMK et al Tribo performance analysis on polymer-based composites. In: Polymer-based composites. CRC Press, pp 115–130
118. Sukur EF, Onal G (2021) Long-term salt-water durability of GNPs reinforced basalt-epoxy multiscale composites for marine applications. Tribol Int 158:106910

119. Tawfik MM (2017) Experimental studies of nanofluid thermal conductivity enhancement and applications: a review. Renew Sustain Energy Rev 75:1239–1253
120. Tekalur SA, Shukla A, Shivakumar K (2008) Blast resistance of polyurea based layered composite materials. Compos Struct 84:271–281
121. Thakur VK, Thakur MK, Gupta RK (2014) raw natural fiber–based polymer composites. Int J Polym Anal Charact 19:256–271
122. Thiagamani SMK, Krishnasamy S, Siengchin S (2019) Challenges of biodegradable polymers: an environmental perspective. Appl Sci Eng Prog x:1. https://doi.org/10.14416/j.asep.2019.03.002
123. Thomas SK, Parameswaranpillai J, Krishnasamy S et al (2021) A comprehensive review on cellulose, chitin, and starch as fillers in natural rubber biocomposites. Carbohydr Polym Technol Appl 100095
124. Thompson RC, Moore CJ, Vom Saal FS, Swan SH (2009) Plastics, the environment and human health: current consensus and future trends. Philos Trans R Soc B Biol Sci 364:2153–2166
125. Thyavihalli Girijappa YG, Mavinkere Rangappa S, Parameswaranpillai J, Siengchin S (2019) Natural fibers as sustainable and renewable resource for development of eco-friendly composites: a comprehensive review. Front Mater 6:226
126. Todor MP, Bulei C, Heput T, Kiss I (2018) Researches on the development of new composite materials complete/partially biodegradable using natural textile fibers of new vegetable origin and those recovered from textile waste. In: IOP conference series: materials science and engineering. IOP Publishing, p 12021
127. Vardhan DH, Ramesh A, Reddy BCM (2019) A review on materials used for marine propellers. Mater Today Proc 18:4482–4490. https://doi.org/10.1016/j.matpr.2019.07.418
128. Verma D, Fortunati E, Jain S, Zhang X (2019) Biomass, biopolymer-based materials, and bioenergy: Construction, biomedical, and other industrial applications. Woodhead Publishing
129. Verma S, Das S, Mohanty S, Nayak SK (2019) Development of multifunctional polydimethylsiloxane (PDMS)-epoxy-zinc oxide nanocomposite coatings for marine applications. Polym Adv Technol 30:2275–2300
130. Vijayakumar K, Prabhu L, Subin BS, et al (2020) Development of hybrid aluminium metal matrix composites for marine applications. In: IOP conference series: materials science and engineering. IOP Publishing, p 12016
131. Wadley HNG (2006) Multifunctional periodic cellular metals. Philos Trans R Soc A Math Phys Eng Sci 364:31–68
132. Xiang B, Jia Y, Lei Y et al (2019) Mechanical properties of microcellular and nanocellular silicone rubber foams obtained by supercritical carbon dioxide. Polym J 51:559–568
133. Xie C, Li C, Xie Y et al (2021) ZnO/acrylic polyurethane nanocomposite superhydrophobic coating on aluminum substrate obtained via spraying and Co-curing for the control of marine biofouling. Surf Interfaces 22:100833
134. Yang L, Kong X, Li F et al (2019) Perovskite lead-free dielectrics for energy storage applications. Prog Mater Sci 102:72–108
135. Yee MS-L, Khiew P-S, Chiu WS et al (2016) Green synthesis of graphene-silver nanocomposites and its application as a potent marine antifouling agent. Colloids Surf B Biointerfaces 148:392–401
136. Youh M-J, Huang Y-R, Peng C-H et al (2021) Using graphene-based composite materials to boost anti-corrosion and infrared-stealth performance of epoxy coatings. Nanomaterials 11:1603
137. Zhao H, Nathaniel GA, Merenini PC (2017) Enzymatic ring-opening polymerization (ROP) of lactides and lactone in ionic liquids and organic solvents: digging the controlling factors. RSC Adv 7:48639–48648

Tensile Behavior of Weft-Knitted Structure for Potential Use in Composite Reinforcement via Factorial and 3D Surface

Thiago F. Santos, Caroliny M. Santos, Emad K. Hussein, Lucas Zilio, Mariana Dias, M. R. Sanjay, Rubens Fonseca, Adriano Amaral, and Marcos Aquino

Abstract This chapter investigated the tensile behavior of weft-knitted structures for their potential use as composite reinforcement. A factorial design and 3D surface analysis were used to explore the effects of knitting parameters, such as yarn type, stitch density, and loop length, on the tensile strength, strain, and modulus of the knitted fabric. Weft knitted fabrics structures are popular for traditional wear because of their elastic and light structures, gentle smoothness, low production costs, and high productivity. Several parameters affect knitted fabrics behavior, the most important factors that determine the fabric properties are the fiber type, length loop, and float stitch. Many types of natural and synthetic fibers be used according to the usage areas and expected performance characteristics of the knitted fabrics. Therefore, it has great importance to know the effects of fiber type, length loop, and float stitch which have different sources, structures, and properties, on the fabric properties. In this chapter, the performance and behavior of tensile properties in three different knitted fabrics structure made from natural, and synthetic fibers (100% CO, 100% PET, 100% PA, and 67% PET/33% CO) were studied. The results showed that the tensile properties of the weft-knitted structure were highly dependent on the knitting parameters, with the yarn type having the most significant effect on the tensile strength and modulus. The study also revealed that increasing the stitch density and loop length resulted in higher

T. F. Santos (✉) · C. M. Santos (✉) · L. Zilio · M. Dias · R. Fonseca · A. Amaral · M. Aquino
Textiles Technologies Study Group (GETTEX), Laboratory of Knitting, Textile Engineering Laboratory, Department of Textile Engineering, Federal University of Rio Grande Do Norte, Natal, Rio Grande Do Norte, Brazil
e-mail: thiagotextilufrn@hotmail.com

C. M. Santos
e-mail: carolinyminely@hotmail.com

M. R. Sanjay
Natural Composite Research Group Lab, Department of Materials and Production Engineering, The Sirindhorn International Thai-German Graduate School of Engineering (TGGS), King Mongkuts University of Technology North Bangkok (KMUTNB), Bangkok, Thailand

E. K. Hussein
Mechanical Power Engineering Department, Mussaib Technical College, Al Furat Al Awsat Technical University, Mussaib, P.O. Box 51006, Babil, Iraq

T. Khan and M. Jawaid (eds.), *Green Hybrid Composite in Engineering and Non-Engineering Applications*, Composites Science and Technology,
https://doi.org/10.1007/978-981-99-1583-5_14

tensile strength and strain, while the tensile modulus remained relatively constant. This research provides valuable insights into the tensile behavior of weft-knitted structures and their potential use as composite reinforcement materials.

Keywords Textile composite · Mechanical properties · Knitted fabric · Natural fiber · Synthetic

1 Introduction

1.1 Textile Knitted Structures

The textile industry has developed the ability to produce distinctive fabrics using highly automated techniques such as weaving, braiding and knitting. Knitted structures, in particular, assume both simple and complex forms, depending on the desired application. When used in engineering applications, especially in the field of composite fabrication, an excellent property is noticed, which is its formability [1–3]. It is possible to distinguish the different fabrics according to their manufacturing method. Knitted are produced by intertwining loops of yarn with knitting needles. The term knitting is related to the technique for the production of knitted by intertwining yarn loops. In this way, a continuous series of loop stitches is formed by the machine needles. The next loop is formed by taking the yarn and pulling it through a previously formed loop [4]. Thus, the next new loop is born. Textile materials impart properties that promote enhanced mechanical potential, and some of the traditional textile technologies have been adopted to manufacture knitted reinforcement for advanced polymeric composites. Knitting is particularly suitable for the rapid fabrication of components with complex shapes due to the high strain of knitted and the flexibility provided by the weft-knitted technology allows the production of a wide range of structures with different properties. Normally, the greatest strain of the knitted weft occurs in the direction of the course (horizontal direction), and, during mechanical stress, there is the contraction of the knitted in the direction of the wales (vertical direction). When it comes to the mechanical properties of the knitted weft, it is noticed that they are related to the properties of the yarn, direction, and structure of the knitted. With this, it is possible to work on these properties, since they be designed according to the needs of a given application. Thus, it is possible to choose the fibers and yarns with the most suitable properties and place them in the structure most suited to the need [2, 3, 5, 6].

1.2 Tensile Behavior of Weft-Knitted Structure

The chapter provides insights into the fundamental mechanics of weft-knitted structures and their potential use as composite reinforcement materials, which help in the development of high-performance composites for various industrial applications. The tensile behavior of these knitted structures is a critical factor in determining their suitability for use as composite reinforcement materials. In this context, this study aims to investigate the tensile behavior of weft-knitted structures under various knitting parameters such as yarn type, stitch density, and loop length. The study utilizes a factorial design and 3D surface analysis to explore the effects of these parameters on the tensile strength, strain, and modulus of the knitted fabric. The mechanical behavior of fabric materials has attracted increasing interest in scientific research, as these materials be used in a wide range of applications. In particular, textile composites in which fabric structures are used as preforms are gaining increasing attention alongside traditional apparel and fashion textile fabrics. Technical textiles that exhibit excellent mechanical behavior at the time of application have attracted the attention of engineers and designers. In this way, the use of these textiles in various applications such as transportation, construction, automotive, aerospace, and clothing has increased.

1.3 Knitted-Reinforced Polymer Composites

Composite materials have gained significant attention in various industrial applications due to their high strength and stiffness-to-weight ratio. Reinforcing these materials with fibers further enhance their mechanical properties. Among different fiber-reinforcement methods, weft-knitted structures are widely used due to their unique properties such as high extensibility and conformability. Several authors have studied the properties of knitted fabrics as reinforcement elements in polymeric composite materials. Duhovic 2011 [7], verified the influence of the types of fibers, yarns, and knitted structures used to produce knitted reinforced composites. In addition, the authors also present the techniques used to combine them with different types of polymer matrices. In this study, some precursor technologies of commercially available textile composites (prepreg) are shown, in addition to manufacturing methods, mechanical properties, and applications. Ruan 1996 [8], in turn, presented experimental and theoretical studies on the elastic behavior of knitted composites, manufacturing two types of preforms of weft knitted based on single stitch and ribbed knitted. Demircan 2015 [9], studied the effect of various knitting techniques on the mechanical properties of thermoplastic composites knitted with biaxial weft, using five types of knitted knitted with biaxial weft (simple, interlock, tuck, tuck-miss and interlock2) and using these structures as reinforcement systems for manufacture thermoplastic composites with polypropylene resin (PP). Finally, the mechanical properties of the composites were investigated with various mechanical tests.

2 Manufacture of Knitted Structures

As exhibited in Fig. 1, three knitted fabrics structures; derivative of double Lacoste was produced from 100% 100% cotton (100% CO with 5.7 Ne), 67% Polyester/33% cotton (67% PET/33% CO with 6.7 Ne), synthetic yarns constituted of 100% Polyester (100% PET with 8.4 Ne) and 100% polyamide (100% PA with 8.4 Ne) all acquired from a textile industry. The knitted fabrics were produced in a Cixing double-fronture (electronic) flat knitting machine, model GE2-52C, which has 346 needles and a 7/inch gauge. Independent variables used in this chapter were the structure (density of float stitch), composition (100% CO, 100% PET, 100% PA and 67% PET/33% CO) and length loop (0.98, 0.84 and 0.71 cm). And all experimental planning was carried out using a factorial design (2^3) of variables with 2 central points and 2 repetitions in the statistical software [10–12].

3 Characterization of Knitted Structures

3.1 Tensile Test

Before of the characterization, all the knitted fabric prepared for this study were conditioned according to ASTM D1776/D1776M-20 [13]. For this study, all the specimens were conditioned in a conditioning room for 24 h at 20 ± 2 ºC temperature and 65 ± 2% relative humidity room condition. the knitted fabric condition is essential as high or low humidity affect the fibers. The knitted fabric exhibited in Fig. 1 were cut to dimensions of 150 mm length and 25 mm wide. Tensile tests were performed with the MESDAN Tensolab 3000 dynamometer, using the ASTM D5034 adaptation [14]. And knitted fabrics for each condition (float stitches, composition and loop length) were tested at a rate of 300 mm/min and distance between grips of 100 mm. Finally, the results were treated in the Origin software [15, 16].

3.2 Statistical Analysis of Knitted Fabrics

The full factorial treatment, analysis of variance (ANOVA) and 3D response surface were performed and the response properties (stress, MOE, strain, and work of energy) were analyzed, and discussed by according to the current literature [17, 18]. Design Expert software was used at this stage to validate the statistical study and correlations between structural characteristics of knitted fabrics [19, 20].

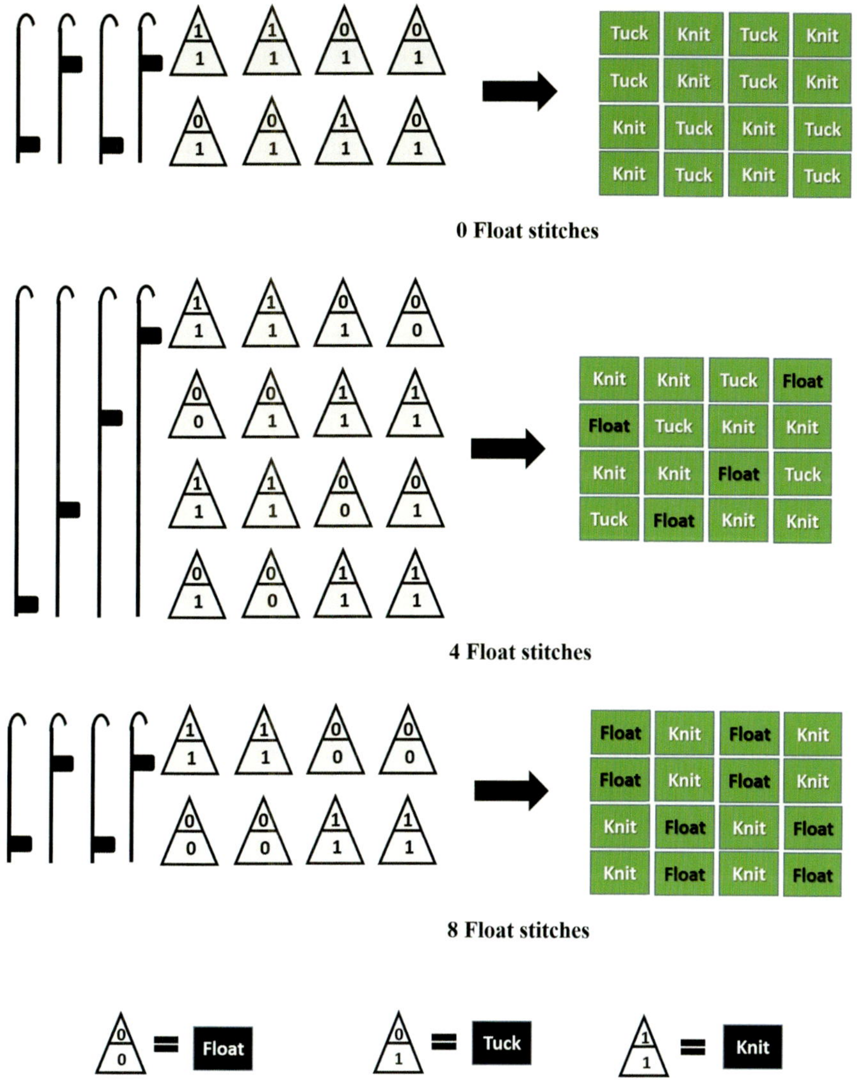

Fig. 1 Illustration of the knitted structures studied

4 Results and Discussion

4.1 Tensile Behavior in the Synthetic Fibers

When a knitted fabric is stretched, the individual yarns in the fabric are pulled in different directions. The tension on the yarns causes them to break, which weaken the overall structure of the fabric. However, by incorporating float stitches into the fabric, the tension on the yarns be distributed more evenly, reducing the risk of individual yarns breaking [21]. Float stitches are created by knitting a yarn over one or more adjacent stitches without looping the yarn through the stitch. This creates a "float" of yarn on the surface of the fabric, which provide additional strength and stability. In the case of synthetic yarns such as PA and PET, which are often used in high-performance fabrics for their strength and durability, the use of float stitches be particularly effective. These yarns are generally more resistant to stretching than natural fibers such as cotton or wool, which make them more prone to breaking under tension as exhibited in Fig. 2. By incorporating float stitches into the fabric, the stress on the yarns be distributed more evenly, reducing the risk of breakage and improving the overall tensile behavior of the fabric [22].

The increase in float stitches in knitted fabric with synthetic yarns such as 100% PA (polyamide) and 100% PET (polyethylene terephthalate) improve the stress behavior of the fabric in several ways. Firstly, increasing the number of float stitches improve the elasticity and flexibility of the fabric. Float stitches are created when the yarn is carried over several stitches without being knitted, which results in longer loops that stretch and contract more easily than tightly knitted stitches. This increased elasticity helps the fabric to better withstand stress and deformation. Secondly, increasing the number of float stitches also improve the strength of the fabric. The longer loops created by float stitches interlock with neighboring loops, creating a more tightly integrated fabric structure. This improves the resistance of the fabric to tearing or breaking under stress. Thus, increasing the number of float stitches also improve the breathability and moisture-wicking properties of the fabric [23]. The longer loops created by float stitches create more open spaces within the fabric, allowing air and moisture to circulate more freely. This help to keep the wearer cool and dry, reducing the likelihood of stress-related discomfort or irritation. Length loops are created by extending the length of the yarn between the needles during the knitting process, resulting in longer stitches that provide greater flexibility and stretch. The longer stitches created by length loops help the fabric better resist stress and deformation [24]. Firstly, longer loops create more surface area for the yarn to interlock and form a stronger bond with neighboring stitches. This result in a fabric that is more resistant to tearing and breaking under stress. Secondly, longer loops increase the elasticity and flexibility of the fabric as exhibited in Fig. 2.

Longer loops stretch and contract more easily than shorter loops, allowing the fabric to better accommodate movements and stresses placed on it. Thirdly, longer loops improve the breathability and moisture-wicking properties of the fabric. The increased surface area of longer loops creates more open spaces within the fabric, allowing air and moisture to circulate more freely. This helps to keep the wearer cool and dry, reducing the likelihood of stress-related discomfort or irritation. the increase in length loop in knitted fabric with synthetic yarns have several benefits for stress behavior, including improved strength, elasticity, breathability, and moisture-wicking properties. However, increasing the number of float stitches, composition and length loop in knitted fabric with synthetic yarns have several benefits for stress behavior, including improved elasticity, strength, breathability, and moisture-wicking

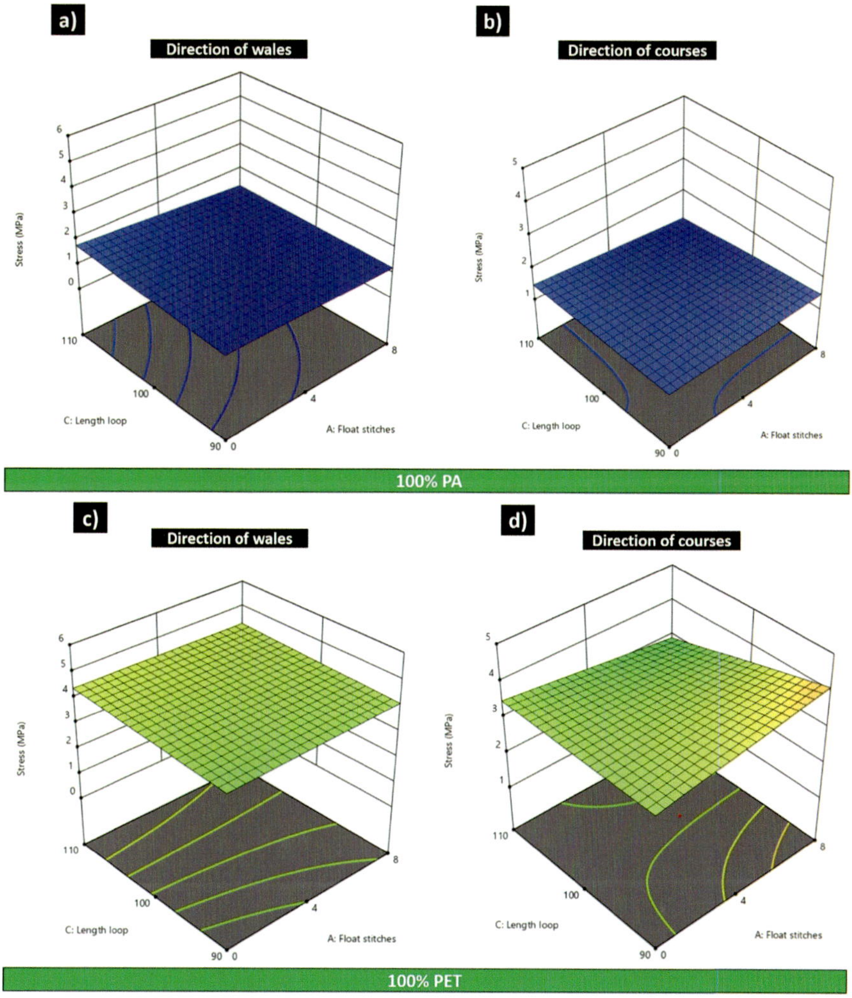

Fig. 2 3D surface of Stress in weft knitted fabrics with synthetics fibers

properties [25]. Incorporating 8 float stitches per inch of fabric in knitted fabrics made from synthetic yarns such as 100% PA (polyamide) and 100% PET (polyethylene terephthalate) improve the stress behavior of the fabric. When a knitted fabric is stretched, the tension on the yarns causes the individual fibers to break, which weaken the overall structure of the fabric. By incorporating float stitches into the fabric, the tension on the yarns be distributed more evenly, reducing the risk of individual fibers breaking [26] as exhibited in Fig. 2. In addition, float stitches help to control the structure of the fabric, which also affect its stress behavior. For example, float stitches help to create a more even distribution of tension across the fabric, which improve the fabric's resistance to stress and deformation. Research has shown that incorporating around 8 float stitches per inch of fabric be effective for improving the stress behavior of knitted fabrics made from 100% PA and 100% PET. This number may vary depending on the specific characteristics of the yarn and the structure of the fabric, but 8 float stitches per inch is a good starting point [26, 27]. It's worth noting that float stitches also affect other properties of the fabric, such as its stretch and recovery, its drape, and its dimensional stability. Therefore, the number and placement of float stitches should be carefully considered in relation to the desired properties of the fabric [26].

The increase in float stitches in knitted fabric with synthetic yarns such as 100% PA (polyamide) and 100% PET (polyethylene terephthalate) improve the strain behavior of the fabric in several ways. Increasing the number of float stitches (8, 4 and 0) improve the elasticity and flexibility of the fabric. Float stitches are created when the yarn is carried over several stitches without being knitted, which results in longer loops that stretch and contract more easily than tightly knitted stitches. This increased elasticity helps the fabric to better withstand strain and deformation. Increasing the number of float stitches also improve the strength of the fabric. The longer loops created by float stitches interlock with neighboring loops, creating a more tightly integrated fabric structure. This improve the resistance of the fabric to tearing or breaking under strain. The increase in length loop in knitted fabric with synthetic yarns such as 100% PA (polyamide) and 100% PET (polyethylene terephthalate) create more surface area for the yarn to interlock and form a stronger bond with neighboring stitches. This result in a knitted fabric that is more resistant to breaking under strain [22]. Then, increase the elasticity and flexibility of the fabric. Longer loops stretch and contract more easily than shorter loops, allowing the fabric to better accommodate movements and strains placed on it. Overall, the increase in length loop in knitted fabric with synthetic yarns have several benefits for strain behavior as exhibited in Fig. 3.

The Modulus of Elasticity (MOE) refers to the stiffness of a material, or its ability to resist deformation when a load is applied. Knitted fabrics made with synthetic yarns such as 100% polyamide (PA) and 100% polyethylene terephthalate (PET) have been found to exhibit improved MOE compared to other materials. Synthetic yarns generally have higher tensile strength than natural fibers such as cotton or wool, which means they withstand greater stress before breaking. This higher tensile strength contributes to increased stiffness and improved MOE. Additionally, knitted fabrics made with synthetic yarns be engineered to have specific properties such as higher density or tighter knit structures, which also contribute to increased stiffness and improved MOE. These fabrics also be treated with various chemical finishes or coatings to enhance their performance properties. the use of synthetic yarns in knitted fabrics lead to improved MOE due to their inherent properties such as higher tensile strength, as well as their ability to be engineered and treated for specific performance properties [27]. An increase in float stitches in knitted fabrics made with synthetic yarns such as 100% PA and 100% PET contribute to improved MOE (Modulus of Elasticity), which is a measure of a material's stiffness as exhibited in Fig. 4.

Float stitches are created when a yarn is carried across the back of a stitch without being knit into the fabric. In knitted fabrics made with synthetic yarns, increasing

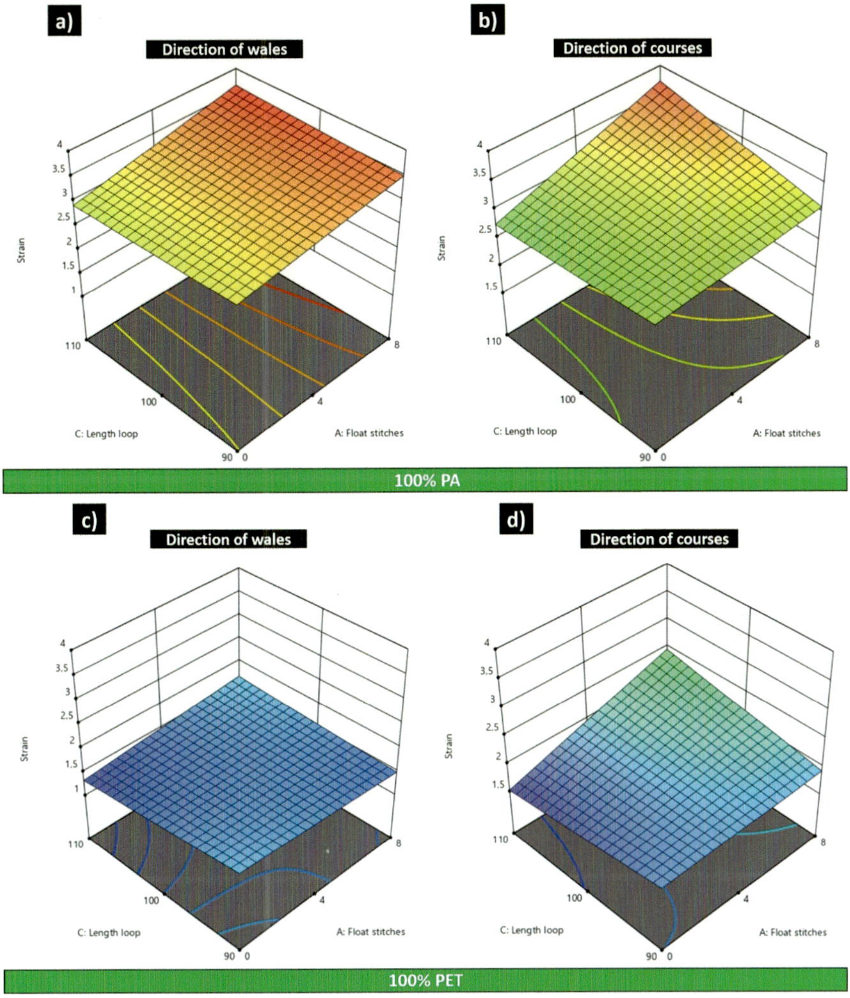

Fig. 3 3D surface of strain in weft knitted fabrics with synthetics fibers

the number of float stitches create a denser, more tightly-knit structure. This denser structure lead to increased stiffness and improved MOE. In addition, the orientation and alignment of the yarns in the fabric also contribute to improved MOE. When synthetic yarns are knitted in a way that aligns the yarns in a single direction, this creates a fabric with greater tensile strength and stiffness [28]. By increasing the number of float stitches, the yarns be oriented and aligned more precisely, leading to improved MOE. Furthermore, the use of synthetic yarns themselves contribute to the improved MOE of knitted fabrics. Synthetic yarns such as 100% PA and 100% PET are known for their high tensile strength and resilience, which further enhance the stiffness and MOE of the fabric.

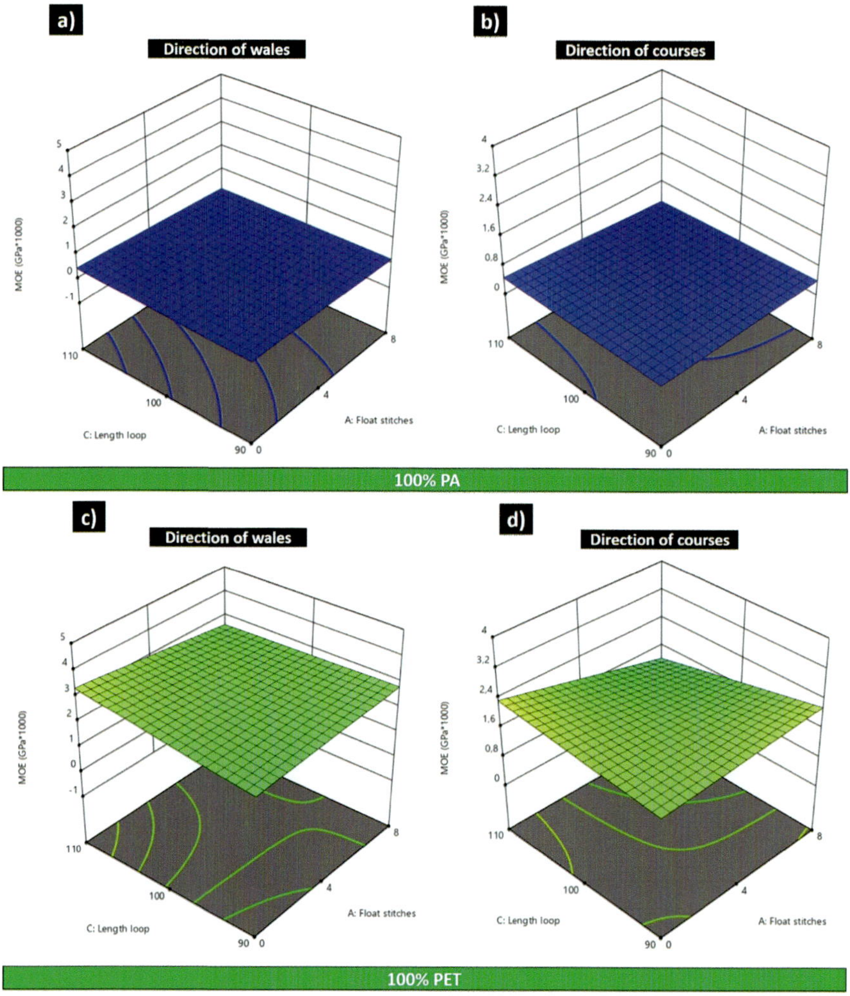

Fig. 4 3D surface of MOE in weft knitted fabrics with synthetics fibers

An increase in the length of loops in knitted fabrics made with synthetic yarns contribute to improved MOE (Modulus of Elasticity), which is a measure of a material's stiffness. Length of loops is an important factor that affects the structure and properties of a knitted fabric. Longer loops create a more open and flexible structure, while shorter loops create a denser and more tightly-knit structure. In knitted fabrics made with synthetic yarns, increasing the length of loops create a more open structure, which lead to increased flexibility and improved MOE. When synthetic yarns are knitted in a way that aligns the yarns in a single direction, this creates a fabric with greater tensile strength and stiffness. By increasing the length of loops, the yarns be oriented and aligned more precisely, leading to improved MOE. Furthermore, the use of synthetic yarns themselves contribute to the improved MOE of knitted fabrics. Synthetic yarns such as 100% PA and 100% PET are known for their high tensile strength and resilience, which further enhance the stiffness and MOE of the fabric. Overall, increasing the number of float stitches, length of loops in knitted fabrics made with synthetic yarns such as 100% PA and 100% PET contribute to a denser, more tightly-knit structure that enhances the orientation and alignment of the yarns and improves the stiffness and MOE of the fabric as exhibited in Fig. 4.

The work of rupture is a measure of the energy required to break a material, which is an important characteristic of its mechanical properties. Knitted fabrics made with synthetic yarns such as 100% PA (polyamide) and 100% PET (polyethylene terephthalate) have been found to exhibit different levels of work of rupture. In general, knitted fabrics made with 100% PA tend to be superior to those made with 100% PET in terms of work of rupture. One reason for this is that polyamide fibers have a higher elongation at break than PET fibers. This means that PA fibers be stretched further before breaking, which result in a higher work of rupture. Additionally, PA fibers have a higher tenacity than PET fibers, which means they are more resistant to breaking under stress. Furthermore, the structure of the knitted fabric also plays a role in the work of rupture. Knitted fabrics made with 100% PA tend to have a more open and flexible structure, which allow for greater elongation and deformation before breaking. On the other hand, knitted fabrics made with 100% PET tend to have a denser and more tightly-knit structure, which may result in a lower work of rupture. It's important to note that the specific properties of the knitted fabric will depend on a variety of factors, including the type of yarn, the knitting process, and any finishing treatments applied to the fabric [29]. Therefore, while knitted fabrics made with 100% PA may generally exhibit superior work of rupture compared to those made with 100% PET, there may be instances where the opposite is true depending on the specific characteristics of the fabric as exhibited in Fig. 5.

The number of float stitches in a knitted fabric affect its mechanical properties, including its work of rupture. A study comparing knitted fabrics made with 8 float stitches and 100% PA (polyamide) to those made with 100% PET (polyethylene terephthalate) found that the 100% PA fabric was superior in terms of work of rupture. One possible explanation for this is that polyamide fibers have a higher elongation at break than PET fibers, meaning they stretch further before breaking. This increased elongation result in a higher work of rupture for the 100% PA fabric. Additionally, polyamide fibers have a higher tenacity than PET fibers, which means they are more resistant to breaking under stress. The presence of 8 float stitches may also contribute to the superior work of rupture in the 100% PA fabric. Float stitches are created when a yarn was carried across the back of a stitch without being knit into the fabric, which create a more open and flexible structure. This increased openness and flexibility may allow for greater elongation and deformation before breaking, leading to a higher work of rupture. It's worth noting that the specific properties of a knitted fabric will depend on a variety of factors, including the length loop, composition and float stitches [30]. Therefore, while the 100% PA fabric with 8 float stitches may have been superior in terms of work of rupture in this particular study.

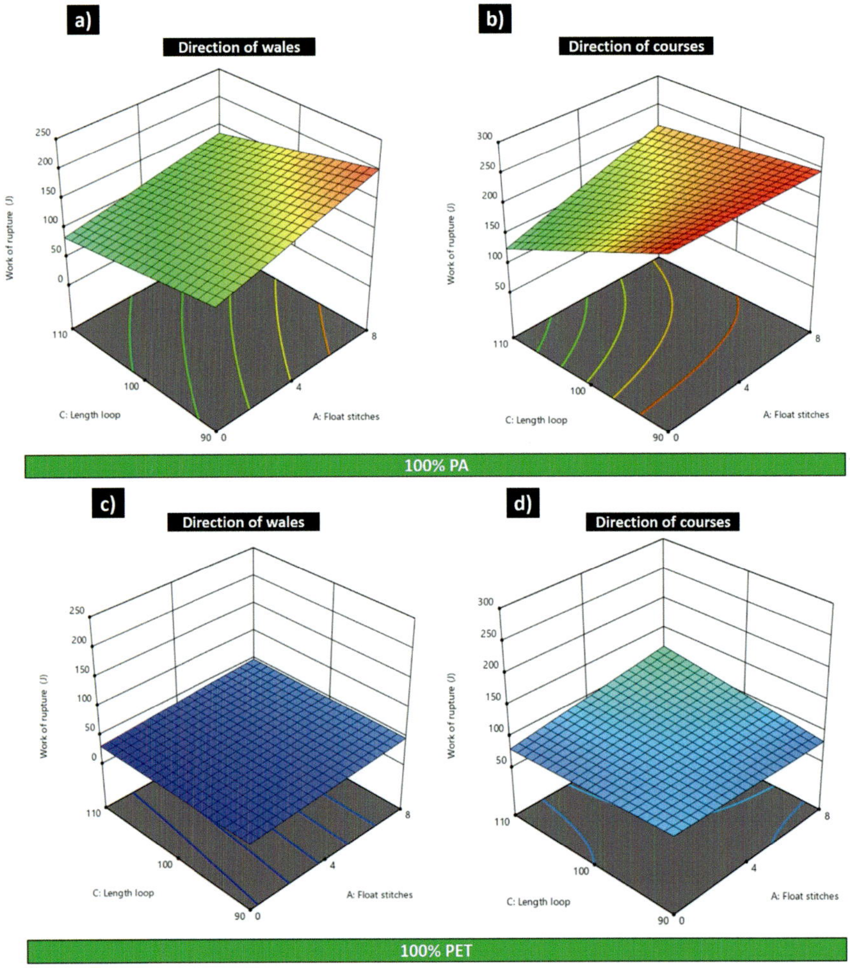

Fig. 5 3D surface of work of rupture in weft knitted fabrics with synthetics fibers

The length of loops in a knitted fabric affects its mechanical properties, including its work of rupture. A study comparing knitted fabrics made with 90 length loops and 100% PA (polyamide) to those made with 100% PET (polyethylene terephthalate) found that the 100% PA fabric was superior in terms of work of rupture. One possible explanation for this is that polyamide fibers have a higher elongation at break than PET fibers, meaning they stretch further before breaking. This increased elongation result in a higher work of rupture for the 100% PA fabric. Additionally, polyamide fibers have a higher tenacity than PET fibers, which means they are more resistant to breaking under stress [25]. Longer loops create a more open and flexible structure in the knitted fabric, which allow for greater elongation and deformation before breaking. This increased flexibility may contribute to the higher work of rupture observed in the 100% PA fabric as exhibited in Fig. 5.

4.2 Tensile Behavior in the Natural Fibers

The stress in a knitted fabric refers to the force required to deform the fabric, and it is an important mechanical property that affect the durability and performance of the fabric. The stress–strain behavior of a knitted fabric be influenced by various factors, including the type of yarn used. In a study comparing knitted fabrics made with synthetic yarns 100% CO (cotton) to those made with 67% CO/33% PET (polyethylene terephthalate), it was found that the 100% CO fabric was superior in terms of stress. This was that cotton fibers have a higher elongation at break than PET fibers. This means that cotton fibers be stretched further before breaking, which result in a higher stress. Additionally, cotton fibers have a lower modulus of elasticity than PET fibers, which means they are more flexible and less stiff. This flexibility may allow for greater deformation before breaking, contributing to the higher stress observed in the 100% CO knitted fabric. The presence of PET fibers in the 67% CO/33% PET fabric may also contribute to its lower stress. PET fibers have a higher modulus of elasticity than cotton fibers, which makes the fabric stiffer and less flexible. This stiffness may result in less deformation before breaking and a lower stress as exhibited in Fig. 6.

Knitted fabrics made with different yarn compositions exhibit different levels of stress. Additionally, cotton fibers have a lower stiffness than PET fibers, which make the fabric more flexible and easier to deform under stress. Knitted fabrics made with 100% CO (cotton) were found to be superior to those made with a blend of 67% CO and 33% PET (polyethylene terephthalate) in terms of stress. Cotton fibers have a higher compressibility than PET fibers. This means that they compress more under stress, which help to absorb and distribute the force, leading to a lower stress level. In contrast, the PET fibers in the blend have a higher stiffness and lower compressibility, which result in a fabric that is less flexible and more resistant to deformation under stress. This increased resistance lead to a higher stress level in the fabric as exhibited in Fig. 6.

The knitted fabrics made with 8 float stitches and 100% CO (cotton) in the courses direction were superior in terms of stress compared to those made with a blend of 67% CO and 33% PET (polyethylene terephthalate). The presence of float stitches in the knitted fabric further increases its compressibility by creating a more open and flexible structure. This allows the fabric to deform more easily and absorb stress more effectively, resulting in a lower stress level. The PET fibers in the blend have a lower compressibility and higher stiffness, which make the fabric less flexible and more resistant to deformation under stress [31]. The presence of float stitches in a knitted fabric affects its mechanical properties, including its stress performance. However, in the case of the study you mention, knitted fabrics made with 0 float stitches and 100% CO (cotton) in the wale direction were found to be superior to those made with

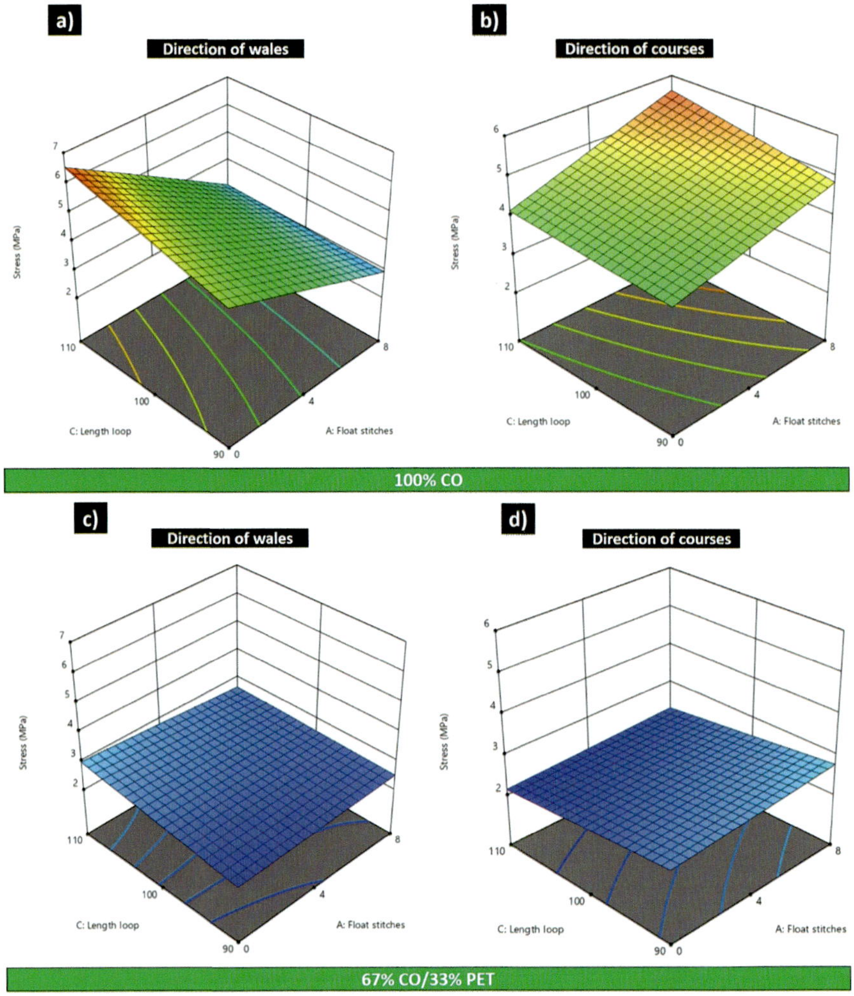

Fig. 6 3D surface of Stress in weft knitted fabrics with natural fibers

a blend of 67% CO and 33% PET (polyethylene terephthalate) in terms of stress as exhibited in Fig. 6.

Knitted fabrics made with 100% CO (cotton) in the wale direction were found to be superior to those made with a blend of 67% CO and 33% PET (polyethylene terephthalate) in terms of strain. Fibers are known to have higher elongation at break than PET fibers, meaning that they stretch more before breaking. This higher elongation contributes to the superior strain performance of the 100% CO fabric. Additionally, the structure of the fabric also plays a role in its strain performance. Knitted fabrics made with 100% CO in the wale direction may have had a more open structure,

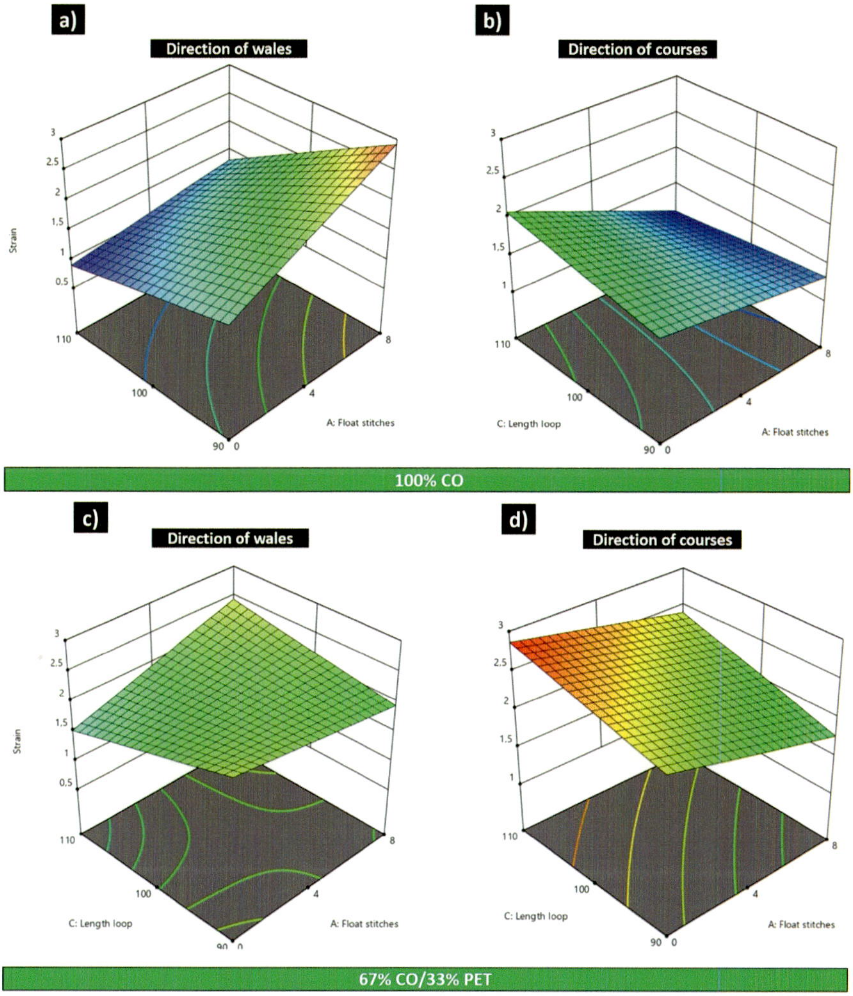

Fig. 7 3D surface of Strain in weft knitted fabrics with natural fibers

which provide better flexibility and allow for more stretching without causing deformation or breakage. In contrast, the blend fabric may have been more rigid due to the presence of PET fibers, which could limit its ability to stretch and cause it to break at a lower strain. However, the composition of 100% CO was the highest value of strain property when compared to the hybrid composition, higher were obtained in the course direction. In Fig. 7 show the strain property of knitted fabric structures composed of 100% CO and 67% PET 33% CO yarns, respectively, in the wale direction. From Fig. 7, it was observed that in the wale direction, the increase of float stitches density and loop length promoted critical changes differently from the course direction in knitted fabric structures composed of yarns as exhibited in Fig. 7.

It was found that knitted fabrics made with 100% CO (cotton) in the wale direction and with either 0 or 8 float stitches were superior to those made with a blend of 67% CO and 33% PET (polyethylene terephthalate) in terms of strain. The absence of float stitches in the fabric creates a more compact and dense fabric structure, which provide better resistance to strain. This be particularly beneficial for cotton fibers, which have a higher elongation at break than PET fibers. The cotton fibers in the 100% CO fabric may have contributed to the superior strain performance by being more compressible and better able to absorb and distribute the force under strain, resulting in less deformation or breakage. The presence of float stitches, on the other hand, create more open spaces in the fabric structure, which reduce its resistance to strain. However, the knitted fabrics with 8 float stitches and 100% CO in the wale direction still outperformed the blend fabric in terms of strain. This may be due to the fact that the presence of float stitches also increases the flexibility and stretchiness of the fabric, which contribute to its strain performance. Therefore, while the 100% CO fabric with 0 or 8 float stitches in the wale direction may have been superior in terms of strain in this particular study as exhibited in Fig. 7.

The knitted fabric made of 100% cotton yarns or a blend of 67% cotton and 33% polyester (CO/PET) with improved MOE (modulus of elasticity) is superior depends on the specific needs and preferences of the application. Knitted fabrics of 100% cotton yarns are known for their softness, breathability, and comfort. They are hypoallergenic, absorbent, and do not irritate the skin, making them an ideal choice for clothing, particularly for those with sensitive skin. Additionally, cotton is a natural fiber that is sustainable and biodegradable, making it an eco-friendly choice. On the other hand, the addition of polyester to cotton yarns improve the fabric's strength, durability, and elasticity, making it more resistant to wear and tear. Polyester is also less prone to wrinkling and shrinking than cotton, and it dries more quickly. However, polyester is a synthetic fiber that is not biodegradable and contribute to environmental pollution. In terms of MOE, a blend of 67% cotton and 33% polyester with improved MOE may be more suitable for applications where elasticity and strength are critical, such as in athletic wear, activewear, or outdoor gear. However, if softness and comfort are more important than strength and elasticity, a knitted fabric made of 100% cotton may be a better choice. An increase in float stitches in knitted fabrics made with natural yarns such as 100% CO and contribute to improved MOE (Modulus of Elasticity), which is a measure of a material's stiffness as exhibited in Fig. 8.

Float stitches are created when a yarn is carried across the back of a stitch without being knit into the fabric. In knitted fabrics made with natural yarns (100% CO), increasing the number of float stitches create a denser, more tightly-knit structure. This denser structure lead to increased stiffness and improved MOE. In addition, the orientation and alignment of the yarns in the fabric also contribute to improved MOE. When natural yarns are knitted in a way that aligns the yarns in a single direction, this creates a fabric with greater tensile strength and stiffness [32]. By increasing the number of float stitches, the yarns be oriented and aligned more precisely, leading to improved MOE. Furthermore, the use of natural yarns themselves contribute to the improved MOE of knitted fabrics. Natural yarns such as 100% CO known for

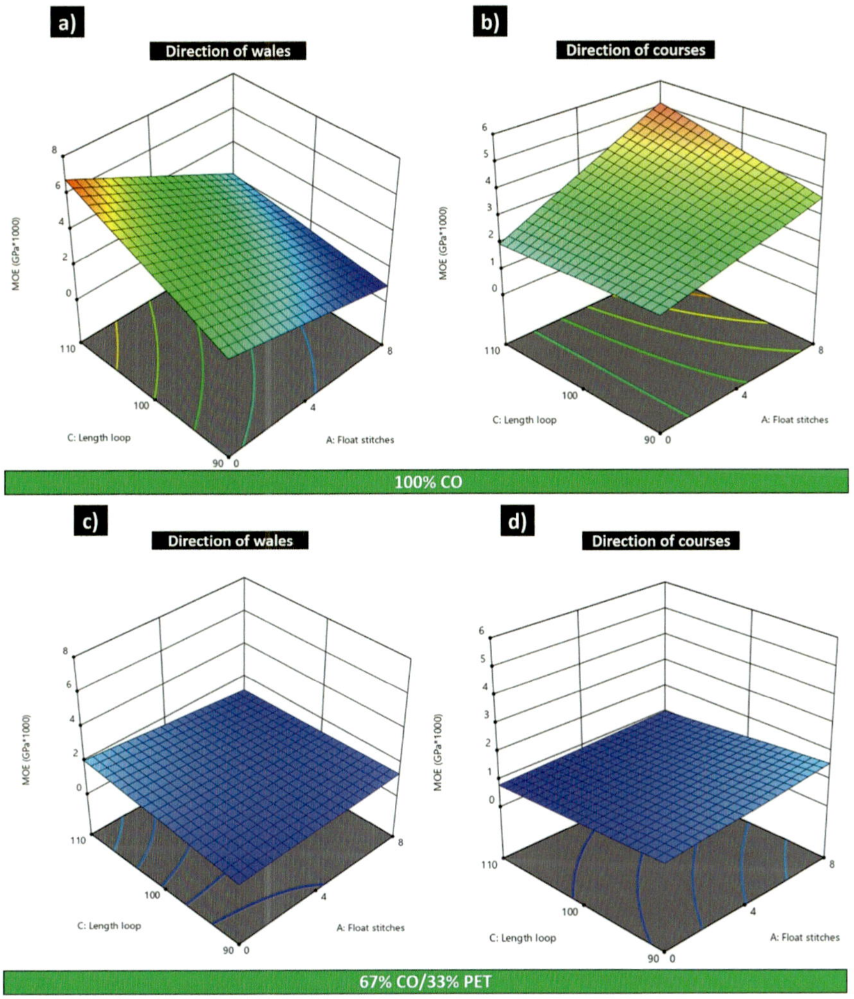

Fig. 8 3D surface of MOE in weft knitted fabrics with natural fibers

their tensile strength and resilience, which further enhance the stiffness and MOE of the fabric. Overall, increasing the number of float stitches, length of loops in knitted fabrics made with natural yarns such as 100% CO contribute to a denser, more tightly-knit structure that enhances the orientation and alignment of the yarns and improves the stiffness and MOE of the fabric as exhibited in Fig. 8.

The work of rupture (energy required to break a material). Knitted fabrics made with natural yarns such as 100% CO (cotton) and 67% CO/33% PET (CO/PET) have been found to exhibit different levels of work of rupture. In general, knitted fabrics made with 67% CO/33% PET tend to be superior to those made with 100% CO in terms of work of rupture. Additionally, CO/PET fibers have a higher tenacity than CO fibers, which means they are more resistant to breaking under stress. Furthermore, the structure of the knitted fabric also plays a role in the work of rupture. Knitted fabrics made with 100% CO tend to have a denser and more tightly-knit structure, which may result in a lower work of rupture. Therefore, while knitted fabrics made with 67% CO/33% PET may generally exhibit superior work of rupture compared to those made with 100% CO, there may be instances where the opposite is true depending on the specific characteristics of the fabric as exhibited in Fig. 9.

The number of float stitches in a knitted fabric affect its work of rupture. Knitted fabrics made with 8 float stitches and 100% CO (cotton) to those made with 67% CO/33% PET (CO/PET) found that the CO/PET fabric was superior in terms of

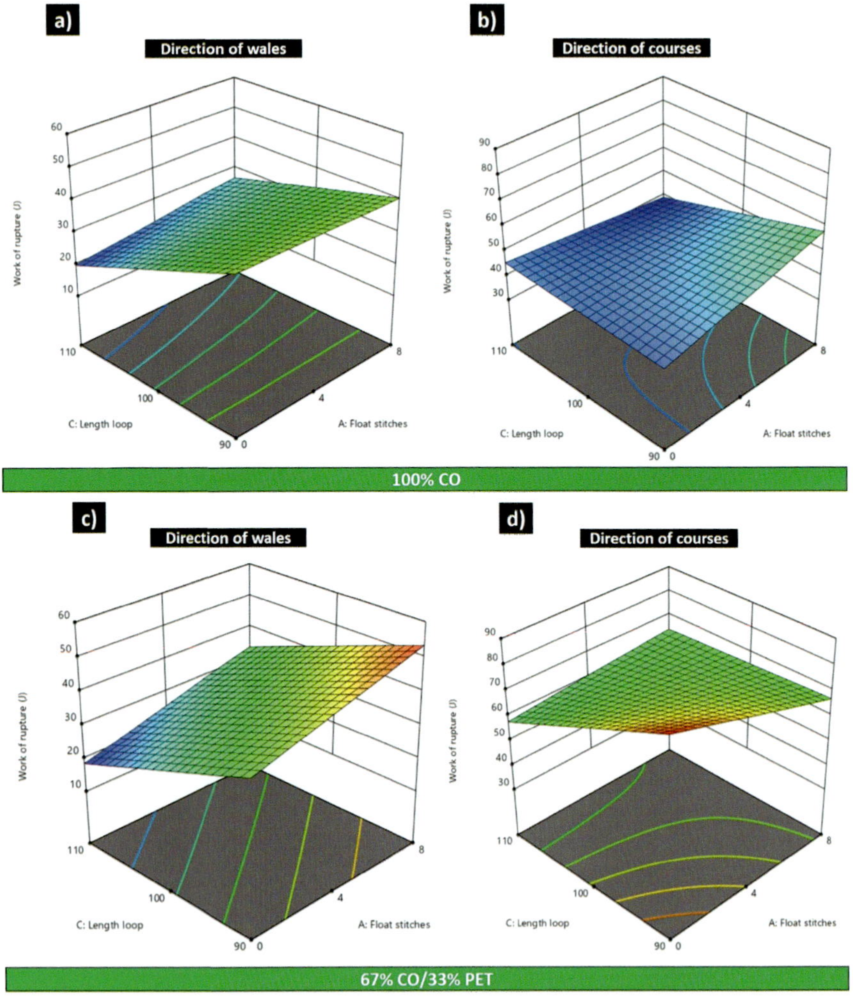

Fig. 9 3D surface of work of rupture in weft knitted fabrics with natural fibers

work of rupture. This increased elongation result in a higher work of rupture for the CO/PET fabric. Additionally, the presence of 8 float stitches contribute to the superior work of rupture in the 67% CO/33% PET fabric. Float stitches are created when a yarn was carried across the back of a stitch without being knit into the fabric, which create a more open and flexible structure [33]. Therefore, while the CO/PET fabric with 8 float stitches may have been superior in terms of work of rupture in this particular study as exhibited in Fig. 9.

Comparing knitted fabrics made with 90 length loops and 100% CO (cotton) to those made with 67% CO/33% PET (CO/PET) found that the CO/PET fabric was superior in terms of work of rupture. Longer loops create a more open and flexible structure in the knitted fabric, which allow for greater elongation and deformation before breaking. This increased flexibility may contribute to the higher work of rupture observed in the CO/PET fabric as exhibited in Fig. 9.

4.3 Analysis of Variance

Analysis of variance (ANOVA) is a statistical method used to determine if there is a significant difference between the means of two or more groups. In the context of tensile behavior, ANOVA can be used to analyze the relationship between the factors float stitches (A), composition (B), length loop (C) and interactions (AB, AC, BC and ABC) and the tensile strength of a material. To conduct an ANOVA analysis of tensile behavior in relation to the factors float stitches (A), composition (B), length loop (C) and interactions (AB, AC, BC and ABC), a researcher would first need to collect data on the tensile strength of the material at different loop lengths. This data could be collected using a tensile testing machine, which measures the amount of force required to stretch a material to its breaking point [19].

4.3.1 Synthetic Yarns

Design of experiments showed that in factorial design of this chapter significant F-values were obtained for all factors float stitches (A), composition (B), length loop (C) and interactions (AB, AC, BC and ABC), therefore, response variables can be adequately expressed by model as shown in Tables 1 and 2. Then, statistical adjustment of data set for stress, strain, MOE and work of rupture in knitted structures (Synthetic yarns) in wales direction (as shown in Table 1). In which, standard deviation obtained from stress (0.42 MPa), strain (0.08), MOE (0.5 GPa) and work of rupture (1.2 J), were significantly low, and shows that level of data dispersion around means (2.9 MPa, 2.4, 1.8 GPa and 86.4 J) and CV% (14.2%, 3.4%, 31.2% and 1.4%) respective, in mechanical responses of knitted structures used as reinforcement in textile composite (as shown in Table 1). Then, statistical adjustment of data set for stress, strain, MOE and work of rupture in knitted structures (Synthetic yarns) in courses direction. In which, standard deviation obtained from stress (0.22 MPa), strain (0.03), MOE (0.26 GPa) and work of rupture (19.3 J), were significantly low, and shows that level of data dispersion around means (2.59 MPa, 1.54, 1.29 GPa and 150.2 J) and CV% (8.34%, 1.18%, 19.9% and 12.8%) respective, in mechanical responses of knitted structures used as reinforcement in textile composite (as shown in Table 2).

4.3.2 Natural Yarns

Once the data has been collected, the researcher can conduct an ANOVA analysis to determine if there is a significant difference in tensile behavior between the different float stitches (A), composition (B), length loop (C) and interactions (AB, AC, BC and ABC). The ANOVA analysis will calculate a p-value, which indicates the probability of obtaining the observed results by chance. If the p-value is less than a predetermined significance level (usually 0.05), then the researcher can conclude that there is a significant difference in tensile strength between the different loop lengths. This information can be used to identify the optimal different float stitches, composition, length loop for maximizing the tensile strength of the material. An ANOVA analysis of tensile behavior in relation to the different float stitches (A), composition (B), length loop (C) provides valuable insights into the mechanical properties of materials and help inform the design and engineering of new materials. Statistical models of tensile behavior (stress, strain, MOE and work of rupture) of knitted structure manufactured using natural yarns become validated when a set of data (residual values) are closer straight line or well distributed straight line along, thus, ρ-values exhibit indices higher or equal to 0.05 ($\rho \geq 0.05$). Consequently, F-value statistically analyzes how much the means in each response variable are oscillating or distinct, so lower F-value means that the mean squares of model statistic are lower than value larger of mean squares of residuals (error higher). If error value is higher, less significant is ρ-values for variables (length loop, structure and composition) studied in this chapter and their interactions (as shown in Tables 3 and 4).

Design of experiments showed that in factorial design of this chapter significant F-values were obtained for all factors structure (A), composition (B), length loop (C) and interactions (AB, AC, BC and ABC), therefore, response variables can be adequately expressed by model as shown in Tables 3 and 4. Then, statistical adjustment of data set for stress, strain, MOE and work of rupture in knitted structures (Natural yarns) in wales direction. In which, standard deviation obtained from stress (0.2 MPa), strain (0.14), MOE (0.3 GPa) and work of rupture (0.9 J), were significantly low, and shows that level of data dispersion around means respective, promotes a lower coefficient of variation (CV%) in mechanical responses of knitted structures commonly used in textile-reinforced composite materials (as shown in Table 3). And, finally, statistical adjustment of data set for stress, strain, MOE and work of rupture in knitted structures (natural yarns) in courses direction. In which, standard deviation obtained from stress (0.3 MPa), strain (0.07), MOE (0.3 GPa) and work of rupture (1.8 J), were significantly low, and shows that level of data dispersion around means respective (3.4 MPa, 1.9, 2.1 GPa and 59.6 J), promotes a lower coefficient

Table 1 ANOVA of tensile properties of knitted structures in Wales direction

	Stress (MPa)		Strain		MOE (GPa*1000)		Work of rupture (J)	
	F-value	p-value	F-value	p-value	F-value	p-value	F-value	p-value
Model	31.9	<0.0001	364.4	<0.0001	15.7	<0.0001	6133	<0.0001
A—float stitches	0.5	0.5	72	<0.0001	0.4	0.5	4242	<0.0001
B—composition	220.6	<0.0001	2409.6	<0.0001	107.0	<0.0001	32.940.4	<0.0001
C—length loop	1.6	0.23	0.4	0.6	0.6	0.5	2123.5	<0.0001
AB	0.7	0.4	39.6	<0.0001	0.001	1.0	1492.3	<0.0001
AC	0.5	0.5	25.6	0.0004	1.2	0.3	63.1	<0.0001
BC	0.02	0.9	1.5	0.3	0.23	0.6	1971.2	<0.0001
ABC	0.02	0.9	2.3	0.2	0.7	0.4	98	<0.0001
Standard dev	0.42		0.08		0.5		1.2	
Mean	2.9		2.4		1.8		86.4	
C.V. %	14.2		3.4		31.2		1.4	
R^2	0,95		0,99		0.91		0.99	

Table 2 ANOVA of tensile properties of knitted structures in courses direction

Stress (MPa)		Strain		MOE (GPa*1000)		Work of rupture (J)	
F-value	p-value	F-value	p-value	F-value	p-value	F-value	p-value
74.2	<0.0001	1577.0	<0.0001	29.5	<0.0001	34.7	<0.0001
0.16	0.69	1399	<0.0001	4.5	0.06	8.5	0.01
484.9	<0.0001	8663.9	<0.0001	184	<0.0001	182.5	<0.0001
11.2	0.007	121.2	<0.0001	1.6	0.24	12.2	0.005
0.9	0.37	1.4	0.3	1.4	0.27	1.8	0.2
15,4	0.0023	842.1	<0.0001	9.7	0.01	13.9	0.003
3.9	0.07	11.7	0.006	0.6	0.44	22.3	0.0006
3.0	0.11	0.1	0.8	4.5	0.06	1.4	0.3
0.22		0.03		0.26		19.3	
2.59		2.45		1.29		150.2	
8.34		1.18		19.9		12.8	
0.98		0.99		0.95		0.96	

of variation (CV%) in mechanical responses of knitted structures commonly used in textile-reinforced composite materials (as shown in Table 4).

5 Conclusions

Knitted fabric structures exhibited significantly high property values under different loop lengths, float stitches and directions of wales and courses. Stress property was strongly influenced by float stitches density, as well as loop length of knitted fabric structures. In general, the specific number of float stitches needed to achieve optimal tensile behavior will depend on the characteristics of the yarn and the structure of the fabric. However, research has shown that incorporating around 8 float stitches per inch of fabric can be effective for improving the tensile behavior of knitted fabrics with synthetic yarns. The use of float stitches in knitted fabric with synthetic yarns (such as PA and PET) can improve the tensile behavior of the fabric due to the way in which these stitches distribute stress across the fabric. The effect of different variable combinations on tensile properties of knitted fabrics. It was also found that the presence of float stitches and length loop for natural and synthetics yarns have a significant effect on stress, strain, MOE and work of rupture mechanical properties. Finally, it was concluded that statistical adjustment of data set for stress, strain, Young's modulus and work of rupture in knitted fabric structures made by natural and synthetic yarns (wale and course direction) were significantly in mechanical responses of knitted fabric structures commonly used in textile-reinforced composite materials.

Table 3 ANOVA of tensile properties of knitted structures in Wales direction

	Stress (MPa)		Strain		MOE (GPa*1000)		Work of rupture (J)	
	F-value	p-value	F-value	p-value	F-value	p-value	F-value	p-value
Model	148.0	<0.0001	48.9	<0.0001	134.9	<0.0001	384.1	<0.0001
A—float stitches	245.7	<0.0001	67.7	<0.0001	222.4	<0.0001	413.5	<0.0001
B—composition	494.1	<0.0001	38.7	<0.0001	286.2	<0.0001	105.6	<0.0001
C—length loop	42.5	<0.0001	94.9	<0.0001	142.8	<0.0001	1933.2	<0.0001
AB	223.4	<0.0001	20.6	0.0008	156.8	<0.0001	187.6	<0.0001
AC	23.8	0.0005	2.5	0.14	56.5	<0.0001	24.9	0.0004
BC	3.6	0.09	51.2	<0.0001	73.7	<0.0001	6.9	0.02
ABC	3.1	0.1	66.5	<0.0001	6.2	0.03	17.1	0.002
Standard dev	0.2		0.14		0.24		0.9	
Mean	3.4		1.9		2.2		32.5	
C.V. %	5.4		7.1		11.3		2.7	
R^2	0.99		0.97		0.99		0.99	

Table 4 ANOVA of tensile properties of knitted structures in courses direction

	Stress (MPa)		Strain		MOE (GPa*1000)		Work of rupture (J)	
	F-value	p-value	F-value	p-value	F-value	p-value	F-value	p-value
Model	36.1	<0.0001	183.3	<0.0001	77.5	<0.0001	183.1	<0.0001
A—float stitches	18.8	0.0012	345.3	<0.0001	103.7	<0.0001	3.6	0.09
B—composition	216.4	<0.0001	742.9	<0.0001	343.98		827.3	<0.0001
C—length loop	0.1	0.74	92.8	<0.0001	0.3	0.5735	195.1	<0.0001
AB	8.9	0.01	4.6	0.0555	56.1	<0.0001	43.9	<0.0001
AC	0.3	0.6	11.3	0.0063	7.6	0.02	2.9	0.12
BC	7.6	0.02	55.02	<0.0001	15.3	0.003	21.4	0.0007
ABC	0.8	0.4	31.2	0.0002	15.3	0.002	187.1	<0.0001
Standard dev	0.3		0.07		0.3		1.8	
Mean	3.4		1.9		2.1		59.6	
C.V. %	9.9		3.4		12.3		3.0	
R^2	0.96		0.96		0.98		0.99	

- Among the independent variables studied (float stitches, composition and the length loop), float stitches was the variable that most influences the mechanical behavior of knitted structures used for textile composites.
- Stress knitted structures behave inversely proportional to the density of float stitches (0, 4 and 8 points).
- Increased density of float stitches promoted higher stress values in both natural and synthetic knitted fabrics.
- The MOE and strain exhibited high anisotropy, as an increase in float stitches density causes changes in MOE and strain.
- Showing that the properties of knitted fabrics improve and intensify regardless of the nature of the fiber (synthetic or natural).

Finally, this proves that float stitches contribute significantly and improve the mechanical behavior of knitted structures used as reinforcement in textile composites.

Acknowledgements This research was supported by our Lord. We thank our colleagues at the Textile Quality Control Laboratory—LABCTEX and the Federal University of Rio Grande do Norte.

Declaration of Conflicting Interests The author(s) declare(s) that there is no conflict of interest.

References

1. Abghary MJ, Nedoushan RJ, Hasani H (2017) Multi-scale modeling the mechanical properties of biaxial weft knitted fabrics for composite applications. Appl Compos Mater 24:863–878. https://doi.org/10.1007/s10443-016-9551-y
2. Dinh TD, Weeger O, Kaijima S, Yeung S-K (2018) Prediction of mechanical properties of knitted fabrics under tensile and shear loading: Mesoscale analysis using representative unit cells and its validation. Compos Part B Eng 148:81–92. https://doi.org/10.1016/j.compositesb.2018.04.052
3. Leong KH, Ramakrishna S, Huang ZM, Bibo GA (2000) The potential of knitting for engineering composites—a review. Compos Part A Appl Sci Manuf 31:197–220. https://doi.org/10.1016/S1359-835X(99)00067-6
4. Sarker ME, Islam MR, Mahmud MS, Islam MZ, Alimuzzaman S, Ahmed ATMF (2022) Effects of yarn, fabric and machine parameters on the dimensional properties of weft knitted single jersey fabric. J Nat Fibers 19:15625–15636. https://doi.org/10.1080/15440478.2022.2131682
5. De Araujo M, Fangueiro R, Hu H (2011) Weft-knitted structures for industrial applications. In: Advances in knitting technology. Elsevier, pp 136–170. https://doi.org/10.1533/9780857090621.2.136
6. Fatkić E, Geršak J, Ujević D (2011) Influence of knitting parameters on the mechanical properties of plain jersey weft knitted fabrics. Fibres Text East Eur 88:87–91
7. Duhovic M, Bhattacharyya D (2011) Knitted fabric composites. In: Advances in knitting technology. Elsevier, pp 193–212. https://doi.org/10.1533/9780857090621.2.193
8. Ruan X (1996) Experimental and theoretical studies of the elastic behavior of knitted-fabric composites. Compos Sci Technol 56:1391–1403. https://doi.org/10.1016/S0266-3538(96)00097-8
9. Demircan Ö, Ashibe S, Kosui T, Nakai A (2015) Effect of various knitting techniques on mechanical properties of biaxial weft-knitted thermoplastic composites. J Thermoplast Compos Mater 28:896–910. https://doi.org/10.1177/0892705713519121
10. Nonato RC, Bonse BC (2016) A study of PP/PET composites: Factorial design, mechanical and thermal properties. Polym Test 56:167–173. https://doi.org/10.1016/j.polymertesting.2016.10.005
11. Melo ABL, Panzera TH, Freire RTS, Scarpa F (2018) The effect of Portland cement inclusions in hybrid glass fibre reinforced composites based on a full factorial design. Compos Struct 202:233–240. https://doi.org/10.1016/J.COMPSTRUCT.2018.01.069
12. Santos TF, Santos CM, Aquino MS, Oliveira FR, Medeiros JI (2020) Statistical study of performance properties to impact of Kevlar® woven impregnated with Non-Newtonian Fluid (NNF). J Mater Res Technol 9:3330–3339. https://doi.org/10.1016/j.jmrt.2020.01.027
13. ASTM International (2020) ASTM D1776/D1776M-20-standard practice for conditioning and testing textiles. https://doi.org/10.1520/D1776_D1776M-20.
14. ASTM D5034-09 (2017) Standard test method for breaking strength and elongation of textile fabrics (grab test). West Conshohocken. https://doi.org/10.1520/D5034-09R17
15. Oliveira LÁ, Santos JC, Panzera TH, Freire RTS, Vieira LMG, Scarpa F (2018) Evaluation of hybrid-short-coir-fibre-reinforced composites via full factorial design. Compos Struct 202:313–323. https://doi.org/10.1016/j.compstruct.2018.01.088
16. de Olveira LÁ, dos Santos JC, Panzera TH, Freire RTS, Vieira LMG, Rubio JCC (2018) Investigations on short coir fibre–reinforced composites via full factorial design. Polym Polym Compos 26:391–399. https://doi.org/10.1177/0967391118806144
17. ASTM D3776/D3776M–09a (2017) Standard test methods for mass per unit area (weight) of fabric. West Conshohocken, PA. https://www.astm.org/Standards/D3776.htm. Accessed 4 Aug 2019
18. ASTM D3887-96 (2008) Standard specification for tolerances for knitted fabrics. West Conshohocken. https://doi.org/10.1520/D3887-96R08

19. Santos C, Santos T, Moreira K, Aquino M, Zillio RFL (2021) Statistical study of the influence of fiber content, fiber length and critical length in the mechanical behavior of polymeric composites reinforced with carica papaya fibers (CPFs). Appl Sci Eng Prog 14:719–726. https://doi.org/10.14416/J.ASEP.2021.07.002
20. Maâtoug N, Sahnoun M, Sakli F (2012) Statistical analysis of surface roughness parameters for weft knitted fabrics measured by the textile surface tester (TST). J Eng Fiber Fabr 7:155892501200700. https://doi.org/10.1177/155892501200700407
21. Sarybayeva E, Kuramysova M, Mukimov M, Shardarbek M, Rakhmanova Z, Makhanbetaliyeva K, Tashmukhamedov F, Jurinskaya I, Kalmakhanova M (2023) Influence of tuck stitches on the physical and mechanical properties of knitted fabrics. Res J Text Appar. https://doi.org/10.1108/RJTA-07-2022-0083
22. Santos TF, Santos CM, Dias M, Zilio L, Melo KM, Cavalcante ME, Castro LF, Hussein EK, Aquino MS (2023) Mechanical behavior of natural/synthetic weft-knitted structures commonly used as reinforcement of hybrid composites via full factorial design: yarn compositions and float stitches. Green Hybrid Compos Eng Non-Eng Appl. https://doi.org/10.1007/978-981-99-1583-5_15
23. Assefa A, Govindan N (2020) Physical properties of single jersey derivative knitted cotton fabric with tuck and miss stitches. J Eng Fiber Fabr 15:155892502092853. https://doi.org/10.1177/1558925020928532
24. Asif A, Rahman M, Farha FI (2015) Effect of knitted structure on the properties of knitted fabric. Int J Sci Res 4:1231–1235
25. Liu S, Yang C, Zhao Y, Ming Tao X, Tong J, Li L (2016) The impact of float stitches on the resistance of conductive knitted structures. Text Res J 86:1455–1473. https://doi.org/10.1177/0040517514555798
26. Zílio L, Dias M, Santos T, Santos C, Fonseca R, Amaral A, Aquino M (2020) Characterization and statistical analysis of the mechanical behavior of knitted structures used to reinforce composites: Yarn compositions and float stitches. J Mater Res Technol 9:8323–8336. https://doi.org/10.1016/j.jmrt.2020.05.089
27. Dos Santos TF, Da Silva Santos CM, Da Fonseca RT, Dos Santos VS, Grosschopf MG, De Aquino MS (2019) Influence of structure and composition in the mechanical properties of textile polymeric fabrics. 27:222–227. https://doi.org/10.1177/0967391118823075
28. Leong KH, Ramakrishna S, Huang ZM, Bibo GA (2000) Potential of knitting for engineering composites—a review. Compos Part A Appl Sci Manuf 31:197–220. https://doi.org/10.1016/S1359-835X(99)00067-6
29. Oner E (2019) Mechanical and thermal properties of knitted fabrics produced from various fiber types. Fibers Polym. 20:2416–2425. https://doi.org/10.1007/s12221-019-9119-1
30. Yip J, Ng SP (2008) Study of three-dimensional spacer fabrics: physical and mechanical properties. J Mater Process Technol 206:359–364. https://doi.org/10.1016/j.jmatprotec.2007.12.073
31. Mikučioniene D, Čepukone L, Milašiene D (2018) Investigation on mechanical and thermal properties of knits from peat fibers and their combination with other natural fibers. Text Res J 88:1660–1670. https://doi.org/10.1177/0040517517705633
32. Hoque MS, Hossain MJ, Rahman MM, Rashid MM (2022) Fiber types and fabric structures influence on weft knitted fabrics. Heliyon 8:e09605. https://doi.org/10.1016/j.heliyon.2022.e09605
33. Yesmin S, Hasan M, Miah MS, Momotaz F, Idris MA, Hasan MR (2014) Effect of stitch length and fabric constructions on dimensional and mechanical properties of knitted fabrics. World Appl Sci J 32. https://doi.org/10.5829/idosi.wasj.2014.32.09.1264

Mechanical Behavior of Natural/Synthetic Weft-Knitted Structures Commonly Used as Reinforcement of Hybrid Composites via Full Factorial Design: Yarn Compositions and Float Stitches

Thiago F. Santos, Caroliny M. Santos, Mariana Dias, Lucas Zilio, Katia M. Melo, Maria Eduarda Cavalcante, Luiz Filipe Castro, Emad Kamil Hussein, and Marcos Aquino

1 Introduction

Understand the behavior of knitted fabrics to apply them in composites. Composite materials are part of materials engineering in general, however, there is a very attractive area in which textile engineers work intensively called textile-reinforced composite materials (TRCM). Textile reinforcement of knitted structures have active mechanical properties and show potential for reduced manufacturing costs and enhanced processability, with more than adequate, or in some cases improved. In this chapter it was reported about the behavior and performance of the mechanical properties of textile knitted commonly used to reinforce hybrid composites materials. Knitted fabric structures were developed using the Cixing Knitting System (CKS) software, then weft-knitted fabrics were manufactured by a double-front (electronic)

T. F. Santos · C. M. Santos · M. Dias · L. Zilio · K. M. Melo · M. E. Cavalcante · M. Aquino
Textile Engineering Laboratory, Department of Textile Engineering, Federal University of Rio Grande do Norte, Natal, Rio Grande do Norte, Brazil
e-mail: carolinyminely@hotmail.com

T. F. Santos (✉) · C. M. Santos · K. M. Melo · M. E. Cavalcante · M. Aquino
Laboratory of Knitting, Textiles Technologies Study Group (GETTEX), Federal University of Rio Grande do Norte, Natal, Rio Grande do Norte, Brazil
e-mail: thiagotextilufrn@hotmail.com

L. F. Castro
Department of Production Engineering, Federal University of Rio Grande do Norte, Natal, Rio Grande do Norte, Brazil

E. K. Hussein
Al Furat Al Awsat Technical University, Kufa, Iraq

T. Khan and M. Jawaid (eds.), *Green Hybrid Composite in Engineering and Non-Engineering Applications*, Composites Science and Technology,
https://doi.org/10.1007/978-981-99-1583-5_15

flat knitting machine. With the objective of evaluating and analyzing the performance and mechanical behavior such as stress, strain and young's modulus of knitted fabric structures commonly used in polymeric matrix textile composites via full factorial design.

1.1 Properties of Knitted Fabrics

Among textile materials, knitted fabric is characterized by having a less stable structure in which parameters change due to various stresses, including elongation. These fabrics are subjected to a variety of stresses and loads/deformations during use. Charges are different according to intensity, direction and duration. The alternation of loading and unloading processes or repose affects the structure of the knitted fabric and can cause changes in the linear dimensions of the knitted fabric or even result in structural deformation of the knitted fabric and ultimately loss of product appearance or of its functional properties [13, 26]. However, knitted fabrics also have special properties and can be used for technical applications. Compared to woven, knitted fabrics have much more extensibility, especially in a transverse direction (courses) as the structure of knitted fabric is made up of knit, unlike woven fabric, which does not use stitches [33]. Total deformation and its components are the most used characteristics in single cycle studies to evaluate the mechanical properties of textile materials. Total deformation consists of these 3 components: first, elastic deformation, which disappears immediately after discharge; then plastic deformation with an additional relaxation period, which occurs in time at low speeds; and, finally, the residual deformation, which does not disappear after unloading in the knitted fabric. During relaxation, due to elastic forces, the stretched knit fabric structure returns to its original position until such forces become equal to the friction forces between the knitted fabric structural stitches (knit, retained and omitted) [26]. When knitted fabrics are subjected to a traction test, a change in knitted fabric dimensions is observed. These changes can be defined by increasing dimensions in one direction as the other direction decreases in dimensions. In many cases, it is important to know the proportion that the knitted fabric will deform in one direction or the other. It is also known that there are different extensibilities (in the direction of courses and wales) in addition to the maximum rupture force for each respective direction. It was also known that the direction of the wales and courses are influenced by the types and amount of specific stitches (knit, retained and omitted) in the structural design of the knitted fabric [33]. Several researchers carried out a study on textile knitted for potential application in textile-reinforced composite materials (TRCM). Zílio et al. [54] specifically investigated the influence of yarn compositions and float stitches in improving mechanical properties and behavior (stress, MOE and strain) for potential application in textile-reinforced composite materials [54]. While Asif et al. [4] evaluated yarn count parameters, stitch length, knitting loom diameter, needle gauge, dyeing parameters and finishing parameters in textile knitted of three weft knitted structures (plain jersey, single lacoste and double lacoste). The parameters were related as a function of several properties, such as dimensional stability, fabric width, area density, spirality, pilling resistance, color fastness to washing, color fastness to light and color fastness to friction in the finished state [4]. A study by Duhovic and Bhattacharyya [14] suggests the use of knitted structures as reinforcing elements in polymeric composite materials [14]. In view of this, the present work proposes to study via full factorial design a combination between fiber composition and float stitches for a more assertive choice in the production of polymeric composites with textile knitted reinforcement.

1.2 Kintted-Reinforced Composites

Textile composites are those whose reinforcements are textiles combined with a matrix. This describes a family of materials used for technical applications in various industrial sectors [37]. The huge variety of textile techniques, in the area of composites, took more than 50 years to be discovered, being basically studied the use of textile fibers for this purpose, and for a long time, textile knitted were not considered in the manufacture of composites. It is important to note that it was only in the early 1990s when the first scientific works appeared reporting systematic studies on knitted composites. The reluctance to use knitted fabrics as a reinforcement for composites was partly based on the thought that their high deformation and instability would not make them suitable for use as an effective reinforcement for composites. This, however, ignores one of the basic laws in composites: once the matrix is rigid enough to immobilize the woven and knitted fabrics, these fabrics are fully operational as reinforcement. Currently, however, it is known that the structure of knitted fabrics allows an extensional deformation of the preform; Complex composite parts can therefore be molded without great difficulty. A second group of arguments was based on facts, since the highly curved nature of yarns in knitted fabric results in mechanical properties that would not exceed those of a nearly isotropic laminate. However, no one had predicted that knitted fabrics would have excellent interlaminar fracture toughness, giving hope for good damage tolerance and impact resistance [51]. It is important to point out that composites reinforced with knitted fabric also have other advantages, such as: excellent elasticity and flexibility, which allows them to combine with any complex shape very well without folds and excellent shape forming ability, that is, the desired shape (even than the double curved shape) can be obtained through knitting by the use of knitting machines. These advantages of knitted fabric result in the following characteristics: flexibility, good impact properties, good interlaminar properties, good moldability, high production rate and short production period, which are important in composite production [37].

1.3 Tests on Textile-Reinforced Composite Materials

Tests on composites of textile knitted are varied and applied to the most diverse structures of knitted fabrics. It is possible to do chemical, physical and mechanical analysis of knitted fabrics [2]. In a study by Hamedi et al. [18], three-dimensional spacer fabrics were investigated regarding their mechanical behavior in terms of flexural performance, and a three-point flex test was performed [18]. In a research related to biaxial knitted fabrics connected in multilayers for carbon fibers, their tensile strength was tested in different types of structures with different volumetric fractions of carbon fiber, using one-way analysis of variance and one-dimensional linear regression [39]. A study on the effect of knitted density, natural fiber type, crack length and fabric direction (wales or courses) on fracture strength and fracture toughness of laminated composites reinforced with natural fiber knitted fabric were investigated. For this, plain fabrics made of mixed yarns 50/50 bamboo/cotton, 50/50 viscose/cotton and 50/50 modal/cotton were studied [50]. A study by Khondker et al. [25] analyzed the tensile, bending and impact properties of polypropylene knitted composites using injection-compression molding [25]. Shekarchizadeh et al. [44] studied investigation of the tensile characteristics of biaxial weft knitted reinforced composites in terms of stiffness, strength and failure mechanism [44]. Alpyildiz et al. [1], in turn, investigated the mechanical performance of composites that are reinforced with glass knitted fabrics composed of tuck stitches. Tensile, compression, impact and compression after impact (CAI) tests were performed and the results of composites reinforced with knitted fabrics derived from full cardigan and 1 $\times$ 1 glass fiber rib knitted fabrics were compared [1]. It is known that as for a composite material, the performance depends not only on the strength of the constituent materials, but also on the internal structural geometry of the constituents. So, it was interesting to study the effects of the fold stitch, which is one of three different types of loops that can be used to build knitted fabrics and has a different geometry than other types of loops. Considering that the type, structure and orientation angle of the reinforcement material used in laminated composite structures directly affect the mechanical behavior of the composite material. Balcıoğlu and Yalçın [8] investigated the influence of these parameters on the mechanical properties and fracture toughness of laminated composites reinforced with knitted fabric. For this, laminated composites were produced by the manual lamination method with fiberglass/epoxy and carbon/epoxy reinforced with 1 $\times$ 1 knitted fabric. The mechanical behavior of the composites produced was determined under tensile, compression and shear loads [8]. This work investigates the influence of wale and course directions, density of float stitches (0, 2 and 4), knitting yarns influence of composition (100% cotton, 50% cotton/50% polyester and 100% polyester) and length loop (0.98, 0.84 and 0.71 cm) on the performance and mechanical behavior of mechanical properties in the knitted fabric structures commonly used to reinforce the textile composites. Using advanced statistical analysis tools, Response Surface Methodology and analysis of variance (ANOVA). Finally, to promote mechanical behavior and mechanical properties in order to understand the remained interactive and attractive effects produced by different structures, compositions and loop length of knitted fabrics.

2 Experimental

2.1 *Knitting Yarns Used in the Manufacture of Knitted Fabrics*

Single pique knitted fabrics were manufactured from of different knitting yarns: composition 1 (100% Polyester—PES), composition 2 (50% cotton—CO/50% Polyester—PES) and composition 3 (100% cotton—CO). Yarns count used in this study were 5.7 Ne (473.13 ± 3.21 μm), 7.3 Ne (415.9 ± 10.12 μm) and 8.4 Ne (781.27 ± 12.30 μm) for the yarn compositions 100% CO, 50% CO/50% PES and 100% PES respectively.

2.2 *Preparation and Experimental Planning of Knitted Fabric*

Single pique knitted fabric shown in Fig. 1 were developed with different densities of float stitches (0, 2 and 4) using Cixing Knitting System (CKS) software to posteriorly manufacture Single pique knitted.

Before carry out experimental planning, because, it provides power or effectiveness of the responses (stress, strain and MOE). Power of the response design, in which, were obtained values of signal-to-noise statistic (2), that is distance between mean values and standard deviation for the independent variables (float stitches, composition and length loop) [36]. Were used density of float stitches (0, 2 and 4),

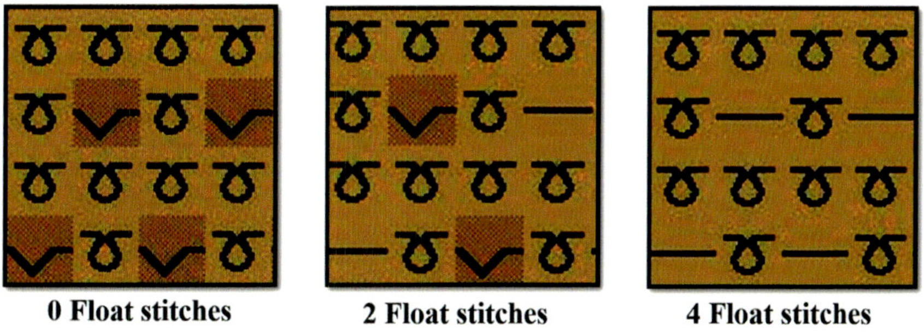

Fig. 1 Stitches diagram of single pique knitted fabric studied

Table 1 Variables and experimental conditions this study

Variables	Type	Low	Middle	High	Power of A, B and C
A—Float stitches	Numeric	0	2	4	96.0%
B—Composition	Numeric	1 (100% PES)	2 (50% CO/50% PES)	3 (100% CO)	
C—Length loop	Numeric	0.71 cm	0.84 cm	m	

composition (1, 2 and 3) and length loop (0.98, 0.84, 0.71 cm), as study variables (Table 1). Planning and experimental design was carried out using a factorial design 2^3 with 2 central points and 2 repetitions (96.0%) in the statistical software Design Expert [54].

2.3 *Manufacturing of Single Pique Knitted Fabrics*

Weft knitted fabrics (single pique) developed in CKS software, were manufactured from knitting yarns (100% CO, 50% CO/50% PES and 100% PES), for a double-fronture electronic flat knitting machine (V-bed) from Cixing (model GE2-52C), which has 346 needles and 7/inch gauge. All single pique knitted fabrics (showed in Fig. 1) have a mean grammage 22.49 ± 2.78 kg/m^2, 4.4 ± 0.5 courses/cm (cpc) and 5.6 ± 1.3 wales/cm (wpc) [5, 6].

2.4 *Tensile Test in Single Pique Knitted Fabrics*

Tensile tests were performed with MESDAN's Tensolab 3000 dynamometer equipment, according ASTM D5034 standard at rate of 30 cm/min and distance between claws 10 cm [7]. Single pique fabrics were cut to 15 cm length and 2.5 cm width for each condition of float stitches (0, 2 and 4), composition (1, 2 and 3) and length loop (0.98, 0.84 and 0.71 cm). And finally, results were treated in the Origin software.

2.5 *Statistical Analysis*

The mechanical behavior of response properties (stress, MOE and strain) were manipulated, analyzed and discussed using design of experiment (DoE), knowing that, the percent contribution ratio was used to show the influence of inputs (A, B and C) and their interactions [31, 32, 34]. That consist of analysis of variance—ANOVA (R^2, value-F, value-ρ, standard deviation, mean and C.V.%), in addition to 3D response surface methodology (RSM) were obtained of Design Expert software [40, 42, 45].

3 Results and Discussion

3.1 Mechanical Behavior in the Direction of Wales and Courses

3.1.1 Statistical Analysis of Stress Mechanical Behavior

Figure 2 show the results of the stress suffered in each of the combinations in the knitted fabrics, in the wales direction. It was generally observed that in all length loops the stresses tend to be higher in composition 3 (100% cotton) and lower in composition 1 (100% polyester). This can be explained by the intrinsic property of cotton to be more rigid than polyester. When the stress/strain curves of more rigid materials have been observed, they are generally able to withstand greater stress before deforming or breaking. In our case, as polyester is more ductile, it deforms earlier before the stress rises. On the other hand, when the variation of float stitches is observed, the mechanical behavior of composition 3 is the opposite of composition 1. It is noticed that, in knitted cotton, as the float stitches are increased, the stress also increases, while in knitted polyester, the stress values are inversely proportional to the number of float stitches. In this way, the mechanical behavior of cotton is predictable, since the more you float stitches, the more stretched and tensioned the yarns are and the structure itself becomes more rigid, which apparently does not happen with knitted polyester [54]. This difference in mechanical behavior is more accentuated in the knitted with the smallest loop (0.71 cm) and a more contorted graph was observed. In Fig. 2 (graph "b"), the highest stress value (knitted cotton with 4 float stitches) and the lowest value (knitted polyester with 4 float stitches) was observed. On the other hand, in Fig. 2 (graph "a") shows a more harmonious situation, which leads us to conclude that the length loop is a determining factor for the observation of this difference in mechanical behavior of natural/synthetic weft-knitted structures commonly used as reinforcement of hybrid composites.

Note that in Fig. 3a, taking into account the float stitches (0, 2 and 4), it was obtained through this investigation that the longer the loop length (0.98 cm) the lower the influence of the float stitches variable, and the greater the influence dominant of the composition variable in the stress results.

It was observed that when increasing the density of float stitches (0 float stitches) for composition 3 (100% cotton), there is a significant increase in stress in relation to all other conditions of the variables (float stitches, composition), except for the condition float stitches (4 float stitches) in composition 1 (100% polyester). The float stitches condition (0 float stitches) for composition 1 (100% polyester) exhibits this complex and exponential behavior, as the length loop variable (0.98 cm) provides

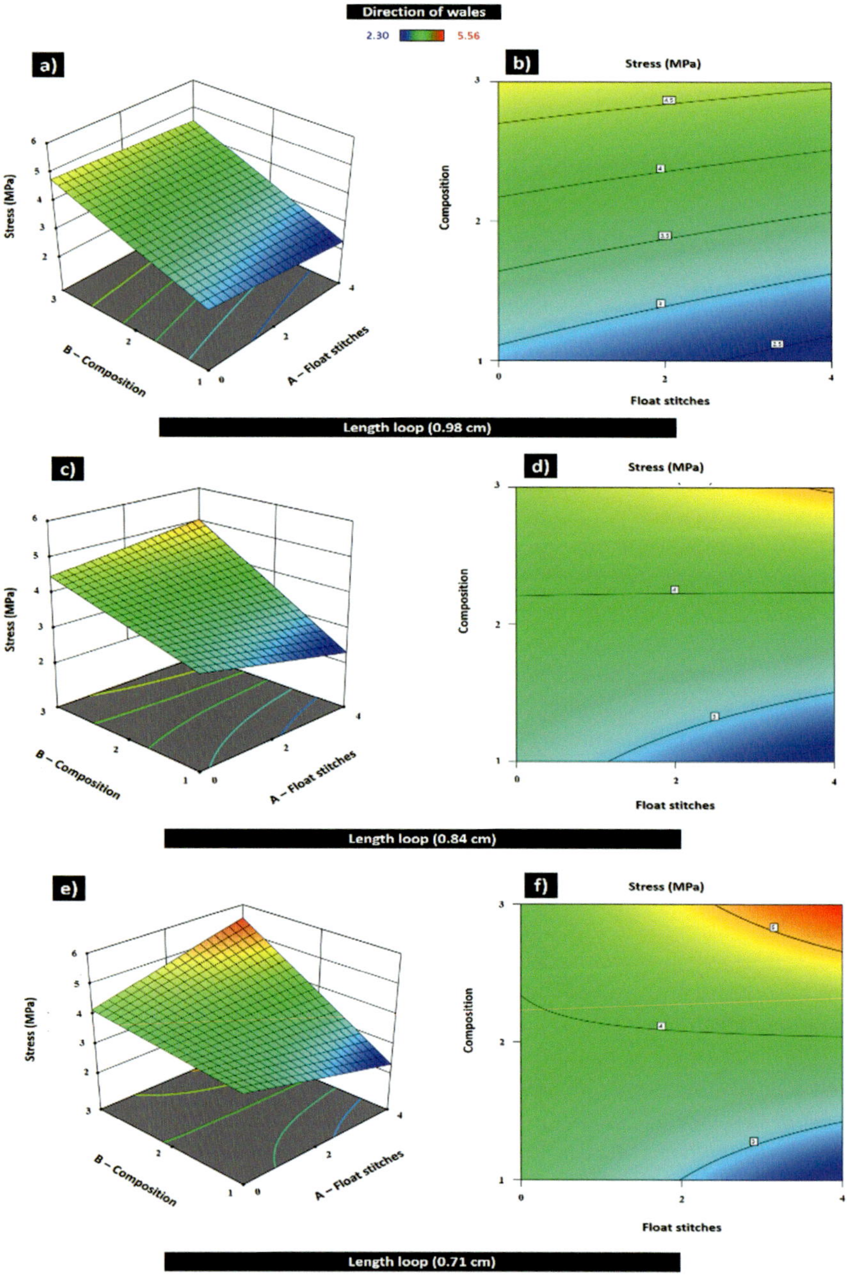

Fig. 2 Response surface and multiresponse contour of stress in the direction of wales

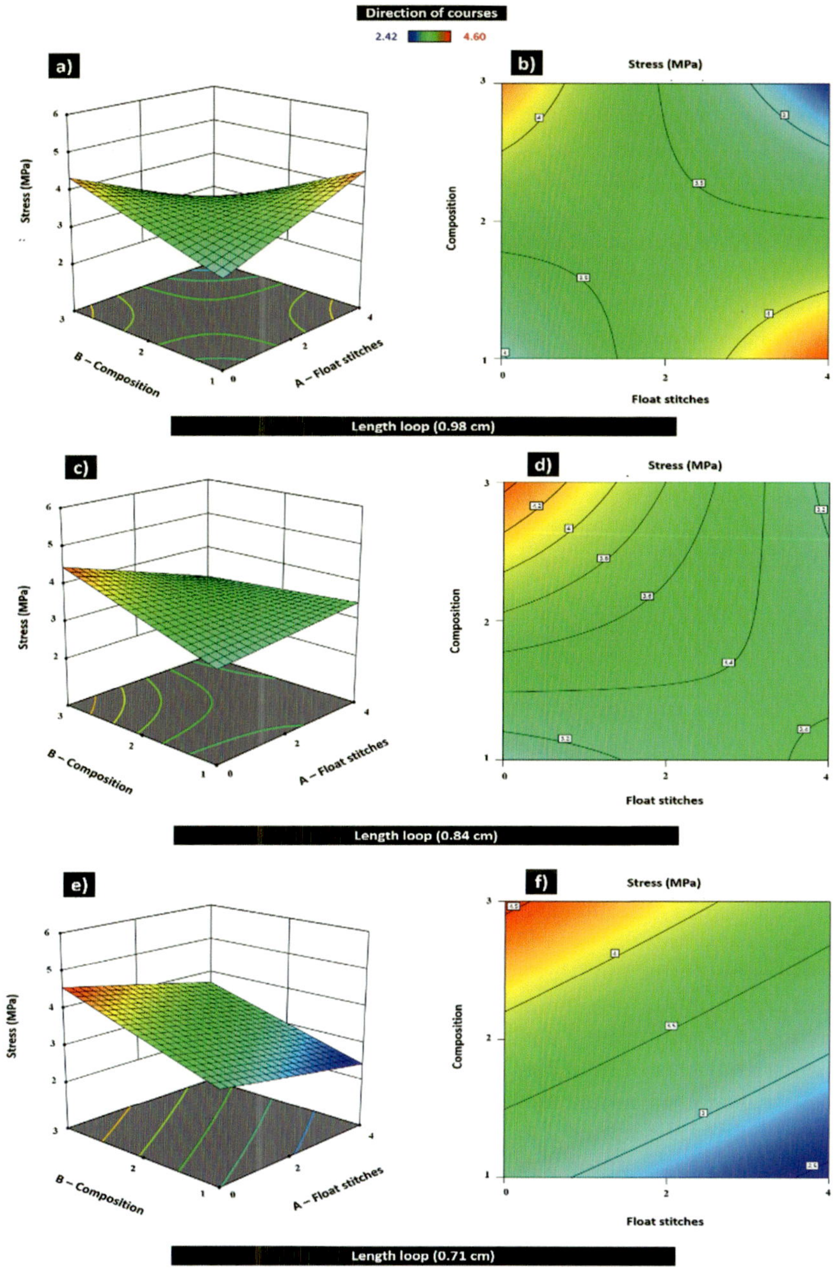

Fig. 3 Response surface and multiresponse contour of stress in the direction of courses

greater evidence of the anisotropy, complexity of the mathematical and statistical model of stress that it describes the curve in Fig. 3a [53]. Note further that in Fig. 3b the conditions of float stitches (0 float stitches) for composition 1 (100% polyester) and of float stitches (4 float stitches) for composition 3 (100% cotton) provide the appearance of a saddle stitch on the curve. These two conditions (0 float stitches and composite 1; 4 float stitches and composite 3) were minimum stress points on the curve in Fig. 3a. Figure 3c, e show that shortening the loop length increases the stability and stress borne by the knitted fabric. By reducing the length loop, the curves in Fig. 3c, e exhibit non-complex and linear behavior of the mathematical and statistical model of stress that provides less evidence of knitted fabric anisotropy with potential use for textile-reinforced composite materials applications [12]. In Fig. 3c, e the condition float stitches (0 float stitches) for composition 3 (100% cotton) exhibit more significant stress when compared to all other conditions and the condition that presented the least significant results was 4 float stitches and composition 1, showing that the increased density of float stitches and the reduction of the length loop provide a minor improvement for this condition [24]. Therefore, it was observed that it is possible to improve the mechanical behavior of knitted structures to enhance their application in weft-knitted structures commonly used as reinforcement of textiles composites.

3.2 *Statistical Analysis of Strain Mechanical Behavior*

Figure 4 shows the strain results of certain combinations between floats stitches (0, 2 and 4), compositions (1, 2 and 3) and length loop (0.98, 0.84 and 0.71 cm) from knitted fabrics in the direction of wales for potential use as textile reinforcement in composite materials. The strain results exhibited in Fig. 4a, strengthen the claim that shortening the loop length increases strain, which is the ability of the knitted fabric to deform to its breaking point [28]. Thus, it is also possible to observe that in Fig. 4a the behavior of the strain is less significant for composition 1 and independent of the floats stitches (0, 2 and 4), this occurred because the longer the length loop (0.98 cm) the smaller the influence of the float stitches variable, and the more significant is the dominant influence of the composition variable on stress outcomes (both to increase and to reduce mechanical behavior) [38]. The curves in Fig. 4c exhibit transitions in the mechanical behavior of knitted fabrics with potential use in textile-reinforced composite materials applications. From Fig. 4a, a behavioral trend in green and yellow colors was observed for composition 1 with 0 floats stitches (absence of floats stitches), in Fig. 4c it becomes clear that that behavior shown in Fig. 4a features a balance with length loop (0.84 cm) for composition 1 (100% polyester) in any density of floats stitches (0, 2 and 4). In the results for the length loop (0.84 cm), Fig. 4c, it is noted that from the combination composition 2 and 2 float stitches in relation to the combination composition 1 and 0 float stitches there were significant results for the strain values, where it can be understood that where there is absence or lesser amount of float stitches, more deformation is achieved due to the low elongation capacity of the float stitches. These results are very significant because it shows that it is possible by changing the length loop to obtain the maximum of behavior and mechanical properties, as well as alternating the composition or density of floats stitches to change the intensity of these properties in the direction of wales to direction of courses and vice versa and versa. Therefore, enhancing or improving the behavior of the strain in the direction of the applied stress in composites reinforced with textile structures. As expected, in Fig. 4e the most effective or most significant result was the combination of composition 1 (100% polyester) with 4 float stitches compared to the other combinations. For the results of a loop length of 0.71 cm, Fig. 4e, f, a significant strain value is concluded for the combination composition 1 and 0 float stitches (absence of float stitches), while for the combination composition 3 with 4 float stitches it presents the lowest value for the strain result. Due to the high elasticity of polyester (PES) and the absence of float stitches (0 float stitches), the higher strain value for this combination is explained, unlike that obtained in the composition condition 3 (100% cotton) with 4 float stitches in which the lowest strain value was obtained in this study, this occurred because cotton (CO) has less elasticity when compared to polyester (shown in Fig. 4f) [49].

Figure 5, the strain results of each combination between floats stitches (0, 2 and 4), compositions (1, 2 and 3) and length loop (0.98, 0.84 and 0.71 cm) of knitted fabrics in the direction of courses for use in composites reinforced with textile materials. In Fig. 5a, the strain results shown with a length loop of 0.98 cm, it is observed that the combination of composition 1 (100% polyester) with 0 float stitches (absence of float stitches) obtained the highest strain result compared to the other combinations (see Fig. 5b). While, all other combinations have moderate or low performance [41]. Because, as the loop length increases, the easier the loops slide over each other [3, 41]. Another important point to highlight is that in the strain composition 1 (100% PES) has high intrinsic elasticity, however, when float stitches were inserted, this property was affected (decrease in the strain). In the loop length of 0.84 cm, Fig. 5c, in the combination composition 2 with 2 float stitches and composition 1 with 0 float stitches there were significant values for strain, where it can be understood that where there is absence or lesser amount of float stitches, it is possible to deform more due to the low elongation capacity of the float stitches [27]. So, when the 0.71 cm length loop results were analyzed, Fig. 5e, f, a significant strain value is concluded for the combination composition 1 with 0 float stitches (absence of float stitches), while for the combination composition 3 with 4 float stitches it has the lowest strain value compared to the other combinations [17]. Therefore, as the length loop decreases, the strain value gradually increases for all combinations, with the exception of composition 3 with 4 float stitches. A greater improvement in the mechanical behavior of the strain was obtained for composition 1 (100% polyester), due to its high elongation capacity when compared to composition 3 (100% cotton) and that, with the decrease in the length loop, the strength of knitted fabrics for potential application in textile composites [16, 37].

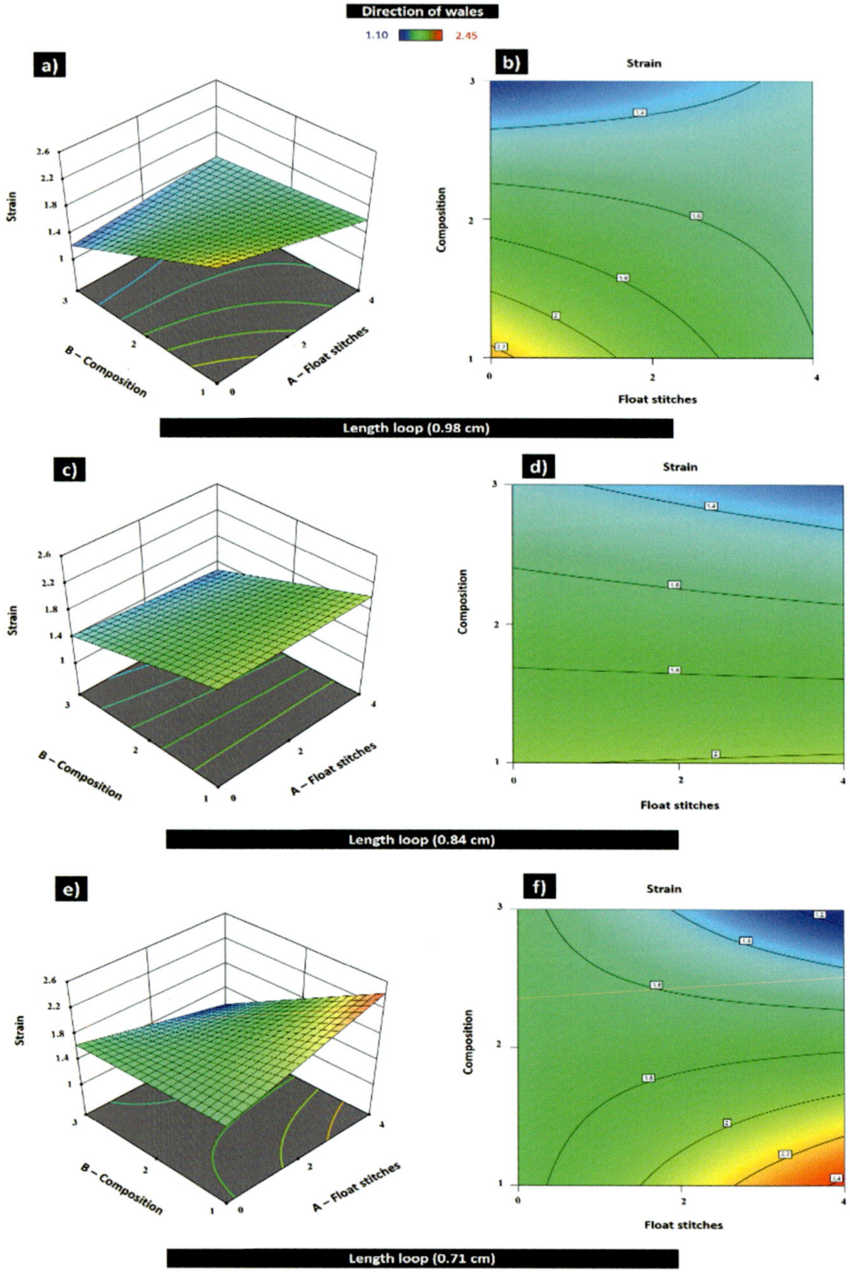

Fig. 4 Response surface and multiresponse contour of strain in the direction of wales

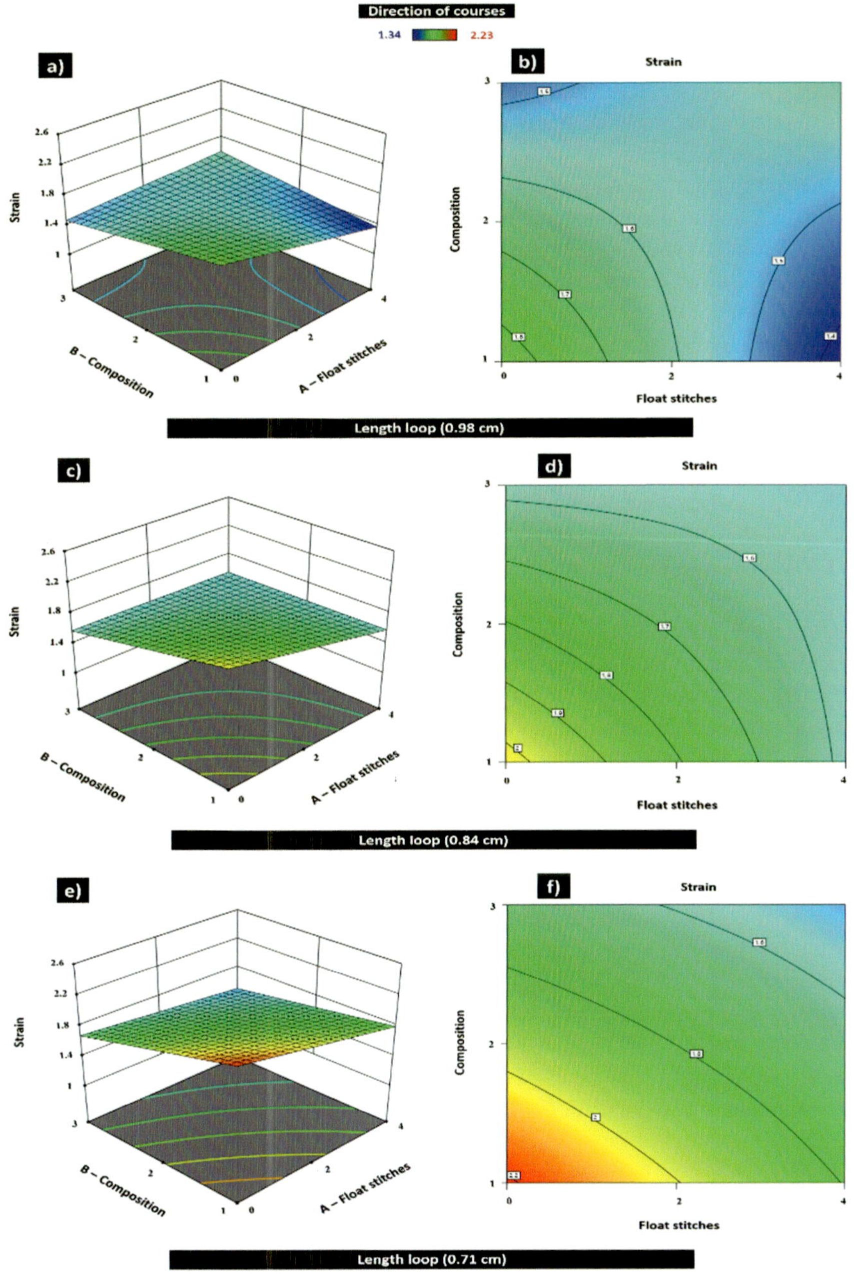

Fig. 5 Response surface and multiresponse contour of strain in the direction of courses

3.3 Statistical Analysis of MOE Mechanical Behavior

Young's Modulus (MOE) is a measure of the stiffness of an elastic material, and it is defined as the ratio of stress to strain (stress/strain). Knitted fabrics with low MOE tend to be ductile, as well as knitted fabrics with high MOE tend to be brittle. Figure 6, represent the response surface and multiresponse contour of MOE in the wales direction of each combinations between floats stitches (0, 2 and 4), compositions (1, 2 and 3) and length loop (0.98, 0.84 and 0.71 cm) of knitted fabrics for use in textile-reinforced composites. In Fig. 6a, the MOE results shown with a length loop of 0.98 cm, was obtained in the combination of composition 3 (100% cotton) and 0 float stitches (no float stitches) had the highest MOE result compared to the other combinations (see Fig. 6b). While, all other combinations have moderate or low performance. As the length loop increased, the strain obtained in Fig. 4a and better mechanical behavior of composition 3 (100% cotton) with 0 float stitches in the wales direction. This was due to the high difficulty of the loops to slide over each other and the low elongation capacity of composition 3 (100% cotton) when compared to composition 1 (100% polyester). Thus, it was very important to note that in MOE, composition 3 (100% CO) exhibited low deformability or a more plastic mechanical behavior compared to composition 1 (100% polyester) and when float stitches were inserted, this property was improved (addition in MOE). In the loop length of 0.84 cm, Fig. 5c, it was observed that intensifying the tightening or reducing the length loop alters the mechanical behavior that promotes and enhances the plastic (fragile) mechanical behavior in the combination composition 3 with 4 float stitches when compared to the other combinations [9, 54]. Thus, exhibiting more satisfactory results for MOE, as it was understood when reducing the length loop (0.84 cm) and increasing the amount of float stitches (4 float stitches), it is possible to reduce the deformability due to the low elasticity capacity of the float stitches. Finally, when the 0.71 cm length loop results were analyzed, Fig. 6e, f and a significant MOE value is concluded for the combination composition 3 with 4 float stitches, while for the combination composition 1 with 4 float stitches it has the lowest MOE value compared to the other combinations. Therefore, as the length loop decreases, the MOE value is gradually improved for all combinations, with the exception of those with composition 1 (100% polyester), mainly (composition 1 with 4 float stitches). Better mechanical behavior of MOE was obtained for composition 3 (100% CO), due to its lower elasticity (deformability) capacity when compared to composition 1 (100% polyester) of knitted fabrics for potential application in textile composites [11, 13].

Figure 7 represent the response surface and multiresponse contour of MOE in the direction of the courses of each combination that presents a behavior directly proportional to the gradual reduction of the length loop for all knitted fabrics. Figure 7a, the MOE results shown with a length loop of 0.98 cm exhibit complex statistical and mathematical behavior, were obtained in the combination of composition 1 (100% PES) with 4 float stitches the highest MOE result followed by composition 3 (100% CO) with 0 float stitches compared to the other combinations (see Fig. 7b) [15, 22].

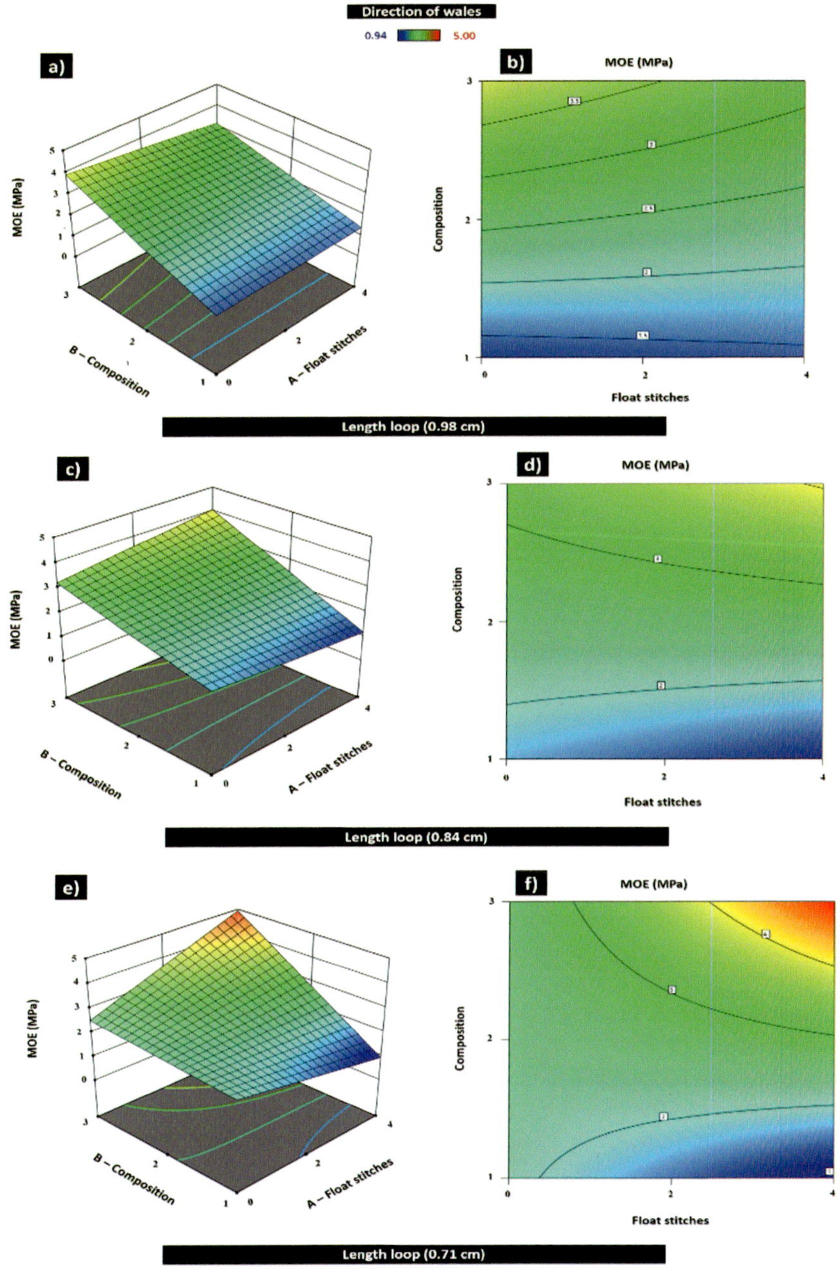

Fig. 6 Response surface and multiresponse contour of MOE in the direction of wales

Thus, when analyzing the mechanical behavior in the length loop (0.98 cm), it was observed that the combination composition 1 (100% PES) with 0 float stitches and composition 3 (100% CO) with 4 float stitches exhibited lower MOE values in the direction of the courses, because in both combinations there was no improvement in the MOE, that is, when the combination composition 1 (100% PES) that has intrinsically good elongation with 0 float stitches (the absence of float stitches) this condition promotes greater deformability, in the same way shape, composition 3 (100% CO) with 4 float stitches (see Fig. 7) [20]. Regarding the length loop 0.84 and 0.71 cm, knitted fabrics was observed that the length loop 0.98 cm exhibited greater prominence in composition 1 (100% PES) with 4 float stitches from knitted fabrics [29]. This feature ensured the best performance in the loop length of 0.98 cm [48]. This is explained by the fact that there is less contraction in knitted fabrics, since float stitches are smaller in length and with the presence of float stitches, the structure becomes uniform, stable, and thus, ensuring a better distribution of forces. Applied making the material more rigid and resistant. Another important aspect to take into account is the greater stiffness presented in the 100% polyester composition with 4 float stitches compared to other float stitches and compositions [19]. This is due to the fact that, as it is a structure formed by threads in the form of filaments, this provides uniformity in relation to the distribution of force applied to the material and thus greater resistance. The high presence of float stitches favors this resistance, since float stitches are free stitches and therefore suffer less stress, traction and wear during the production of knitted fabrics and even when it is already manufactured, making it suitable for potential application in textile composites [23, 35].

3.4 *Analysis of Variance in Single Pique Knitted Mechanical Behavior*

Analysis of variance (ANOVA) is a conceptually simple, powerful, and popular way to perform statistical testing on experiments that involve two or more groups [46]. The results exhibited in Table 2 correspond to each response studied using design of experiments (DoE) towards the wales. The values found for both R^2 and adjusted R^2 were above 0.6, suggesting that the goodness-of-fit of the model is consistent and compatible. The ANOVA Table 2 indicates that the regression model is significant because $p < 0.05$ [52]. Table 2 study indicates that behavior and mechanical properties of knitted fabric can be predicted on the basis of float stitch density (0, 2 and 4), loop length (0.98, 0.84 and 0.71 cm), yarn compositions (100% cotton, 50% cotton/50% polyester and 100% polyester) [52]. Then, in ANOVA perceived that the properties that describe the studied mechanical behavior. Thus, the mean obtained from stress (3.77 ± 0.04 MPa), strain (1.68 ± 0.03) and MOE (2.51 ± 0.05 MPa), and a significant standard deviation due to the low dispersion of the data. Therefore, the coefficient of variation (CV%), stress (1.16%), strain (2.04%) and MOE (2.14%) of this statistical model is significant [30, 47].

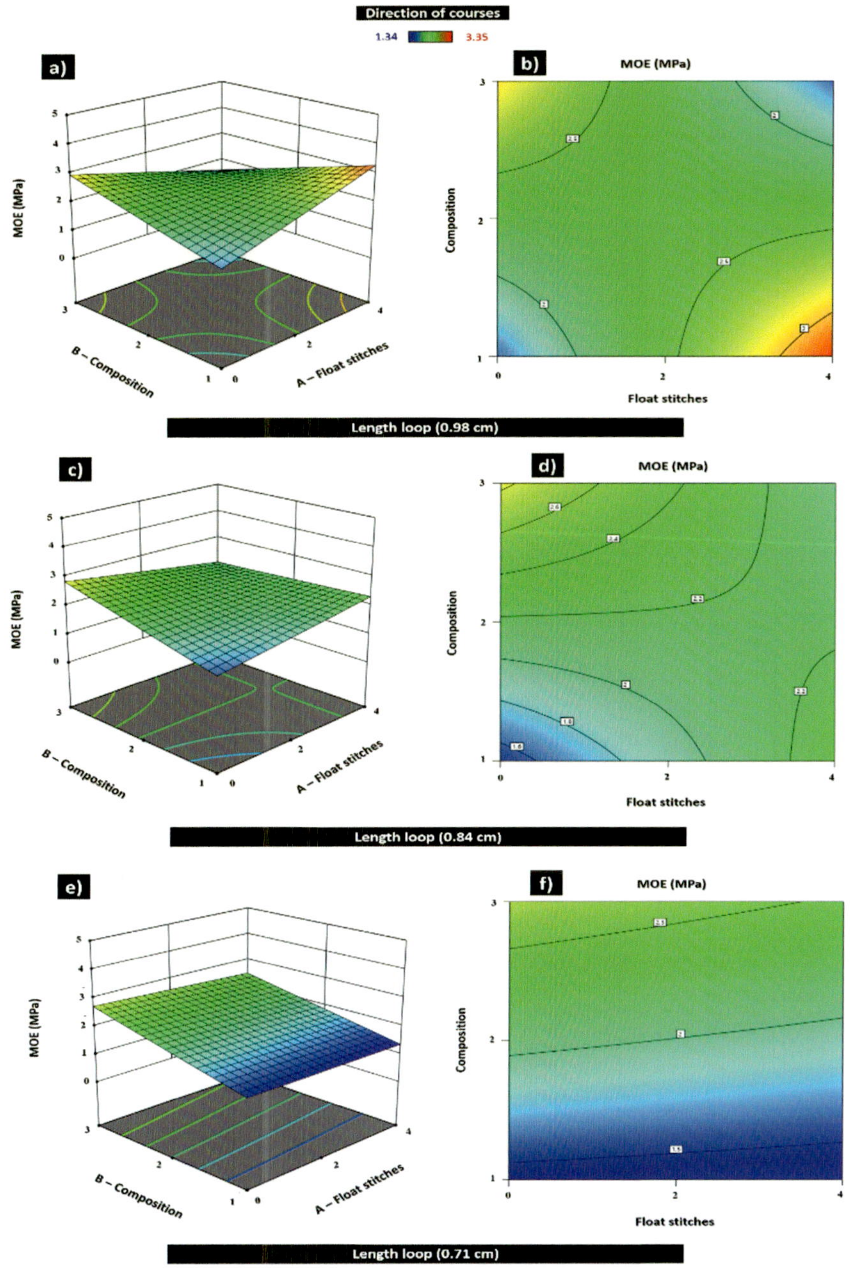

Fig. 7 Response surface and multiresponse contour of MOE in the direction of courses

Table 2 ANOVA of mechanical properties and behavior of single pique knitted in wales direction

	Stress (MPa)		Strain		MOE (MPa)	
	F-value	p-value (Prob > F)	F-value	p-value (Prob > F)	F-value	p-value (Prob > F)
Model	1493.8	<0.0001	360.22	<0.0001	1348.8	<0.0001
A—Float stitches	83.62	<0.0001	12.28	0.0067	34.48	0.0002
B—Composition	8076.6	<0.0001	1452.8	<0.0001	6715.8	<0.0001
C—Length loop	179.45	<0.0001	39.43	0.0001	45.37	<0.0001
AB	1262.6	<0.0001	29.20	0.0004	626.13	<0.0001
AC	96.44	<0.0001	74.24	<0.0001	300.79	<0.0001
BC	20.88	0.0013	8.53	0.0170	0.0979	0.7615
ABC	737.00	<0.0001	905.03	<0.0001	1719.9	<0.0001
Standard dev	0.04		0.03		0.05	
Mean	3.77		1.68		2.51	
C.V.%	1.16		2.04		2.14	
R^2	0.99		0.99		0.99	
Adjusted R^2	0.99		0.99		0.99	

In Table 3 the effect of float stitch density (0, 2 and 4), loop length (0.98, 0.84 and 0.71 cm), yarn compositions (100% cotton, 50% cotton/50% polyester and 100% polyester) on knitted fabric were evaluated, using ANOVA analysis was employed.

Table 3 ANOVA of mechanical properties and behavior of single pique knitted in courses direction

	Stress (MPa)		Strain		MOE (MPa)	
	F-value	p-value (Prob > F)	F-value	p-value (Prob > F)	F-value	p-value (Prob > F)
Model	1614.4	<0.0001	119.64	<0.0001	408.18	<0.0001
A—Float stitches	929.06	<0.0001	186.30	<0.0001	0.0039	0.9515
B—Composition	1268.9	<0.0001	198.25	<0.0001	389.46	<0.0001
C—Length loop	69.81	<0.0001	167.20	<0.0001	190.12	<0.0001
AB	3127.4	<0.0001	149.14	<0.0001	889.59	<0.0001
AC	471.65	<0.0001	12.90	0.0058	35.55	0.0002
BC	2802.7	<0.0001	95.45	<0.0001	652.77	<0.0001
ABC	2630.9	<0.0001	28.26	0.0005	699.77	<0.0001
Standard dev	0.03		0.04		0.05	
Mean	3.47		1.73		2.09	
C.V.%	0.88		2.01		2.55	
R^2	0.99		0.98		0.99	
Adjusted R^2	0.99		0.98		0.99	

Stress (3.47 ± 0.03 MPa), strain (1.73 ± 0.04) and MOE (2.09 ± 0.05 MPa), if you get a large value-F, it is understood that parts of the data are significant, that is, the input values, do not describe the behavior of the response variables in totality [30]. While a small value-ρ means all your results are significant. Because, that value-F corresponds to level of discrepancy between the mean values of the data studied and that higher the value-F the more significant value-ρ ($\rho \geq 0.05$) becomes for the variables float stitches (A), Composition (B) and Length loop (C) and their respective interactions (AB, AC, BC and ABC) exhibited in Table 3 [21, 32]. The coefficient of variation obtained were, stress (0.88%), strain (2.01%) and MOE (2.55%). Thus, the value-F and value-ρ for the float stitches (<0.0001), composition (<0.0001) and length loop (<0.0001), as well as for the interactions (AB, AC, BC, ABC and AC) of all response variables (stress, strain, MOE) were significant [42, 54]. That is, studied variables (Float stitches, Composition and Length loop) and their respective interactions, promote a significant and intense improvement on the mechanical properties and behavior of knitted fabric. This model is significant and can adequately describe the response variables [10, 43]. All the response variables studied showed high compatibility (between 0.98 and 0.99) comparing the experimental data and statistical model in direction of courses. Finally, since the P-value in the ANOVA exhibited Tables 2 and 3 is less than or equal to 0.05, there is a statistically significant relationship between the variables (float stitches, composition and length loop) at the 98–99% confidence level in both knitted fabric directions (wales and courses).

4 Current and Future Trends

Among the independent variables studied (float stitches, composition and the length loop). The length loop is the independent variable that most influences and alters the mechanical behavior of single pique knitted fabrics commonly used in textile composites. It is noted that in the direction of courses and direction of wales were high for composition 1 (100% PES) and composition 3 (100% CO), as well as the results obtained revealed significant evidence of anisotropy compared to those composed of (50% CO/50% PES). Finally, exposing that the properties of single pique knitted fabrics improve and intensify regardless of the nature of the fiber (synthetic or natural). Finally, this proves that float stitches significantly contribute and improve the mechanical behavior of knitted structures used in knitted-reinforced composite materials. Thus, future investigations focus on new knitted fabrics and matrix systems, and the structural design of knitted fabrics, in addition to improving existing technologies. Successful implementation of mono-, bi- and triaxial textile composites for textile reinforcement in composite materials requires further research development into the production economics, engineering design and structural behavior of composites reinforced with knitted fabrics to create a cost-effective production technology and also sustainable. For this, it is currently necessary to combine fabric geometry model (FGM) with computer-aided design (CAD)—computer-aided manufacturing (CAM) or (CAD-CAM) and advanced statistical analysis (ASA) using tools such as analysis of variance (ANOVA) and 3D response surface methodology (RSM) can create a useful framework that can help integrate the structural design of knitted fabrics, as well as the processing parameters in the structural analysis (composition, length loop, yarn count, and etc.) and ultimately efficiently design the properties needed for a specific end use in composite materials reinforced with knitted fabrics. Future research will investigate the phenomenon of adhesion and knitted fabrics-matrix interface in textile-reinforced composite materials. For the high performance of the properties obtained is the result of the quality of adhesion at the knitted-matrix interface in a textile composite material. The great advantage of knitted fabrics continues to be the high deformability, which allows efficient use in Composite products of complex format. However, it is clear that the potential of knitted fabrics as a reinforcement applied to textile composites has not yet been fully explored, and that knitting technology awaits the imagination and creativity of textile, materials and composites engineers.

Acknowledgements This research was supported by our Lord. We thank our colleagues at the Textile Quality Control Laboratory—LABCTEX and Textile Engineering Post Graduate Program—PPgET at the Federal University of Rio Grande do Norte.

References

1. Alpyildiz T, Icten BM, Karakuzu R, Kurbak A (2009) The effect of tuck stitches on the mechanical performance of knitted fabric reinforced composites. Compos Struct 89:391–398. https://doi.org/10.1016/j.compstruct.2008.09.004
2. Aquino MS, Holanda SM, Santos TF et al (2019) Analysis of polymeric knitted fabrics on properties for telecommunications devices application. Fibers Polym 20:2348–2354. https://doi.org/10.1007/s12221-019-8724-3
3. Arumugam S, Kandasamy J, Venkatesan S et al (2022) A review on the effect of fabric reinforcement on strength enhancement of natural fiber composites. Materials (Basel) 15:3025. https://doi.org/10.3390/ma15093025
4. Asif A, Rahman M, Farha FI (2015) Effect of knitted structure on the properties of knitted fabric. Int J Sci Res 4:1230–1235
5. ASTM D3776/D3776M—09a (2017) Standard test methods for mass per unit area (weight) of fabric. West Conshohocken, PA
6. ASTM D3887-96 (2008) (2008) Standard specification for tolerances for knitted fabrics. West Conshohocken
7. ASTM D5034-09(2017) (2017) Standard test method for breaking strength and elongation of textile fabrics (grab test). West Conshohocken
8. Balcıoğlu HE, Yalçın D (2020) The determination of fracture characterization of knitted fabric reinforced composites using arcan test. Fibers Polym 21:849–863. https://doi.org/10.1007/s12221-020-9619-z
9. Bhattacharya P (Dr) SS, Agrawal MSA (2017) Textile reinforced structure: a review. Int J Eng Res Appl 07:84–86. https://doi.org/10.9790/9622-0707088486
10. Çeven EK, Günaydın GK (2021) Evaluation of some comfort and mechanical properties of knitted fabrics made of different regenerated cellulosic fibres. Fibers Polym 22:567–577. https://doi.org/10.1007/s12221-021-0246-0

11. de Holanda SM, da Silva JP, de S Queiroz I et al (2017) Microstrip antenna based on knitted fabrics with biodegradable synthetic fibers. J Mater Sci Mater Electron 28:15118–15126. https://doi.org/10.1007/s10854-017-7387-3
12. Döbrich O, Gereke T, Cherif C (2016) Modeling the mechanical properties of textile-reinforced composites with a near micro-scale approach. Compos Struct 135:1–7. https://doi.org/10.1016/j.compstruct.2015.09.010
13. dos Santos TF, da Silva Santos CM, da Fonseca RT et al (2019) Influence of structure and composition in the mechanical properties of textile polymeric fabrics. Polym Polym Compos 27:222–227. https://doi.org/10.1177/0967391118823075
14. Duhovic M, Bhattacharyya D (2011) Knitted fabric composites. In: Advances in knitting technology. Elsevier, pp 193–212
15. Duhovic M, Bhattacharyya D (2006) Simulating the deformation mechanisms of knitted fabric composites. Compos Part A Appl Sci Manuf 37:1897–1915. https://doi.org/10.1016/j.compositesa.2005.12.029
16. Gommers B, Verpoest I, Houtte P (1998) Analysis of knitted fabric reinforced composites: part II. Stiffness and strength. Compos Part A Appl Sci Manuf 29:1589–1601. https://doi.org/10.1016/S1359-835X(98)00096-7
17. Gommers B, Verpoest I, Van Houtte P (1996) Modelling the elastic properties of knitted-fabric-reinforced composites. Compos Sci Technol 56:685–694. https://doi.org/10.1016/0266-3538(96)00053-X
18. Hamedi S, Hasani H, Dibajian SH (2017) Numerical simulating the flexural properties of 3D weft-knitted spacer fabric reinforced composites. J Compos Mater 51:1887–1899. https://doi.org/10.1177/0021998316665240
19. Hasan KMF, Horváth PG, Alpár T (2021) Potential fabric-reinforced composites: a comprehensive review. J Mater Sci 56:14381–14415. https://doi.org/10.1007/s10853-021-06177-6
20. Hasani H, Hassanzadeh S, Abghary MJ, Omrani E (2017) Biaxial weft-knitted fabrics as composite reinforcements: a review. J Ind Text 46:1439–1473. https://doi.org/10.1177/1528083715624256
21. Hoque MS, Hossain MJ, Rahman MM, Rashid MM (2022) Fiber types and fabric structures influence on weft knitted fabrics. Heliyon 8:e09605. https://doi.org/10.1016/j.heliyon.2022.e09605
22. Huang ZM, Ramakrishna S (2000) Micromechanical modeling approaches for the stiffness and strength of knitted fabric composites: a review and comparative study. Compos Part A Appl Sci Manuf 31:479–501. https://doi.org/10.1016/S1359-835X(99)00083-4
23. Kalaimathi TRP, Shanmugam B (2022) A review paper on mechanical properties of flexural and impact test on textile reinforced engineered cementitious composites. Mater Today Proc 50:2160–2169. https://doi.org/10.1016/j.matpr.2021.09.443
24. Karaduman NS, Karaduman Y, Ozdemir H, Ozdemir G (2017) Textile reinforced structural composites for advanced applications. In: Textiles for advanced applications. InTech
25. Khondker OA, Yang X, Usui N, Hamada H (2006) Mechanical properties of textile-inserted PP/PP knitted composites using injection–compression molding. Compos Part A Appl Sci Manuf 37:2285–2299. https://doi.org/10.1016/j.compositesa.2005.12.032
26. Kyzymchuk O, Melnyk L (2018) Stretch properties of elastic knitted fabric with pillar stitch. J Eng Fiber Fabr 13:155892501882072. https://doi.org/10.1177/1558925018820722
27. Leong KH, Ramakrishna S, Huang ZM, Bibo GA (2000) The potential of knitting for engineering composites—a review. Compos Part A Appl Sci Manuf 31:197–220. https://doi.org/10.1016/S1359-835X(99)00067-6
28. Li Y, Ma P, Tian M, Yu M (2022) Dynamic equivalent resistance model of knitted strain sensor under in-plane and three-dimensional surfaces elongation. Polymers (Basel) 14:2839. https://doi.org/10.3390/polym14142839
29. Liu Y, Hua H, Li B (2022) Exploration and design of knitted composites for architectural application: the MeiTing project. Front Archit Res 11:594–608. https://doi.org/10.1016/j.foar.2022.01.004

30. Maâtoug N, Sahnoun M, Sakli F (2012) Statistical analysis of surface roughness parameters for weft knitted fabrics measured by the textile surface tester (TST). J Eng Fiber Fabr 7:155892501200700. https://doi.org/10.1177/155892501200700407
31. Mark H, Workman J (2018) Experimental designs, PART 2: one-way ANOVA. In: Chemometrics in spectroscopy, 2nd. Elsevier, pp 63–68
32. Melo KM De, Dos Santos TF, Santos CMDS et al (2019) Study of the reuse potential of the sisal fibers powder as a particulate material in polymer composites. J Mater Res Technol 8:4019–4025. https://doi.org/10.1016/j.jmrt.2019.07.010
33. Mikučioniene D, Čiukas R, Mickevičiene A (2010) The influence of knitting structure on mechanical properties of weft knitted fabrics. Medziagotyra 16:221–225
34. Molugaram K, Rao GS (2017) ANOVA (analysis of variance). In: Statistical techniques for transportation engineering. Elsevier, pp 451–462
35. Muralidhar BA (2013) Tensile and compressive behaviour of multilayer flax-rib knitted preform reinforced epoxy composites. Mater Des 49:400–405. https://doi.org/10.1016/j.matdes.2012.12.040
36. Oliveira LÁ, Santos JC, Panzera TH et al (2018) Evaluation of hybrid-short-coir-fibre-reinforced composites via full factorial design. Compos Struct 202:313–323. https://doi.org/10.1016/J.COMPSTRUCT.2018.01.088
37. Pamuk G, Çeken F (2008) Manufacturing of weft-knitted fabric reinforced composite materials: a review. Mater Manuf Process 23:635–640. https://doi.org/10.1080/10426910802316484
38. Pham, Döbrich, Trümper et al (2019) Numerical modelling of the mechanical behaviour of biaxial weft-knitted fabrics on different length scales. Materials (Basel) 12:3693. https://doi.org/10.3390/ma12223693
39. Qi Y, Li J, Liu L (2014) Tensile properties of multilayer-connected biaxial weft knitted fabric reinforced composites for carbon fibers. Mater Des 54:678–685. https://doi.org/10.1016/j.matdes.2013.08.051
40. Rossi F, Mirtchev V (2016) Analysis of variance (ANOVA). In: Statistics for food scientists. Elsevier, pp 19–29
41. Ruan X (1996) Experimental and theoretical studies of the elastic behavior of knitted-fabric composites. Compos Sci Technol 56:1391–1403. https://doi.org/10.1016/S0266-3538(96)00097-8
42. Santos TF, Santos CM, Aquino MS et al (2020) Statistical study of performance properties to impact of Kevlar® woven impregnated with Non-Newtonian Fluid (NNF). J Mater Res Technol. https://doi.org/10.1016/j.jmrt.2020.01.027
43. Seboka N (2021) Statistical analysis of effects of fabric thickness, loop shape factor, fabric tightness factor and aerial weight on water vapor permeability of single jersey polyester knitted fabric. Int J Sci Res Comput Sci Eng 9:55–62
44. Shekarchizadeh N, Jafari Nedoushan R, Dastan T, Hasani H (2021) Experimental and numerical study on stiffness and damage of glass/epoxy biaxial weft-knitted reinforced composites. J Reinf Plast Compos 40:70–83. https://doi.org/10.1177/0731684420938446
45. Smalheiser NR (2017) Reproducibility and robustness. In: Data literacy. Elsevier, pp 3–15
46. Smalheiser NR (2017) ANOVA. In: Data literacy. Elsevier, pp 149–155
47. Smalheiser NR (2017) Experimental design: design strategies and controls. In: Data literacy. Elsevier, pp 65–85
48. Stolyarov ON (2009) Mechanical properties of polymer composites reinforced with knit fabrics made of high-strength aramid fibres. Fibre Chem 41:53–55. https://doi.org/10.1007/s10692-009-9110-5
49. Sulochani RMN, Jayasinghe RA, Nilmini AHLR, Priyadarshana G (2022) Recent developments in textile reinforced polymer composites. Asian J Chem 34:487–496. https://doi.org/10.14233/ajchem.2022.23504
50. Tiber B, Balcıoğlu HE (2019) Flexural and fracture behavior of natural fiber knitted fabric reinforced composites. Polym Compos 40:217–228. https://doi.org/10.1002/pc.24635
51. Verpoest I, Gommers B, Huysmans G et al (1997) The potential of knitted fabrics as a reinforcement for composites. Compos Appl Des 1:108–133

52. Wong ASW, Li YI (2006) Prediction of clothing sensory comfort. In: Li Y, Wong ASW (eds) Clothing biosensory engineering. Elsevier, pp 178–188
53. Zamanov AD (2005) Natural vibrations of a strip made of a composite material with a locally curved structure. Mech Compos Mater 41:49–54. https://doi.org/10.1007/s11029-005-0006-z
54. Zílio L, Dias M, Santos T et al (2020) Characterization and statistical analysis of the mechanical behavior of knitted structures used to reinforce composites: yarn compositions and float stitches. J Mater Res Technol 9:8323–8336. https://doi.org/10.1016/j.jmrt.2020.05.089

GPSR Compliance

The European Union's (EU) General Product Safety Regulation (GPSR) is a set of rules that requires consumer products to be safe and our obligations to ensure this.

If you have any concerns about our products, you can contact us on ProductSafety@springernature.com

In case Publisher is established outside the EU, the EU authorized representative is:

Springer Nature Customer Service Center GmbH
Europaplatz 3
69115 Heidelberg, Germany

Batch number: 10756376

Printed by Printforce, the Netherlands